"*Moravian Soundscapes* is a stunning achievement that deftly crosses disciplinary boundaries to offer a compellingly immersive journey into eighteenth century Moravian communities as experienced by German and Native peoples. Woven throughout is Eyerly's own family story, which reminds readers that all history writing gains its fuel in our own more recent pasts." —Rachel Wheeler, Indiana University

"*Moravian Soundscapes* brings a compelling and necessary new approach to the study of music, sound, space, and colonial encounter in early America. Combining historical research, sound mapping, and autobiographical reflection, Eyerly reveals the way in which listening and singing were integral to European and Native Moravians' understanding of their environments, experiences of faith, and construction of community. In doing so, she offers an intimate exploration of how family, place, and music intertwine." —Glenda Goodman, University of Pennsylvania

"*Moravian Soundscapes* is an important contribution to our understanding of the musical dimension of European religious subcultures in colonial-era North America. . . . Eyerly positions Moravian song and sound at the center of this history and shows how its creators used it to impose order on their social and natural worlds." —Olivia Bloechl, author of *Native American Song at the Frontiers of Early Modern Music*

MUSIC, NATURE, PLACE

Sabine Feisst and Denise Von Glahn

MORAVIAN SOUNDSCAPES

A Sonic History of the Moravian Missions in Early Pennsylvania

Sarah Justina Eyerly

INDIANA UNIVERSITY PRESS

Published with financial assistance from the AMS 75 Publication Awards for Younger Scholars Fund of the American Musicological Society, supported in part by the National Endowment for the Humanities and the Andrew W. Mellon Foundation. Publication of this book also was supported in part by a grant from the H. Earle Johnson Fund of the Society for American Music.

This book is a publication of

Indiana University Press
Office of Scholarly Publishing
Herman B Wells Library 350
1320 East 10th Street
Bloomington, Indiana 47405 USA

iupress.indiana.edu

© 2020 by Sarah Eyerly

All rights reserved

No part of this book may be reproduced or utilized in any form or by any means, electronic or mechanical, including photocopying and recording, or by any information storage and retrieval system, without permission in writing from the publisher. The paper used in this publication meets the minimum requirements of the American National Standard for Information Sciences—Permanence of Paper for Printed Library Materials, ANSI Z39.48-1992.

Manufactured in the United States of America

Cataloging information is available from the Library of Congress.

ISBN 978-0-253-04766-3 (hardback)
ISBN 978-0-253-04769-4 (paperback)
ISBN 978-0-253-04775-5 (ebook)

1 2 3 4 5   25 24 23 22 21 20

*For my mother,
Mary Ann Erickson Eyerly,
and my father,
Raymond Werner Eyerly.*

Do you feel the words that you sing in your heart?
Ktammachtammen neen aaptonawaganan anekanawojanne ktahak?
Fühlest du die Worte, die du singest, im Herzen?

JOHANN JACOB SCHMICK,
MISCELLANEA LINGUAE NATIONIS INDICAE MAHIKAN DICTAE (C. 1753–1767)

CONTENTS

ABOUT THE COMPANION WEBSITE XI

ACKNOWLEDGMENTS XIII

NOTE ON NAMING, TERMINOLOGY, AND ARCHIVAL SOURCES XV

Prologue: The Pennsylvania Wilds 1

Introduction: Sounding New Histories of the Moravian Missions 7

Peale 49

1 Penn's Woods 51

Bethlehem 105

2 Friends & Strangers 107

Herrnhut 151

3 Sound & Spirit 154

Moravian Run 186

4 1782 191

Epilogue: Petquotting 223

GLOSSARY: A MORAVIAN VOCABULARY 227

BIBLIOGRAPHY 229

INDEX 261

ABOUT THE COMPANION WEBSITE

Moravian Soundscapes

https://doi.org/**10.33009**/moraviansoundscapes_music_fsu

Moravian Soundscapes: A Sonic History of the Moravian Missions in Early Pennsylvania features a companion website. Readers are encouraged to visit the website to view interactive and static maps, archival materials, and pictures, and to listen to sound samples that illustrate the content of the book. Icons located throughout the book indicate online content 🌐. For assistance in navigating the website and determining the location of the online content, please see the following "List of Audiovisual Materials." To learn more about navigating the interactive maps, please see the online user guide. More information about this project can be found at the "About This Project" section of the website.

The initial maps and sound examples for this book were created by Sarah Eyerly, Mark Sciuchetti, and Andy Nathan using ArcGIS 10.3 and ArcGIS Online, courtesy of the Florida State University, and Logic Studio. Over time, we will update the content of the website in response to new technologies and new modes of presentation. We would like to express our thanks to the American Council of Learned Societies, the American Musicological Society, the Society for American Music, the Council for Research and Creativity at FSU, and the Lucille P. and Elbert B. Shelfer Professorship in Music for funding the first stage of the development and production of the recordings, maps, and website.

List of Audiovisual Materials

https://doi.org/**10.33009**/moraviansoundscapes_music_fsu

Introduction: Sounding New Histories of the Moravian Missions

1. Timeline: "Moravian Missions in North America, 1740–1794"
2. Static Map: "The Moravian Atlantic"
3. Map Collection: "Mapping Pennsylvania"

4. Picture Collection: "Modern-Day Pictures"
5. Interactive Sound Map: "Moravian Soundscapes"

Chapter 1: Penn's Woods

1. Static Map: "Early Moravian Missions in Pennsylvania and Ohio"
2. Interactive Map: "The Pennsylvania Frontier"
3. Interactive Sound Map: "The Great Shamokin Path"
4. Timeline: "Zinzendorf's Pennsylvania Journey"
5. Interactive Sound Map: "Zinzendorf's Journey to Shamokin and Wyoming"

Chapter 2: Friends & Strangers

1. Interactive Sound Map: "Bethlehem in 1758"
2. Interactive and Static Maps: "Sound Boundaries of Bethlehem"

Chapter 3: Sound & Spirit

1. Sound Recordings: "Mohican-Moravian *Singstunde*"
2. Soundscape Recording: "Gnadenhütten, Pennsylvania"
3. Interactive Map: "Spiritual Singing in Bethlehem"

Chapter 4: 1782

1. Interactive Map: "The Pennsylvania Frontier"
2. Interactive Sound Map: "Journeys of the Native *Gemeine*, 1763–1772"
3. Interactive Map: "The Journey of the Native *Gemeine* from Friedenshütten II to Friedenstadt"
4. Static Map: "The Ohio Country, 1782"
5. Historic Document and Map Collection: "Ohio"

ACKNOWLEDGMENTS

Books are journeys. Their narrative maps have the potential to transport us to new places and perspectives. Sometimes, they allow us to slip deftly between the past, present, and future, the unexpected or the familiar. They can disappoint or delight. Regardless, these are journeys we undertake in fellowship with other travelers—our readers, our research subjects, our families and friends, our professional colleagues and students, and the institutions and organizations that support our work. In writing this book, I have come to value the act of journeying and the transhistorical, storytelling work of history. I have come to appreciate that our work as historians is informed by how we understand and honor the relationships that bind us together in the past and the present. These are networks as wide as the Atlantic and as intimate as "home." So, it is with great pleasure that I thank my fellow travelers. I am grateful to have undertaken this journey with you.

To my father, Raymond Eyerly, and my mother, Mary Ann Erickson Eyerly. It was your passion for Pennsylvania's history, environment, and people that inspired this book.

To my husband, Andy Nathan, and my children, Jesse and Ian. You have patiently endured this long journey with love and devotion. Thank you.

To my wonderful research assistants at the Florida State University: Mark Sciuchetti, Miranda Penley, Rebekah Franklin, Laura Clapper, Rachel Bani, Alexandra Taggart, and Joseph Cramer. The countless hours you have contributed to this project have enriched it beyond what I could have ever hoped to achieve alone. And to the many students at FSU who have shared their ideas and creativity—I owe you a debt of gratitude.

To my research collaborator, Rachel Wheeler. Thank you for sharing the experiences of motherhood and academic life, and for a friendship that carries far beyond our shared interest in the history of the Moravian missions.

To my musicology colleagues at FSU: Valerie Arsenault, Michael Bakan, Charlie Brewer, Michael Broyles, Laura Gayle Green, Frank Gunderson, Margaret Jackson, Panayotis League, Eduardo Lopez-Dabdoub, Douglass Seaton, and Denise Von Glahn. You are an inspiring and extraordinary community of scholars and I am grateful to have undertaken this journey in your company. To Debbie Whitaker, whose wise council helped me through the darkest of times. Thank you for believing that I could finish this book. And many thanks to Dean Patricia Flowers of the FSU College of Music for her unwavering support of my scholarship.

To the many scholars and archivists who have assisted with this project: Paul Peucker and Thomas McCullough of the Moravian Archives in Bethlehem; Nola Reed Knouse, Gwyneth Michel, and David Blum of the Moravian Music Foundation in Bethlehem and Winston-Salem; Charlene Donchez Mowers, Tavia Minnich, and Philip Trabel of the Historic Bethlehem Partnership; Olaf Nippe and Claudia Mai of the Unity Archives of the Moravian Church in Herrnhut, Germany; Kaitlyn Pettengill of the Historical Society of Pennsylvania; Aaron McWilliams of the Pennsylvania Historical and Museum Commission and the Pennsylvania State Archives; and Janet Johnson, Curator of Archaeology at the State Museum of Pennsylvania, I am extraordinarily grateful to all of you for your assistance. To Katie Faull, Patrick Erben, Paul Peucker, Olivia Bloechl, John Corrigan, David Bodenhamer, and Rachel Wheeler. Your generosity of intellect and time have shaped this book in many unexpected and beautiful ways.

To my wonderful editorial team at the Indiana University Press: series editors Denise Von Glahn and Sabine Feisst, acquisition editors Janice Frisch and Allison Chaplin, executive director Gary Dunham, project managers Darja Malcolm-Clarke and Carol McGillivray, copyeditor Ann Aubrey Hanson, indexer Kathy Bennett, and marketing and publicity manager Rachel Rosolina, thank you for believing in this project and shepherding it toward publication. And to the Press's anonymous reviewers, thank you for generously donating your time to this project, and for providing careful readings and thoughtful critiques that helped to strengthen the manuscript.

To the institutions and organizations who tangibly supported the development of the book and website. This project would not have been possible without generous funding from the American Council of Learned Societies (Collaborative Research Fellowship), the Society for American Music (Sight & Sound Subvention, H. Earle Johnson Publication Subvention), the American Musicological Society (publication subventions from the AMS 75 Publication Awards for Younger Scholars Fund, supported in part by the National Endowment for the Humanities and the Andrew W. Mellon Foundation), the Council for Research and Creativity at FSU (First Year Assistant Professor Award, Planning Grant), and the Lucille P. and Elbert B. Shelfer Professorship in Music.

NOTE ON NAMING, TERMINOLOGY, AND ARCHIVAL SOURCES

Naming and Terminology

Native, Native American, Native Moravian: Tribal names have been used whenever possible in this book and on the companion website, or "Native" when referring to individuals or groups of multiple tribal heritages. Due to the particular pressures of colonization, the Moravian missions were home to people from various geographic locations and tribal affiliations. Therefore, I have also adopted the term "Native Moravian" to refer broadly to Indigenous people who affiliated with the Moravian Church.

European, European Moravian: I have chosen to use the terms "European" or "European Moravian" in recognition of the fact that the Moravian missions, and also Pennsylvania in general, were home to people from various parts of continental Europe and Britain. Since this study involves multiple ethnic groups, I have used terms such as English Moravian or German Moravian only where it applies to particular people or groups.

Moravian Church: In the eighteenth century, the Moravian Church had many names: Unitas Fratrum, *Brüdergemeine*, Ancient Unity, *erneuerte Brüdergemeine*, Herrnhuter, and the Brethren's Congregation, among others. For the purposes of this study, I have adopted the common English language name, Moravian. It is also important to note that the term "Moravian" was a spiritual designation in the eighteenth century, and did not denote a particular race, class, or geographic origin.

Pennsylvania or *Penn's Woods*: There is not a single Native word for the region of North America that would become the colony of Pennsylvania in the seventeenth century, so indigenous toponyms have been used when possible for local geographies, and the process of naming is discussed throughout the book and in the accompanying maps as a feature of colonization and decolonization.

Archival Sources

Primary source materials from the Moravian Archives in Bethlehem, Pennsylvania, and the Unity Archives of the Moravian Church in Herrnhut,

Germany, are the backbone of this study. It is my intention to present these sources as authentically as possible. Quotes preserve the original spelling, errors, and eighteenth-century conventions. I have avoided "*sic*," due to the sheer number of anomalies deemed incorrect by modern English or German standards. For instance, the written form of German permitted capitalization of the first two letters of words such as *HErrn* or *JEsu*, or capitalization of entire words, to add emphasis. All translations were prepared specifically for this study, unless attribution is given.

All pictures and photographs of archival materials are printed with permission from the Unity Archives of the Moravian Church, Herrnhut, Germany; the Moravian Archives, Bethlehem, PA; the Moravian Historical Society, Nazareth, PA; and the Historical Society of Pennsylvania. The abbreviation "UA" refers to the Unity Archives of the Moravian Church in Herrnhut; "MAB" refers to the Moravian Archives in Bethlehem. The conclusions of this study apply principally to the American communities of the Moravian Church. For the purposes of this study, I did not review archival records from mission communities outside of North America and Germany.

Additionally, I have sometimes used newer versions of spelling and orthography for the Mohican language, where these differ from Moravian missionary transcriptions, according to current language revival guidelines being adopted by a descendant community on the Stockbridge–Munsee Reservation in Wisconsin.

MORAVIAN SOUNDSCAPES

PROLOGUE

The Pennsylvania Wilds

IN THE EARLY SUMMER OF 1794, MY ANCESTOR, the German missionary Johann Jacob Eyerly Jr., walked across Pennsylvania via the Allegheny, Raystown, and Venango Trails.[1] His journey took him from the Delaware and Mohican mission town of Bethlehem, founded by the Moravian Church along the eastern border of European settlements between the Susquehanna and Delaware Rivers, to Fort Pitt in the west. From there, he journeyed northward to survey Native lands granted to the Moravian Church by the United States government in the early 1790s along the shores of Lake Erie at Presqueisle. As he walked through the forest to Presqueisle and back, he kept a diary.[2] He wrote about the massive trunks of chestnut trees, six or seven feet in diameter and of "an amazing height," towering over the forest floor. Lower in the canopy he observed shellbark, hickory, black and white oak, beech, maple, poplar, sugar maple, and ash, spreading their tangled boughs in a dense cathedral-like ceiling that blocked the undergrowth. In the deep thickets near French Creek—in the Shawnee territory surrounding Fort Le Boeuf—he was struck by the richness of the soil, and the variety of trees and plants: "[this] is very good rich land, with many clearings where, from all appearances, the Indians used to dwell. Where these bottoms are not cleared, they are densely overgrown with White walnut, wild cherries, and the like. I have seen hawthorns here that were from 12 to 15 inches in diameter. There are all sorts of trees on the uplands."[3] He traveled through "woods and glades, wading through streams and through grass half as high as a man."[4] The path, he wrote, was difficult to locate, especially after a rainstorm. Often, he oriented himself by the sun or by sound: listening for the distant roar of the Susquehanna or the Juniata River, the lapping of waves on the shores of Lake Erie, or the silent spaces that signaled the densest parts of the forest. Along the way, he paused to study some of the unknown plants that grew along the trailside, noting the "sassaparill, ginseng, and nettles [that] grow here in abundance, large and juicy."[5] Most of all, he chronicled the sheer human endurance required for the journey—it was so wet that his clothing began to rot off of his back—and his joy at hearing again the sounds of his home community. As he made his return journey, he listened intently for the soundscapes of Bethlehem, heard but not seen through the forest.

It was this chronicle of how my ancestor experienced the landscapes and soundscapes of early Pennsylvania that first inspired me to write this book. I was fascinated by his detailed descriptions of plants and trees, and by the fact that he listened so carefully to the acoustic ecology of the forest that surrounded him. He wrote of Pennsylvania's natural environment with such joy that I could not help but envision a man who walked through the forest with his eyes open and ears unstopped. Like my ancestor, I grew up in Pennsylvania, or Penn's Woods. The farm where I spent my childhood was located in a sparsely inhabited part of the state known as "the Pennsylvania Wilds"—a two-million acre tract of forest that is home to numerous stands of Longfellow pines, the tallest trees in the eastern United States, and some of the only remaining areas of virgin forest in the mid-Atlantic. As a marketing website for the region proudly states, this is a place with night skies so dark that "the Milky Way casts a shadow," and where travelers and residents alike can be rejuvenated by "crisp, mountain air," "nature," and "thousands of miles of forest trails."[6]

The bookshelves of our house sported an array of Audubon field guides to Pennsylvania's plants, trees, insects, birds, reptiles, and amphibians. I spent many summer afternoons in the fields and forests near my house, trying to puzzle out the names of particular ferns or hardwoods, or looking vainly under logs for elusive salamanders with red, white, and brown spots. I clearly remember the day when I first heard the song of a male wood thrush in the marshes near the Moshannon Creek and Moravian Run.[7] As it had Henry David Thoreau, the thrush's song struck me as one of the most musical sounds I had ever heard: "Whenever a man hears it [the wood thrush] he is young, and Nature is in her spring; wherever he hears it, it is a new world and a free country, and the gates of Heaven are not shut against him. . . . The thrush alone declares the immortal wealth and vigor that is in the forest."[8] Listening to the vibrant songs of the wood thrush and the other birds that populated the nearby meadows and woods, I learned to orient myself by sound to the environment around my home. But, in the sparsely populated spaces of the Pennsylvania Wilds, it was also possible to hear and to respond to quieter sounds: wind, water, insects, and the rustling of trees and grasses mixed with the faraway sounds of vehicles and homes.

These sounds oriented me to the structures of forest and farmland, imparting a sense of meaning, home, and place within what might otherwise have been a formidable tract of wild land. It was possible to get lost in these wild spaces, so my mother installed an iron bell on the corner of our house. Its ringing signaled dinnertime to me and to the horses and sheep who roamed on the pastures to the east of our house, out of sight of barns and buildings. Within the sound of the bell was home. Like my ancestor, I listened intently for those sounds of human place and geography heard but not seen through the forest.

* * *

Almost twenty years after I moved away from my family's farm near the Moshannon Creek and Moravian Run in Cooper Township, Pennsylvania, I embarked on the writing of this book. I can only describe the process itself as an unexpected journey that has caused me to consider deeply the imaginative and storytelling work of history, and the ways in which we are influenced by networks of material objects, manuscripts, sounds, people, and places that connect us with our research subjects in the past and the present. In June 2017, as I was preparing the first draft of the manuscript and simultaneously preparing for my mother's funeral, I found a picture of her as a teenager. It was taken at our family's farm, her childhood home, which I happened to be writing about in this book. In the picture, she was smiling radiantly while leading her pet sheep on a leash. In the background, I could see the corner of the house and the farm fields behind it. In the distance rose the edge of the Allegheny Front, falling away toward the deep cleft of the Moshannon Creek where it joined Moravian Run. My mother looked perfectly at home. I sensed that the picture had somehow captured a deeply revealing image of her as not just the woman who had loved and cared for me for over forty years, but also the gentle person for whom the beauty and solace of life were centered in that place. I thought about the times we sat quietly together on the porch of our home, just looking at the leaves as they turned in the wind, or observing the flight of bluebirds along the fence rows, or listening to the choruses of small frogs in nearby puddles and ponds as they rejoiced in the spring and the thawing of the once-hardened earth.

My mother loved that place—the farm and plot of land where she had many happy childhood memories. Caring for it was one of her greatest joys. But like many things, that joy came with great sadness when she had to leave it, and when she could no longer care for the place she loved in the silent, steadfast way that she valued most of all. Toward the end of her life, her body was almost completely immobilized, and she struggled to find new meaning for a life that had been spent in doing, rather than in imagining. For the past three years, she had done what she could. She watched the birds at the feeder outside of her window at the nursing home, and she looked at the pictures of her grandsons and thought of all of the things they might someday accomplish in life. And she talked with my dad and me on the phone, many conversations both joyful and heartbreaking. And then, she lost even those things. She was no longer able to make a phone call, and the narcotics she had been given to ease the pain of two broken legs took away her mind. At the very end, she returned to the place she had always loved—in her mind she came back to her home. When I visited her and sat by her bedside at the nursing home, her thoughts were far away at the edge of the Moshannon and the Moravian.

As my mother's life ebbed away, I struggled to finish this book. The gradual loss of my mother coupled with the loss of the farm that I had also deeply loved made the prospect of writing an academic book seem remote and almost unimaginable. On March 17, 2017, I made the painful decision to sell the farm. The rights to use that land passed to a new family. And, on June 10, my mother passed away quietly in her sleep. As I struggled to cope with what seemed like interminable loss and little gain, I wondered how I could possibly write this book. Surprisingly, I found that creating the book became an unforeseen journey of healing and discovery—the telling of two stories, separate but intimately intertwined. Through the writings of my Moravian ancestor, Johann Jacob Eyerly, and other songs, diaries, maps, and community records kept and maintained by the Moravian Church since the eighteenth century, I gained the historical perspective necessary to craft the narrative of this book. What I did not sense was that in crafting that narrative, in studying those documents, in seeking to hear those sounds again, I would also come to a deeper understanding of the history of my own family and the very place that my mother and I had both loved. In telling that history, I discovered my own.

Writing this book has also challenged me to consider how my own family participated in the history of the Moravian missions in the eighteenth century. It has challenged me to consider what it means to be a descendant of a European Moravian. History is never objective. It is dependent on personal experience, and individual and collective interpretations of written and musical sources, material culture, land usage and property rights, and the legacy of human encounters over many generations. Understanding the legacy of my own ancestors involved getting out into Pennsylvania's places and understanding them as place-worlds that held meaning in the present and in the past. It involved learning about historic musical traditions, songs, and soundscapes, and why those sounds had mattered to the people who made or heard them. It involved coming to terms with stories of heartbreak, terrible pain, and loss, and acknowledging stories of hope and survivance. In the process, I learned the many histories of the place my family called "home."

Every farm, every community, every town and city in Pennsylvania is built on indigenous lands. Pennsylvania's places have long and complex living histories embedded with relationships and networks, memories, sounds, and ideas. It was important to me to consider how I would deal with these histories as a scholar descended from European settlers. How would I address the legacies of colonialism in the history of Pennsylvania, in my own home community, and indeed within my own family? The work we do as historians can have ramifications that may affect Native communities who have a stake in these histories. We are all bound by relationships and networks, by territories, by treaties, and by the actions of the past. I have been especially inspired by the work of Lisa Brooks and Daniel Heath Justice to create a new narrative of Pennsylvania's history, and indeed of Pennsylvania's soundscapes, that acknowledges past legacies and traumas, and that moves forward to chart new stories that are inclusive and honest. I cannot claim to have absolutely accomplished that task in this book, but as Justice has argued, the process of scholarship is permissive of many paths of inquiry and allows for a process of "becoming" rather than simply "arriving."[9] This book is not a final answer, but a journey to understand Pennsylvania in the past and the present through the auditory experience of sound.

The journey of writing the book has also simultaneously been a process of coming to terms with the grief of losing my mother and the childhood home that we shared. My mother and I were not the first to love the land at the confluence of the Moravian and the Moshannon, and we would not be the last. With that knowledge came the first signs of healing. Rather than a final end to the story of my own family's relationship with this place, I sensed a narrative that stretched compellingly into the past and into the present and future—a narrative that was embedded within the many stories and songs that had been and would be told in this place. Where I hoped to perceive traces of soundscapes long lost, I found instead the ongoing and ever-changing soundscapes of the Pennsylvania Wilds.

Notes

1. For a discussion of the complex network of trails that Eyerly followed, see Paul A. W. Wallace, *Indian Paths of Pennsylvania* (Harrisburg, PA: Commonwealth of Pennsylvania, Pennsylvania Historical and Museum Commission, 2005).

2. Johann Jacob Eyerly, Jr., "Ein Bericht von der Reise der Brüder Jacob Eyerly jun. und Johann Heckewälder zur Ausmessung des Landes am Lake Erie, welches von der General Assembly in Pensylvanien der Societät der Brüder zur Ausbreitung des Evangelii unter den Heyden geschenkt worde, im May und Juny 1794 [A report of a trip taken by Brothers Jacob Eyerly, Jr. and Johann Heckewälder to survey the lands on Lake Erie which the Pennsylvania General Assembly gave the Society of Brethren to spread the gospel among the heathen, in May and June of 1794]," Records of the Moravian Missions to the American Indians (hereafter cited as MissInd) 213.10, Moravian Archives, Bethlehem (hereafter cited as MAB). Translated as Wallace, "Jacob Eyerly's Journal, 1794: The Survey of Moravian Lands in the Erie Triangle," *Western Pennsylvania Historical Magazine* 45, no. 1 (March 1962): 5–23.

3. Wallace, "Jacob Eyerly's Journal, 1794," 21.

4. Wallace, "Jacob Eyerly's Journal, 1794," 19.

5. Wallace, "Jacob Eyerly's Journal, 1794," 20–21.

6. Pennsylvania WILDS, www.pawilds.com, accessed May 1, 2017.

7. The modern name of Moshannon Creek is likely a derivative form of an earlier Delaware name, Mos-hanna-nk (Elk River Place).

8. Henry David Thoreau, *Thoreau On Birds: Notes on New England Birds from the Journals of Henry David Thoreau*, ed. Francis H. Allen (Boston: Beacon Press, 1993), 429.

9. Daniel Heath Justice, *Why Indigenous Literatures Matter* (Waterloo, Ontario, Canada: Wilfrid Laurier University Press), 33–41.

INTRODUCTION

Sounding New Histories of the Moravian Missions

Denk an sie und ihre müh, Heiland,	Think of them and their efforts, O Lord,
sie haben den rechten paß.	For they have the true knowledge.
Wo sie gehn, laß gnade wehn,	Wherever they go, let your mercy flow,
und der verklä[r]ger verliere was.	And those who complain will lose.
In sankt Thomas und Barbies,	In Saint Thomas and Barbados,
Capo, Ceylon, Acra, Crüs,	Capo, Ceylon, Accra, St. Croix,
Pensilvanien, Algier, Grönland,	Pennsylvania, Algiers, Greenland,
Surinam, und hier.[1]	Suriname, and here.

THE WINTER OF 1782 WAS PARTICULARLY BITTER. At the Moravian mission of Gnadenhütten along the Tuscarawas River in the Ohio Country, little remained of the carefully planted corn crops from the previous summer. Wind blew incessantly through the missions' abandoned fields and homes, flattening the cornstalks and whistling through the cracks of walls. Frost lay thick on gardens, roofs, and pathways. The well was frozen solid. But in early March, more than one hundred Delaware and Mohican Moravians returned to the mission after their forced removal in the fall of 1781 by British and Wyandot soldiers.[2] Throughout the winter, the Native Moravian congregation had lingered in captivity along the Sandusky River in the Wyandot-controlled Captives' Town. The German missionaries and three Native elders who served their community were imprisoned by the British at Fort Detroit, suspected of sending military intelligence to the Americans at Fort Pitt. Caught on the western boundary of the American Revolution, a global conflict that stretched from Ohio to India, the Moravians were left to eke out a meager existence far from their agricultural plantations on the Tuscarawas. After several months of near starvation, the desperate refugees risked a return to Gnadenhütten in early March to harvest what little remained of their corn.

As the Moravians worked in the cornfields near Gnadenhütten on the morning of March 7, there was little conversation. Ice crystals filled the air, refracting and amplifying the sounds of the wind. Overhead, red-tailed hawks, mourning doves, and black-capped chickadees spun and called, eager for corn kernels or insects frightened in the passage of the harvesters. The river ran rough despite the ice blocks that littered its surface, and the splashes of otters and muskrats mixed with the dampened biophony of water striders, sauger, bass, and walleye. Flathead catfish and crawfish dredged the riverbed. The reeds rustled in the damp marshes on the river's bank. The harvesters did not hear the militia coming. Despite the size of the company, 160 men from Washington County, Pennsylvania, emerged unheard from the forest. Their muster had been sounded to raze Gnadenhütten and its sister missions of Schönbrunn and Salem to the ground in retaliation for Delaware and Wyandot attacks and the killing of Scots-Irish settlers along the Pennsylvania–Ohio border. No longer would the Moravians be allowed to trade secrets with the Delaware chief, Netawatwees, at Gekelmukpechunk (Newcomerstown). Although the militiamen likely knew that the Delaware and Mohican Moravians were not responsible for the recent raids on Scots-Irish homesteads, their desire for retribution was strong. For them, the stakes were high—the very safety and security of the homes and farmsteads of Washington County and other settler communities in western Pennsylvania.

By the evening of March 7, the little mission settlement of Gnadenhütten resounded with a cacophony of voices in English, German, Delaware, and Mohican. This clamor of tongues was recorded in diary records, letters, and conversations for decades to come: the songs, speeches, and prayers of the Moravians as they begged for mercy from their captors; and the arguments of their captors as they debated whether to take the Moravians as hostages to Fort Pitt, or to kill them and burn their villages. In the end, the most vocal militiamen persuaded all but eighteen of their number to herd the Moravians into "killing houses." The following morning, twenty-eight men, twenty-nine women, and thirty-nine children were bludgeoned, tomahawked, scalped, and burned. No one cared to write down or remember the terrible thud of a cooper's mallet on bone, or the dripping of blood through floorboards; the murmurs and cries of the dying, crushed together in piles in the mission buildings; or the heavy breathing of the Pennsylvanians as they methodically killed ninety-six unarmed people. But some sounds were remembered: the weeping of Nathan Rollins as he tomahawked nineteen people and still felt no relief from having lost his father and brother to Wyandot raiders. The sounds of the dying Moravians subsumed into the crackling fires of burning buildings and the drunken yells of the militia, as remembered by two boys, Jacob and Thomas, who escaped. Militiaman Obadiah Holmes would later write of the quiet remorse of the eighteen abstaining Pennsylvanians who sat huddled on the riverbank, as far away from the atrocities as they dared to creep.[3] And Jacob and Thomas would later recount to the missionary David Zeisberger at Sandusky the jubilant shouts of those who spent the night singing and telling stories about the destruction of the mission and the justice they had exacted from the Moravians.[4]

For those who escaped, perpetrated, or heard tales of the massacre in the days, months, and years following "Gnadenhütten," it was the sounds of that day that were especially remembered. As the years passed, and the horrors of the massacre dimmed, smaller and perhaps lesser sounds were forgotten. But the singing of the Moravian Christians as they prepared for death lingered persistently in legends of the massacre. When the first reports of the event were sent to the North American headquarters of the Moravian Church at Bethlehem, Pennsylvania, they all related a sonic story of the martyrdom of the Delaware and Mohican Moravians, who had "*die ganze Nacht Hymns u. Psalms gesungen* [sung hymns and psalms for the whole night]."[5] It was through their songs that the Native martyrs had achieved, in the Moravian cosmology, a good death. Their sung prayers had allowed them passage from the mortal world and their own physical suffering into the spiritual realms beyond all pain. The fact that the Native martyrs spent their final hours singing Christian hymns, as was customary in German Moravian communities at the point of death, was proof of their sincere adherence to Christianity, in the opinion of church elders. Zeisberger was certain that the Native Moravians had died as true Christians, since they "began to sing hymns and spoke words of encouragement to another until they were all slain."[6] Even in 1792, ten years after the massacre, when the missionary Johann Heckewelder lodged for the evening in Washington County, he found that the Scots-Irish communities where many of the perpetrators still lived remembered the hymns of the Moravians: "I . . . was invited to supper by Mr. van Sweringen, Esq. The good man spoke about the massacre of our Indians, threw his hands together over his head & said: 'I have heard from the lips of the murderers themselves that they killed them while they were praying, singing, and kissing,' & he was not surprised that . . . great blood-guilt lay upon the land and must be atoned for.'"[7]

For the lawyer, Mr. van Sweringen, and others in his community, it was not just the murders that had brought "great blood guilt" upon their communities. Gnadenhütten was not the first or the only massacre to occur during the American Revolution. However, it was set apart from similar atrocities for one reason: the perpetrators had refused to recognize the distinction between Christian and non-Christian Native Americans.[8] Their urgent desire to extact retribution for their own murdered family members and friends had impelled them to heedlessly kill innocent men, women, and children even as they "began to sing hymns and spoke words of encouragement and consolation to each other."[9] Despite the European-style houses and spatial design of the Ohio missions, and the prospect of Native Christians dressed in very much the same manner as the members of the militia, the Pennsylvanians were not interested in believing these were "peaceful Indians." In the years following the massacre, both the fledgling United States government and nativist movements among Native communities in Ohio and Indiana would agree that the blood guilt of the murders at Gnadenhütten rested squarely on the fact that Christians had killed Christians. Two decades later, the Shawnee chief Tecumseh would remind future president William Henry Harrison: "You recall the time when the Jesus Indians of the Delawares lived near the Americans, and had confidence in their promises of friendship, and thought they were secure, yet the Americans murdered all the men, women, and children, even as they prayed to Jesus?"[10] The truth was that the Christian songs and prayers of the Moravians had not saved them from death or tremendous suffering.

Song and Sound in the Moravian Missions

The massacre at Gnadenhütten raises a number of difficult, but important, issues. The ninety-six people who perished there died partly as a result of their tenuous existence at the boundaries between cultures, ethnicities, nations, and religions. What had begun in the 1740s as an attempt by German missionaries to create new multicultural Christian utopias in New York, Connecticut, and Pennsylvania had ended tragically in the death of many of these same Delaware and Mohican Moravians along the Ohio frontier. Although singing had for a brief time in the early history of the Moravian missions created a space for exchange of spiritual and cultural ideas between Delaware and Mohican communities and Moravian missionaries, ironically it was those very hymns that would linger most powerfully in memories of the massacre.

Working forward in time from the founding of the first Moravian settler community at Bethlehem, Pennsylvania, in 1741, to the massacre at Gnadenhütten, Ohio, in 1782, this book positions song and sound at the center of interactions between German Moravian missionaries and Native communities in eastern North America. It is no coincidence that the singing of the Native Moravian congregation at Gnadenhütten persisted in memories of the massacre. Hymns were central not only to the Moravians' missionary philosophies, but also to daily Christian practice and life-ways in mission communities. Hymns and rituals involving singing served as sonic markers of history, place, and identity. The role of hymns in eighteenth-century Moravian life accomplished what Gary Tomlinson has termed "songwork," which he defines as the place and efficacy of song in given societal circumstances.[11] Moravian hymns did cultural work. They were sung at weddings, funerals, and baptisms; they accompanied manual labor; they served as forms of greeting and celebration; they comforted the sick and dying; and they regulated personal mental health. Hymn singing was not confined to the sacred space of a worship hall, but integrated into daily life.[12] Moravians envisioned hymns as powerful tools to integrate people into their communities and to communicate core Moravian values both inside and outside of their communities. Study of Moravian hymnody yields a deeper understanding not only of the relationships between missionaries and Native Christians, but also of the connections of Moravian mission communities with the wider world they inhabited.[13]

Studying hymnody as it is embedded within historical cultures of hearing and listening is important to understanding concepts of social and religious identity and place both for European and Native Moravians. Just as religious experiences so often happened through ordinary day-to-day, person-to-person exchanges, experiences of sound were similarly intimate and heard within the confines of a meeting space, worship hall, or bedroom, or were bounded by the wider soundscapes of communities or the acoustic ecologies of the natural environment. This is as true of Moravian missions as it is of other historic and modern communities. In the context of a time period in American history when the border between settler and Native communities and nations was a shifting spatial and cultural space, sound mattered. People listened carefully to each other and the world around them. These cultures of hearing and listening encompassed and also went beyond musical traditions such as song and hymnody. The natural and human environments of early Pennsylvania were comprised of complex biophonic, geophonic, and anthrophonic acoustic soundscapes. Study of these acoustic environments is important to understanding the social, religious, and spatial relationships that characterized life in both Native and settler communities. The closer we can come to comprehending how early Americans heard their world, the closer we will be to critically understanding not only the history of the Moravian missions but also the difficult and often violent histories of the emergence of the modern American nation on Native soil during the Seven Years' War (1756–1763) and the American Revolution (1775–1783).[14]

Yet, despite the importance of sound and song to the Moravians, there have been no comprehensive attempts to study their mission communities and missionary practices from that perspective. While there is a vast and growing literature on the Moravian missions and encounters between Indigenous peoples and missionaries in early America, only recently have scholars begun to incorporate information about musical practices in Moravian mission contexts.[15] Of particular note are studies by Walter Woodward, as well as Rachel Wheeler and me, on the indigenization of Moravian hymnody and the role of song as a mediator of cultural interactions between Moravian missionaries and Native Christians.[16] These studies also intersect with recent publications on music and Christian missions in colonial contexts more broadly, including Glenda Goodman's work on Native-language psalmody in New England and the soundscapes of colonial encounters, and studies by Kristin Dutcher Mann and Geoffrey Baker on music in Spanish mission contexts.[17] Although music and sound are often relegated to the margins of history, or remain under the purview of musicological inquiries, sensory perceptions and cultural practices surrounding sound are important ways of understanding Indigenous responses to colonialism. This is especially true because hymns and other musical forms often embed traces of Native agency even when contained within the archival records of settler communities that are often dominated by non-Native voices.[18]

Recovering the sonic history of the Moravian missions also restores a part of American history that is often overlooked. In early America, sounds and silences possessed the power to unite and divide, to produce understandings and misunderstandings, and to constitute adaptive or destructive strategies for navigating an unprecedented period of cultural shift and physical copresence between European settlers and Native nations and communities. Studies by Richard Cullen Rath, Peter Charles Hoffer, Geoffrey Baker, and Sarah Keyes have demonstrated that colonial efforts to remodel the landscapes of the Americas after traditional European settlement patterns also included transformation of the soundscapes, or aural landscapes, of the Americas to resemble the familiar soundscapes of European places.[19] As Sarah Keyes has argued, building fences, felling trees, and effecting other physical changes to particular ecological environments went hand in hand with transforming the American landscape into a civilized soundscape of ringing axes, lowing cattle, and clanging bells. Indigenous peoples reacted against this physical and aural encroachment, and the sounds of ritualized speech and music became crucial in encounters between colonial settlers and Native communities.[20]

Soundscapes also defined the nature of both settler and Native communities and their geographic boundaries. The ability to control the aural dimension of landscapes, places, and communities was inextricably intertwined with struggles for power and access to natural resources. When Alexis de Tocqueville toured America to observe new systems of government, he also observed a process of sonic colonization of America's natural environment, including its fauna: "As soon as a European settlement forms in the neighborhood of territory occupied by the Indians wild game takes fright. Thousands of savages wandering in the forest without fixed dwelling did not disturb it; but as soon as the continuous noise of European labor is heard in the vicinity, it begins to flee and retreat toward the west, where some instinct teaches it that it will find limitless wilderness."[21] This sonic space of colonization, according to de Tocqueville, stretched for almost two hundred miles west of continually advancing eighteenth-century colonial settlement.

Moravian communal, environmental, and religious soundscapes, and their attendant hymn traditions, can be understood as colonial structures that attempted to standardize, indeed to colonize, indigenous soundscapes, musical practices, and religious traditions. Moravian Christianity and the Moravian missions were intertwined with the process of colonial settlement, in the way that Matthew Hunter Price has framed Methodism as a perpetuator of colonial settler networks.[22] Religious networks, such as the transatlantic missionary enterprise of the Moravian Church, were used for the economic and social gain of the worldwide Moravian Church. While these mission networks sometimes advanced the purposes of Native Christians in surviving the damaging effects of colonization, European Moravians and the church government certainly also received distinct economic advantages from their connections with the British and Danish empires that were not easily accessible to Native Christians, or which harmed them. The church purchased lands taken from Indigenous peoples through these colonial networks in places as geographically diverse as Greenland and Suriname, and thrived commercially on a global scale. In Pennsylvania, the Moravians acquired the land to build Bethlehem as a consequence of the Walking Purchase—the deceitful stripping and repurposing of Delaware traditional lands by the colonial government in Philadelphia. As George Tinker has shown, even with good intentions, missionary encounters often equated or presaged acts of cultural genocide and forced relocations of Native people.[23]

However, it is also important to note that there has never been a consistent or monolithic Christian missionary practice. Rather, missions are entirely dependent on the particular Christian sect and also the local cultural, social, and political context in which the missions operate. In the case of the Moravian missions, if we leave the narrative at the point of cultural genocide and seizure of Native lands, we may risk discounting the ways that individuals and indeed whole Native communities adapted and found meaning in new and changing traditions and soundscapes, despite the imposed structures of the Moravian Church or colonial agendas. Although Moravian hymn singing and the soundscapes of mission communities were a form of colonialism, on an individual and community level, people were modifying hymns and adapting them. Native Moravians were not passive actors enmeshed in colonial processes. In studying the Moravian missions, we might take some inspiration from Sarah Rivett's reexamination of missionary transcriptions of Native American languages. Rather than emphasizing a process of language erasure, Rivett has sought to highlight the adaptive power and survivance of Native languages. Even as missionaries sought to convey Christian theology through new linguistic mediums, Native languages often resisted simple acts of translation, instead preserving and encoding different theologies and religious worldviews in their very structures and grammars. While many Native American languages ceased to be spoken languages during several centuries of colonial contact, Rivett argues that the essential grammatical elements of those languages survived in missionary transcriptions to encode the cultures and religions they represented.[24]

So, what did the soundscapes of Moravian missions encode? Did Moravian hymns embed the survivance of cultural ways? Did these hymns represent the concerns of Moravians from Delaware and Mohican backgrounds, as well as from German backgrounds? Both Native and European Moravians accessed Christianity through music and ritual. It should not be assumed, though, that Native Christians accepted or even simply mimicked German Moravian hymns and rituals under duress or because of Moravian impositions. Music, hymns, processions, choirs, feast days and celebrations, baptisms, and other Christian rituals certainly created spiritual connections and fostered understandings of commonality, but as Jane Merritt has argued, "Moravian theology [also] influenced the development of a distinctive native Christian religion."[25] Recent scholarship on the history of Christian missions in the eighteenth-century Northeast has turned toward studying Native Christianity not from an either-or-paradigm of conversion versus nonconversion, but from a perspective that seeks to uncover how Native peoples appropriated elements of Christian theology or practice while also negotiating the drastic changes in their communities and the natural environment caused by colonization. This type of approach places Native stories at the center, rather than European American conceptions of religious conversion or environmental transformation.[26] In Moravian communities, Native Christians were active in both music production and performance. They became vocalists and instrumentalists, they learned to build musical instruments, and they copied music. As they helped to create and shape the musical repertories of their communities, they added their own touches to musical manuscripts, instruments, and compositions. While missionaries may have simply desired to preach the Gospel through their own particular style of sung Christian community articulated through hymns, the process of becoming Moravian allowed Native Christians considerable space to develop indigenized forms of Christianity and music-making.[27]

The legacy and history of Moravian hymnody is also the legacy of a musical tradition that represented Native culture and the value of adaptation and the building of relationships and cultural ties even in the face of colonialism. This process can be seen in many different modern-day Native-language hymn traditions, such as the Catholic and Protestant hymns of the Anishinaabe and Kiowa, or the adapted Christian repertory of Inuit Moravians. Although in both instances, these musics were originally introduced by missionaries as a strategy to extinguish Native music and worship, for many Anishinaabe and Inuit people today, singing these pieces in Native languages, and adapting their performing practices, has become a way to maintain traditions and a sense of communal integrity in the face of rapid globalization and cultural instability.[28] As Native scholar Lisa Brooks has commented on historic traditions of Native Christianity in New England:

> In many indigenous communities, the practice of Christianity in Native New England was syncretic, combining indigenous and European spiritual practices, taking on its own character in relation to particular brands and movements of Christianity, and becoming a staple of life for many families, thus part of the fabric of communal identity and history. Now, we might not *like* that so many of our ancestors sought refuge in Christianity, and we may be able to see clearly in retrospect the damaging impact of such choices, but we should not deny our own histories and what we might learn from them or fall into the illusion that those choices made them somehow less Indian.[29]

Both Brooks and Native literary scholar Craig Womack argue for the inclusion of possibilities, such as Christian piety and knowledge, as ways to expand our historical understandings of Native individuals and communities, instead of limiting our understandings of Native experience.[30] Christianity was intertwined with colonization. But the story does not end there. In the case of the Moravian missions, the particular story of both Native and European attempts to create what Juanita Little, a Mescalero Apache Catholic sister, has described in the present day as a "world-wide [Christian] family," became intertwined in the eighteenth century with the geographic position of mission communities on the relentless frontier boundary of European colonization and Native settlements.[31] As Daniel Richter, James Merrell, and Katherine Faull have argued, early Pennsylvania was a place where it was still possible for shared communities to exist, and where Native and settler interests were not mutually exclusive, but still in flux. The strict marking of people by racial categories was a consequence of the eighteenth-century wars that shattered both settler and Native communities. This book tells the story of a Pennsylvania that existed before, during, and following those wars. The history of the Moravian missions intersected with this unprecedented period of political, environmental, and cultural instability, and weaving music and sound back into these well-known historical events in early American history gives us a new perspective not only on the Moravian missions, but also on an important time period in the mid-eighteenth century when peaceful coexistence still seemed possible despite the pressures of colonization.[32]

A Brief History of the Moravian Missions

Pennsylvania's Moravian communities existed within a network of settler and Native communities in North America, as well as a wider, transnational network of Moravian places and people. Both of these networks would have important implications for the fate of the Pennsylvania missions. Beginning in the 1730s, the German Moravian Church had established mission communities across the Atlantic World—the largest Protestant missionary enterprise of the eighteenth century. ⊕ **See website Intro.1, Timeline: "Moravian Missions in North America, 1740–1794."** In constructing their mission network, Moravian Church leaders were adept at capitalizing on the existing trade and political structures of the British and Danish empires. The first missionaries left Germany in 1732 to minister to West African slaves in the sugar plantations of the Danish West Indies, and the following year a mission ship sailed to Danish-controlled Greenland to work with Inuit communities. Church elders, many of whom were members of the European nobility, also established connections in London that allowed them to send missionaries to the British colony of Georgia. The first Moravians in North America arrived in Georgia in 1735 with the aim of converting the local Creek and Cherokee population and slaves who worked in rice plantations of the low country. In 1752, a mission ship arrived on the rugged coast of Labrador. By the late eighteenth century, the Moravian missions in North America extended over a vast geographic distance from coastal Labrador to North and South Carolina (map I.1).[33] ⊕ **See website Intro.2, Static map: "The Moravian Atlantic."**

In the northern American colonies, the founder of the Moravian Church, Count Nikolaus Ludwig von Zinzendorf, outlined a plan for working with Indigenous communities in New York, Connecticut, and Pennsylvania, entitled the *Heiden Collegia* (*Plan for the Heathen*) (fig. I.1a–b).[34] When he created this plan in 1742, Zinzendorf and other church elders had already established a central mission community in 1741 at Bethlehem in eastern Pennsylvania. In the summer and fall of 1742, Zinzendorf traveled from Bethlehem north of the Kittatinny Mountains into New York and the river valleys of the Susquehanna, to meet with the leaders of various Haudenosaunee, Delaware, Shawnee, and Mohican settlements.[35] Based on these meetings, his mission plan focused on several communities he believed would be receptive to Christianity. Two of the most important Native communities in Pennsylvania—Shamokin, a Haudenosaunee-controlled

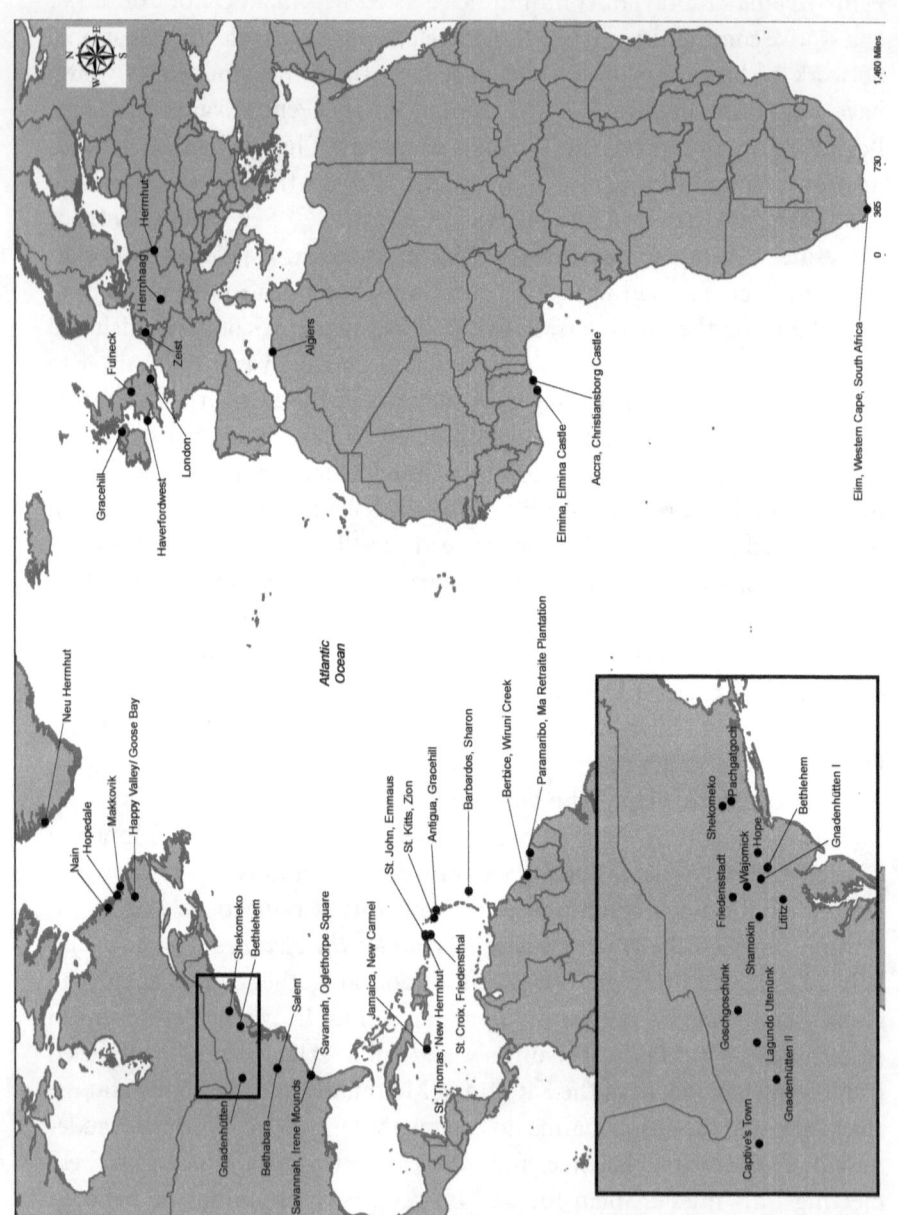

Map I.1 The Moravian Atlantic (created by Mark Sciuchetti and Sarah Eyerly).

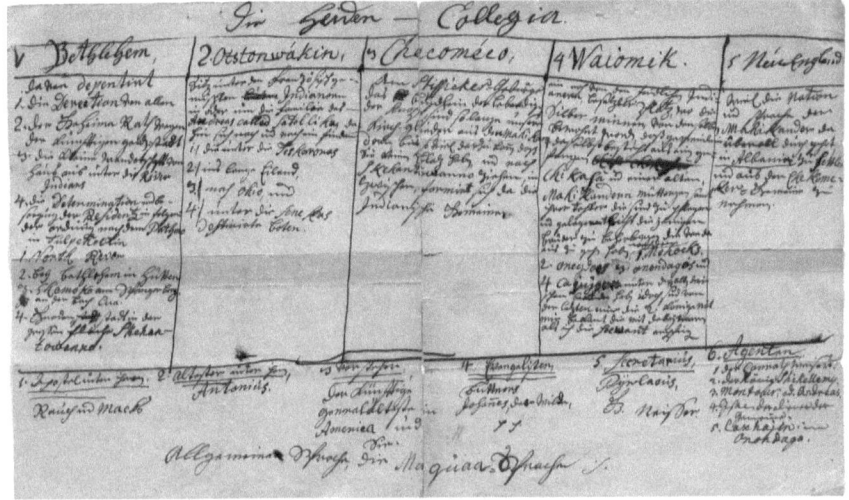

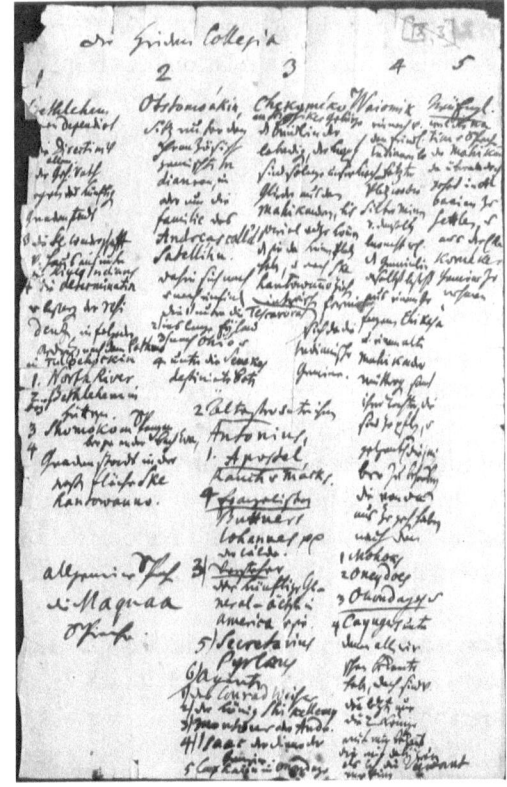

Fig. I.1a–b Copies of Nikolaus Ludwig von Zinzendorf, "Die Heiden Collegia," 1742. Personal papers of Zinzendorf, PPZdf 48, and MissInd 217.12b, MAB.

town at the confluence of the western and northern branches of the Susquehanna, and Wyoming (called by the Moravians, Wajomick), a primarily Shawnee and Delaware settlement in the Wyoming Valley—were to become central hubs for a Native Christian network beyond the borders of European settlements. Zinzendorf's mission plan also included Otstonwakin on the West Branch of the Susquehanna, a settlement led by a French Canadian and Algonquin woman named Madame Montour. A mission center at Otstonwakin would allow the Moravians access to the Great Shamokin Path that led across the Allegheny Mountains to Native communities in the Ohio and Beaver River watersheds, and the Great Warrior's Path that led north to the Council Fire of the Six Nations at Onondaga near Lake Ontario. The existing Mohican mission at Shekomeko in Dutchess County, New York, was to serve as a central hub from which Mohican Christians could be sent to form new mission villages in the Upper Colonies.[36]

This detailed vision for Christianizing Native Americans living in eastern North America may have been conceived by Zinzendorf, but its success rested in the hands of Moravian missionaries from Europe—men and women who traveled by ship across the Atlantic to live in Native American communities. It was ultimately the day-to-day relationships formed by people working, eating, singing, and speaking together that would create the Moravians' transatlantic network of people and places.[37] The Moravian missions were successful because they employed a two-pronged strategy that operated well on both a micro and macro level. In the larger sense, the Moravian Church had created links with the Danish and British Empires that allowed them to quickly build new communities or to establish a presence in existing communities already under the umbrella of larger colonial governing structures. But, on a smaller level, the Moravians' missionary strategy skillfully deployed a network of hundreds of young men and women who ardently believed in the cause of spreading the Christian message to people living in every country and region. They were the cornerstone of the Moravian mission plan, and the success of the church's agenda rested on them.

The first Moravian missionary to live in a Native American community in North America was Christian Heinrich Rauch. Despite having never left the small geographic area in Germany where he was born, Rauch was inspired at the age of nineteen to immigrate to America by a letter sent from a German (non-Moravian) settler already living in Pennsylvania. In late 1737, Rauch had sat on a long wooden bench in the *Saal* (worship hall) of the Moravian community of Marienborn, Germany, eagerly awaiting the reading of a letter from Moravian leader August Spangenberg to Christian David. It was dated November 19, 1737, and contained an account by Conrad Weiser, a German from the Palatinate who now lived in the Tulpehocken Valley of Pennsylvania. Thanks to his ability to speak several Native languages, Weiser now served as the principle messenger to the Haudenosaunee on behalf of the Pennsylvania and Virginia governments. Earlier that year, Weiser had experienced a remarkable encounter in the forest as he journeyed north to Onondaga. This is Weiser's story, as told in his own words and heard by Rauch:

> In the year 1737, I was sent the first time to Onondaga, at the desire of the governor of Virginia. I departed in the latter end of February very unexpectedly for a journey of 500 English miles, through a wilderness where there was neither road nor path, and at such a time of the year when animals could not meet with food. There were with me a Dutchman and three Indians. On the 9th of April I found myself extremely weak, through the fatigues of so long a journey with cold and hunger which I had suffered. There having fallen a fresh snow about twenty inches deep, and we being yet three days' journey from Onondaga in a frightful wilderness, my spirit failed, my body trembled and shook, and I thought I should fall down and die. I stepped aside, and sat down under a tree, expecting there to die. My companions soon missed me. The Indians came back and found me sitting there. They remained awhile silent; at last the old Indian [Shikellamy, Haudenosaunee leader at Shamokin] said, "My dear companion, thou hast hitherto encouraged us; wilt thou now give up? Remember that evil days are better than good days, for when we suffer much we do not sin; *sin will be driven out of us by suffering*, and God cannot extend his mercy to the former; but contrary-wise, when it goeth evil with us, God has compassion on us." These words made me ashamed. I rose up and traveled as well as I could.[38]

Weiser's fortitude earned him the name *Tarachiawagon* (Holder of the Heavens) among the Haudenosaunee. In Marienborn, his account also stirred admiration, but not for Weiser. It was the words of Shikellamy that moved several of the young brethren to tears. Surely, here was a person upon whom the Holy Spirit was already acting. Christian Heinrich Rauch was inspired. By 1740, he had sailed to the New York Colony, and requested permission from community elders to live in the Mohican village of Shekomeko. He did not preach, but instead offered basic medical care to the community, and set about living his usual daily life, including speaking prayers and singing hymns in their midst. Eventually, Rauch was able to report to the church in Europe that three people had requested Christian baptism. More followed and Shekomeko became the first Moravian mission in America.[39]

Early Moravian missionaries such as Rauch adapted to lifeways in Native American communities. They learned to boil maple sugar, interpret wampum, build and use bark canoes, and hunt using Native methods. They lived in bark homes and studied the languages of the communities they lived among. Rauch himself would learn to speak Mohican, Munsee, Unami, and Mohawk. According to an early Moravian Church history: "They [missionaries] earned their own bread, chiefly by working for the Indians, though the latter were not able to pay much for the produce of their labor. They lived and dressed in the Indian manner, so that in travelling to and fro they were taken for Indians. But whenever they could not subsist by the work of their own hands, they were provided with the necessaries of life by the Brethren [Moravians] at Bethlehem."[40] When Rauch eventually built a Christian worship space in Shekomeko, it was a bark structure like the other buildings in the village. As Richard Pointer and Rachel Wheeler have both stressed, Rauch was seemingly a willing observer and practitioner of Mohican culture.[41] He may originally have been interested in Mohican culture because he wanted to make sense of Native religious customs in order to present his Christian teachings in a more effective way, but by participating in local cultural practices he also participated in Native religious life. Missionaries such as Rauch listened to, recorded, and interpreted dreams, blessed hunters and hunting lodges, dispensed medicines, performed rituals for the dead and dying, and offered personal spiritual power through the blood of Christ.

When Christian hymns were sung, or sermons preached, they were usually done in Native languages. Because song, especially as a means of communicating with the divine, was an important part of religious life in Native and European Moravian communities alike, missionaries stressed this connection. In his foreword to his hymnal in the Delaware language, missionary David Zeisberger stated: "As the singing of psalms and spiritual songs has always formed a principal part of the divine service of our Church, even in congregations gathered from among the heathen . . . all our converts find much pleasure in learning verses with their tunes by heart, and frequently sing and meditate on them at home and abroad."[42] Hymns were a part of daily life for Native Christians, just as songs had always been a part of daily life in Native communities: they were sung to and by the sick and the dying. They were sung at gravesides. They were sung by men while hunting. They were sung by women in the travails of childbirth. They were sung to bring comfort, to call spiritual power, and to create and fortify community. They cemented treaties and sounded bravery in battle. And hymns became gifts that could be received from spirit helpers or *manitou*, continuing aspects of sacred song practices in Native communities. Hymns fulfilled similar functions in European Moravian communities, too, forging commonalities in the way that both groups understood the social, cultural, and spiritual purposes of song.[43]

Rauch's missionizing efforts were based on the Christian customs, theologies, and cultural ways that he had carried with him across the Atlantic. What theologies of song had he shared with residents of Shekomeko, and how had these theologies influenced their responses to Moravian Christianity? Rachel Wheeler has argued that there was a distinct difference between the missionizing efforts of the Moravians and the version of Christianity promulgated by English Congregationalist missionaries at the Christian Indian town in Stockbridge, Massachusetts. While Native Christian practice at Stockbridge was shaped around literacy—learning to read from the Bible and the hymnal and interpret Scripture—Moravians emphasized an embodied sense of theology that was principally conveyed through hymns and worship rituals, connecting with previous uses for sacred songs in Mohican communities. Moravians also emphasized teachings about the divine that connected with already existing spiritual understandings of *manitou* or spirit beings. Jesus was presented primarily as a God who had become a man, a great warrior who was killed, yet whose wounds and blood held redemptive, life-giving power. Rather than relying on biblical texts and exegesis, Moravian missionaries presented a distinctly embodied version of Christianity that particularly resonated with Mohican communities.[44] Moravian rituals and songs formed the core of missionary practices in North American mission contexts.

When Rauch arrived in the New York Colony in 1740, the *Brüdergemeine* (Brethren's Community), known in English as the Moravian Church, was a new church body and its religious practices were still very much in development. The first Moravian community of Herrnhut (The Lord's Watch) had been established in 1722 in southeastern Saxony. In the early 1720s, the young German nobleman Count Nikolaus Ludwig von Zinzendorf had purchased a large acreage along the Zittauer Straße, a road connecting the nearby towns of Löbau and Zittau, following his marriage to fellow noblewoman Erdmuthe Dorothea Reuss. At the time, the young couple could scarcely have foreseen the consequences of that purchase. By 1722, together with religious refugees from Bohemia and Moravia, members of the persecuted Protestant denomination of the Unitas Fratrum, they had established a small Christian town on their lands. Although the immediate aim of the community was to provide a sanctuary for Protestants persecuted by the Hapsburg Empire, in 1727, the community developed an entirely new mission. At a worship service on August 13, 1727, Zinzendorf and his fellow "Herrnhuters" sensed a particularly strong presence of the Holy Spirit in the church building, mirroring the biblical tale of the day of Pentecost. Although the worshippers that day hailed from many different Protestant backgrounds—Lutheran, Pietist, Unitas Fratrum—this experience of religious revival impelled them to come together as an entirely new church: the *erneuerte Brüdergemeine* (Renewed Brethren's Community). Their religious fervor also encouraged them to move beyond the boundaries of their small community to spread the Christian message. Within the next twenty years, the people of Herrnhut had founded a transcontinental and transoceanic network of Christian missions stretching from Tibet to Suriname.

According to Zinzendorf, he had foreseen the development of this mission network in a dream in 1723. In this dream, the Holy Spirit had revealed to him a landscape with many Christian towns similar to the community of Herrnhut.[45] It was this specific method of planned spiritual landscapes and towns—communities that resembled each other in both worship and spatial construction—that would fuel the development of more than thirty newly constructed international settlements, in addition to numerous outlying missions such as Shekomeko that were established in already extant towns and communities. The planning of each new Moravian community's physical structure was approved in Europe, and then constructed with natural materials available on site. Most settlements were built around a central town square, and contained communal houses and worship spaces, as well as gardens, trade buildings, and agricultural fields. Lifeways in Moravian communities were also predictably replicated based on a system of communal living called the *Oeconomie* (Economy), in which community members contributed their earnings directly to the church. Moravians were divided into gendered "choirs" based on age and marital status. The desired goal was to create a shared, spiritual space where each person could live solely for the purpose of serving Christ without fear of monetary poverty. Christ, and not a human church official, was the elder of all Moravian communities. At meetings, his presence was signified by an empty chair and his will was ascertained by casting of lots (*das Los*).⁴⁶ Lots were an important arbiter of not just communal decisions but also individual choices. Moravians carried pieces of paper or "lot chips" in their pockets that could be cast to invoke a randomized answer or "Christ's will." Lots settled disputes, interpreted Scripture, and sanctioned marriages, missionary activities, and social customs. The entire social organization of Moravian communities depended on the lot as an arbiter of social and spiritual will (fig. I.2a–d).

Lots also governed worship practices, including music. All Moravians learned to publicly demonstrate their improvisational abilities in daily musical worship services, called *Singstunden* (singing meetings). In a Singstunde, individual hymn-verses and phrases of chorale melodies, from a memorized repertory of several thousand preexisting hymns, were extemporaneously combined by a community member called a *Liturg* (worship leader, liturgist), and repeated by other participants to create a *Liederpredigt* (hymn-sermon). The *Liederpredigt* was itself a sounded explication of a particular scriptural passage called the *Losung* (watchword) that was chosen from a set of hymn verses and scriptural passages selected by Zinzendorf.⁴⁷ If no suitable verses could be drawn from the memorized repertory to suit the Losung, then the liturgist would improvise a new hymn. The entire practice was called "singing from the heart" (*aus dem Herzen gesungen*). The soundways of Moravian communities were therefore as planned as the buildings and communal living practices. The choir system immersed community members in a daily cycle of religious hymns and prayers, directing the entirety of life to spiritual contemplation and a close relationship with God. As these practices were replicated in each community, they constituted a worldwide community based on divinely communicated sound.⁴⁸

Transmitting the Gospel through singing was therefore an important part of Moravian missionary practice. Moravians did not rely on literacy to impart Christian teachings. Instead, rituals (baptism, communion, Singstunden), songs, and prayers were the main methods of connecting with potential Christians.⁴⁹ New Moravians were taught to improvise hymns as a way of channeling the Holy Spirit, a practice that required extensive training and study to learn. Moravians were also encouraged to sing in their own languages, and Moravian hymns in German were sometimes macaronic (sung in two or more languages) or incorporated "loan" words from various languages such as Hebrew, Mohawk, Mohican, Arawak, and Latin. Multilingual Singstunden and other worship services were characteristic forms of worship in Moravian missions, a practice that Joanne van der Woude

Fig. I.2a Box of lots in German and English belonging to the Provincial Elders Conference. OC 210, MAB. Photograph by author.

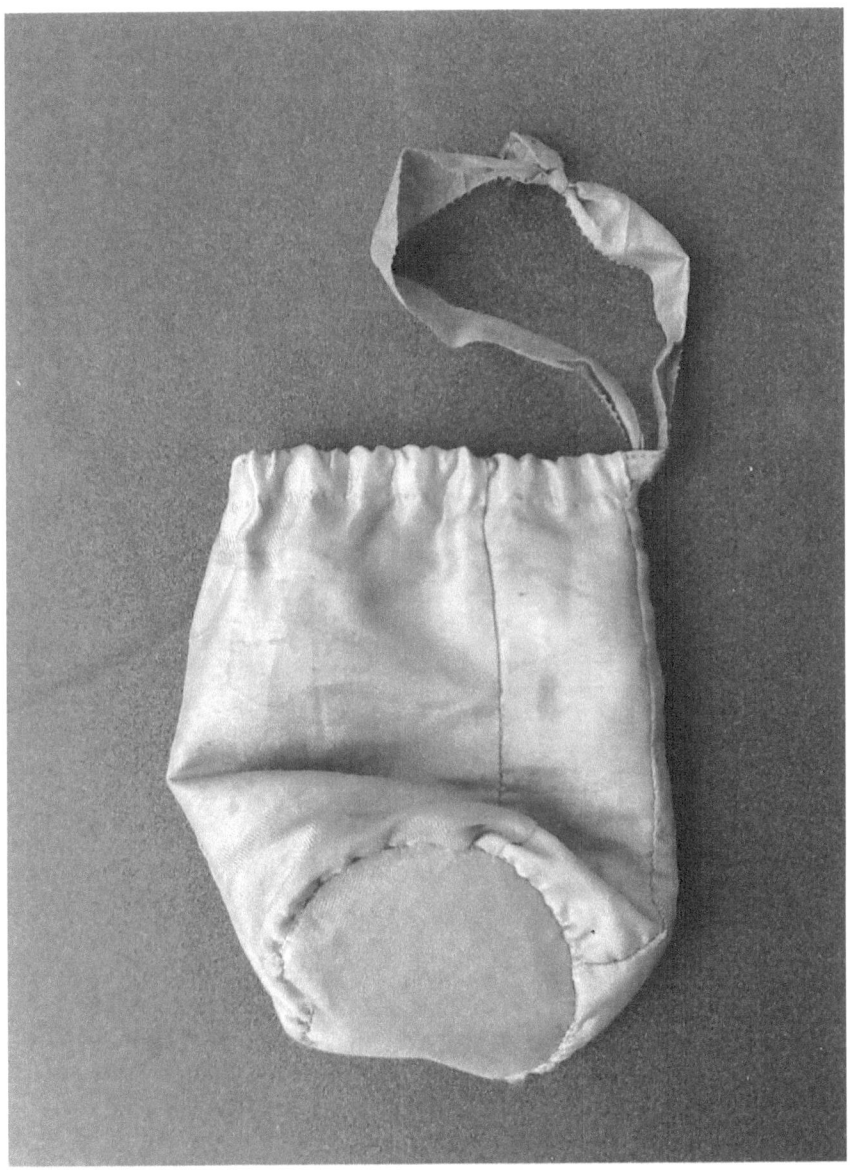

Fig. I.2b Gold silk drawstring lot bag, with gold cord and tassels, containing 51 scrolls with daily watchword texts. OC 510, MAB. Photograph by author.

Fig. I.2c–d Lot boxes and chips. M.20, M.26, and M.28, UA Herrnhut.

has termed "polyglot harmony."⁵⁰ Polyglot singing reflected the Moravian belief that singers were pre-sounding the voices of the multiethnic Christian community that would gather around the throne of God at the end of the world. In the end-time, the cacophony of the tower of Babel would be silenced and all people would sing with one voice.⁵¹ On Earth, in anticipation of that great awakening at the end of time, Moravians could improvise and sing multilingual hymns as a way of channeling their future divinity through the singing body.

Multilingual hymns were also a way to reflect a localized and indigenized version of Christian practice, and Moravians believed they further demonstrated the presence of the Holy Spirit in potential Christians. Missionaries were taught that the Holy Spirit must first be active in the hearts of those "who would hear the message."⁵² Moravian missionaries should not concern themselves with converting large groups of people, but rather with seeking those who freely responded to the Christian message: "We are like servants at their master's door who scratch softly so that those who will want to hear will hear, while others not so inclined can ignore us."⁵³ Zinzendorf claimed that conversion was only accomplished through the Holy Spirit, and encouraged missionaries to only baptize those who they sensed had an intuitive feeling of the presence of the Holy Spirit in their lives. But the work of conversion did not happen without missionaries themselves and the transmission of their particular beliefs and cultural norms. It was the distinctive worship practices of the Moravians that attracted a wide variety of people to join their church communities. While other Protestant groups, such as the Methodists, remained predominantly British and Anglo American, Moravian Church communities boasted an almost globally representative membership.⁵⁴

The founding of the Moravian missions in Pennsylvania began in 1740 with the arrival of a Moravian group from Savannah, Georgia. In 1735, Zinzendorf and the leadership of the Moravian Church had established a settlement in the new British colony of Georgia. That year, several Moravians had traveled to the colony, where they occupied a house in the center of Savannah. But while the missions in the Danish Caribbean thrived, the outbreak of war between England and Spain in 1739 destroyed the Moravians' hopes for further missionary work in the southern British colonies. Fleeing Georgia, they sought refuge in the northern British colonies, arriving in Philadelphia in 1740 on the sloop of George Whitefield, a notorious and celebrated English revivalist. Whitefield had met the Moravians in

Savannah while working with John Wesley, founder of the Methodist Episcopal Church, on behalf of the Society for the Propagation of the Gospel. Whitefield's destination in Pennsylvania was "Nazareth," a 5,000-acre tract of Delaware land including the village of Welagameka that he had recently purchased from the Penn family at the Forks of the Delaware River. He intended to build a school for orphans and the children of slaves from the surrounding European settlements. Whitefield himself soon departed for England, but he appointed the Moravians to build the school in his absence, and allowed them to plan their own settlement on his land. However, within a year, theological disagreements about styles of prayer and Scripture reading erupted between landlord and tenants and the Moravians were evicted. It was then that the Moravian Church purchased its own parcel of Delaware land eleven miles south of Whitefield's tract along the Lehigh River at the abandoned village of Menagachsuenk. There, in 1741, they began to construct their first newly built North American community, Bethlehem.

This new town was to be a geographical and spiritual center that would provide people, materials, and financial support for an expansive mission program. Its location on what was at the time the western boundary of the Pennsylvania Colony was ideal for a variety of missionary purposes. The area stretching from Bethlehem and the Lehigh River Valley southward to Philadelphia, including the Tulpehocken Valley, and the towns of Lancaster and Germantown, was the center of an active German religious diaspora. Lutherans, German Reformed, Arminians, Socinians, Schwenkfelders, German Old Tunkers, New Tunkers, New Lights, Inspired, Sabbatarians, Hermits, Independents, and Free Thinkers had all settled in southeastern Pennsylvania. For Zinzendorf, this offered an unprecedented opportunity to unite these disparate sects under one German church. Although Zinzendorf's ardent plan for a united German church community never materialized, Pennsylvania also offered the chance to live in close proximity to many different Native American communities. By 1748, in accordance with Zinzendorf's *Heiden Collegia*, at least 132 missionaries had been sent out from Bethlehem to work in communities from Pachgatgoch, Connecticut, to Meniolagomeka and Shamokin, Pennsylvania. Rauch's small Christian congregation at Shekomeko, New York, would also eventually relocate to the area around Bethlehem, where they would build several new communities with other Native Christians from various tribal affiliations and backgrounds.

At the height of the Moravians' presence among Delaware, Wampano, Haudenosaunee, Shawnee, and Mohican communities in the 1740s and 50s, the number of Native Moravians living at any one time in mission communities numbered at more than 250 individuals. A catalog of Native baptisms from Pachgatgoch lists 471 men, women, and children who were baptized from the beginning of the mission era to 1769.[55] But this early period of Moravian mission history was relatively brief. Surrounded by escalating conflicts between Native Americans, European settlers, and the European colonial empires of France and England during the Seven Years' War, Native Moravian communities in eastern Pennsylvania were eventually destroyed or abandoned, and many members of the Native Moravian congregation fled westward into the Ohio Country. There, Native Moravians would struggle to maintain viable communities along new spatial, political, and cultural frontiers that would arise in the wake of the American Revolution. After the massacre at Gnadenhütten in March 1782, this once-vibrant community of Native Christians was reduced to a handful of families and individuals. Although the Moravian missions continued into the nineteenth century, with the establishment of new communities in Ohio, Michigan, Indiana, and Ontario, the flourishing numbers of Native Christians who had chosen to live as Moravians, and the vital musical practices that had sustained the first forty years of the missions, no longer characterized life in these newer communities.

Re-Sounding the Moravian Missions

In writing this book, it was my desire to create a rich and inclusive narrative of sound and musical practices in the first forty years of the Moravian missions in North America. But I wondered whether a written book could truly capture the complex affective geographies of past soundscapes. Eventually, I came to the realization that if I wanted to create a new type of history—a sonic history—I would need to consider the possibility that communicating in sound might be an even more effective way to tell the story of the Moravian missions. It was the desire to incorporate sound into the book that led me to available digital technologies, such as mapping and sound design software (ArcGIS and Logic Studio), and the creation of the 🌐 *Moravian Soundscapes* website, the online companion to this book. As you read this book, I invite you to simultaneously study and listen to the recordings on the website to explore what can be learned when a variety of approaches (sound studies and audible history, composition, historical performance, mapping and spatial humanities) are brought together to "sound" the history of the Moravian missions. The narrative of each chapter of the book is expressed on the website through interactive sound maps that contain soundscape compositions, field recordings, and historically informed recordings of spoken texts and hymns in Delaware, Mohican, English, and German. These interactive maps also contain the Global Positioning System (GPS) coordinates for each location, making it possible to take the book and the maps to the places that are being discussed and to literally listen and learn in place.[56]

The *Moravian Soundscapes* website follows the book's structure, which is chronological, beginning with the founding of Bethlehem in 1741 and moving forward to 1782. The book and website also follow a geographic trajectory. Chapter 1 starts broadly with the travels of early Moravian missionaries through the natural environments of Pennsylvania as they encounter Native communities beyond the boundaries of colonial settlements. Chapters 2 and 3 narrow in scope to focus on the internal social and spiritual geographies of the Bethlehem community during the 1740s and 1750s. The final chapter (chap. 4) returns to a broader geographic perspective, detailing the end of Bethlehem's communal Economy in 1762 and the forced migration of the Native Moravian community into northern Pennsylvania and Ohio during the 1760s and 1770s, leading to the massacre at the Ohio community of Gnadenhütten in 1782. Each chapter reveals the spatial, social, and spiritual structures of Moravian communities through detailed discussions of sound and musical practices that link with sound maps or sound examples on the *Moravian Soundscapes* website. It is my hope that the audible history components of the website will serve as helpful frameworks for interpreting the historic acoustic environments of Moravian communities.[57]

As you consider these digital elements, the process of creating them is worth discussing. The process itself has greatly informed my own understanding of sound in historic Moravian contexts. It has been more than 275 years since Bethlehem was founded, and more than 250 years since the Native Moravian communities at Nain, Friedenshütten, and Gnadenhütten, Pennsylvania, were destroyed or abandoned. So, before I could recreate the acoustic environments and singing practices that characterized life in eighteenth-century Moravian communities, I had to first establish the physical locations of many of the nonextant communities, as well as buildings and spaces that had been built over or destroyed since the mid-eighteenth century. This process involved more than two years of fieldwork, mapping, and spatial reconstruction based on archival documents, archaeological data, and georectification of historical maps against modern satellite data. I am happy to say that this was a project that I did not undertake alone. It was a collaborative and interdisciplinary effort involving Mark Sciuchetti, then a doctoral student in geography at the Florida State University, and my husband, sound designer and composer Andy Nathan.

We began the project by conducting fieldwork to collect the GPS locations for all of the sites that existed in Bethlehem in 1758, and all known Moravian mission locations in Pennsylvania, in addition to other neighboring settler and Native communities.[58] Some of these settlements, buildings, and places were quite easy to locate. Bethlehem's Gemeinhaus, and other communal buildings such as the Single Brothers' House, the blacksmith shop, and pottery shop are either extant or have been rebuilt based on data gleaned from archaeological excavations and other historical evidence. Locating other places, though, often required extensive archival and historical research, as well as some degree of informed speculation based on the location of known geographical features such as streams or hills, or spatial data gleaned from a combination of historic and contemporary maps. 🌐 **See website Intro.3, Map collection: "Mapping Pennsylvania."** For instance, after the Native Moravian community at Gnadenhütten, Pennsylvania, was abandoned in 1755, Benjamin Franklin purchased the site and built Fort Allen on the opposite side of the Lehigh River. Over the intervening two and a half centuries, the settler community of Lehighton was gradually built around the fort and over the site of the mission. Using historic eighteenth- and nineteenth-century maps, we were able to locate the original position of the mission on the southern edge of the modern-day Lehighton cemetery.[59] There have also been recent archaeological excavations to locate the Native community at Nain along the Monocacy Creek north of Bethlehem. We were able to obtain the archaeologists' reports and examine artifacts from Nain, in addition to visiting the site and one of Nain's original homes that had been moved into central Bethlehem in 1765.[60]

The most important part of the process of locating eighteenth-century places was "learning in place"—walking, driving, and studying the natural and built topography of historic Moravian places by experiencing them in the present. At the time, I didn't fully appreciate the value of the many days and miles of driving and walking my collaborators and I undertook. When

Andy and I, along with our two sons, hiked through the forests along the Moshannon Creek in central Pennsylvania to map and photograph places visited and traversed by eighteenth-century Moravians along the Great Shamokin Path, I didn't truly comprehend the ways that my historical understanding of those places would be informed by experiencing them with my family in the present. The days that Mark and I spent trekking Bethlehem's streets, ducking under railroad bridges, and venturing through fields and nearby creeks, provided invaluable knowledge that we would not have gained without experiencing those places in person.[61] 🌐 **See website Intro.4, Picture collection: "Modern-Day Pictures."** It was through these "place-visits," literally reading the land as an archive, that my collaborators and I learned to look for traces of vanished places that were still written onto the landscape: the contours of the Moravian grain mill that had ceased operation more than two hundred years ago, but whose trace was still visibly carved into a meadow near the Monocacy Creek; the early King's Roads that had once connected Bethlehem with the colonial Pennsylvania government at Philadelphia, simply paved with asphalt and renamed; and the ancient Minisink Path that linked Native American communities in the area for centuries before the arrival of German missionaries, now a gravel pathway beside a spring that had provided water to travelers for hundreds of years.

We learned that modern place names often revealed layers of history that had since disappeared from human memory or archival documents. As we searched for the location of the mission of Friedenshütten, we found that the place where the mission had once stood along Wyalusing Creek was still marked as "Moravian Street." In our frustrating search for the location of the Rose Inn, when eighteenth-century maps were confusing at best and contradictory at worst, we discovered that a modern street built over the former site of the inn was fortuitously labeled "Rose Inn Avenue." Nearby, almost completely covered with grass, a memorial stone in the yard of a family home still marked the inn's location. Even venturing beyond the bounds of Pennsylvania, we discovered, for instance, that the Moravian mission on the Ma Retraite Plantation in Suriname had not completely vanished. The plantation itself had now become a neighborhood bearing the same name in the capital city of Paramaribo. While buildings and memories may have disappeared, historic places were often recorded on the contours of the landscape, or in the names of streets, rivers, or neighborhoods. Modern roads still followed older pathways. Despite the efforts of modern engineers in the 1960s to carve straight pathways through the challenging terrain of the Allegheny Mountains, Interstate 80 conforms for the most part to the ancient contours of the Great Shamokin Path. These historical and modern intersections were discoverable with the right data and with a curiosity to seek the right places.

In the process of looking for past places, though, my collaborators and I also learned that while the mill trace, the King's Road, the Minisink Path, and the site of the Rose Inn still existed, if you knew where to look for them, the sounds of eighteenth-century Bethlehem had vanished. How could we represent the acoustic environments that had once characterized life in Moravian communities? Were those soundscapes irrecoverable? Sometimes, when faced with a seemingly insurmountable problem, it helps to return to where you started. If we could map eighteenth-century Bethlehem, for instance, by using extant historic buildings, then we could also enter those buildings to learn more about how they operated as acoustic spaces. In addition, there are twelve museums in modern-day Bethlehem dedicated to historic preservation and education about the city's Moravian past. Their collections house historic tools and implements, musical instruments, and other elements of Moravian material culture that once contributed to its soundscapes. And, every Thursday through Saturday, Bethlehem's current blacksmith, Philip Trabel, utilizes the forge in the rebuilt blacksmith shop to make horseshoes, nails, and other metal products. Perhaps the sounds of Bethlehem had not entirely vanished. It would be possible, at least, to conduct acoustic studies and produce field recordings in some locations.

We began by recording the plantation bell at Burnside Plantation, a former Moravian farm that is now a museum on the outskirts of Bethlehem. We also obtained permission to do field recordings of hymn singing inside of the Old Chapel, the worship spaces in the Single Sisters' House and Gemeinhaus, and to record Trabel's work in the blacksmith shop.[62] During the process of field recording, we also collected decibel readings at approximately five-foot intervals from the sound sources (singers, bells, forge). Then, we processed the readings through a mathematical formula for understanding sound decay over distance. The particular formula we chose was capable of taking into account various types of landscapes in Bethlehem over which sound might have traveled, including agricultural fields, coniferous and deciduous forests, grass lands, shrub lands (low trees and bushes), water, and urban or built environments. Our assessment of the historical topography around Bethlehem was based on a 1758 map by the Moravian cartographer Christian Gottlieb Reuter. Reuter helpfully designated varying terrains by type on the map, including particular species of trees. The resulting "sound boundary" maps, which can be found on the website under chapter 2, allowed us to understand what the geographic limits of Bethlehem's soundscapes might have been, and how people might have understood the boundaries of their community by listening.[63] 🌐 **See website chap2.2, Interactive map: "Sound Boundaries of Bethlehem."**

We also wanted to represent sounds and places that were no longer present, and for which it was not possible to create field recordings. In seeking to replicate the diversity of acoustic environments and perceptions of sound that once existed in eighteenth-century Bethlehem, we took advantage of the spatial frameworks provided by Geographic Information Systems (GIS) software and aural cartography or sound mapping.[64] We had already collected the spatial data necessary to reconstruct the built environment of Bethlehem, so it was only one further step to add sound to the maps. We felt that sound maps held great potential for offering the multisensorial approach to historic space that characterized daily life in Bethlehem. They could also allow us to reconstruct some sense of the Moravians' cultural, social, and aesthetic perceptions of sound. Placing sounds within a spatial framework, such as a map could, we hoped, permit readers to explore new experiential and interpretive frameworks for understanding past soundscapes. However, it perhaps goes without saying that there were inherent challenges involved in this idea of reconstructed and experiential soundscapes, not least of which was the problem of how exactly to represent acoustic environments that no longer existed.[65] In the case of the sound maps created for the *Moravian Soundscapes* website, we turned to the work of electronic composers and sound designers for inspiration. Barry Truax and Hildegard Westerkamp's "imaginary soundscapes" or "virtual or simulated soundscapes" became the framing methodology for creating the sounded portions of the project.[66] In their work as electronic composers, both Truax and Westerkamp have simulated past acoustic environments through "soundscape compositions" created from digitally layered field recordings or prerecorded sound samples, generating what Truax has termed a "representation of acoustic environments."[67]

In the case of our sound maps, we created soundscape compositions from field recordings of available industrial and agricultural machinery recorded in Bethlehem. Some of the hymns that are layered into the soundscapes are also field recordings. Other hymns, and the Mohican and German dialogues and sermons that are also represented on the website, were recorded in a studio. However, since the soundscapes of Bethlehem itself have changed dramatically since the eighteenth century, we also used digitally sampled environmental and historical sounds from the sound libraries of the British Broadcasting Corporation (BBC). These sound libraries were originally recorded by BBC recording engineers for film and radio broadcasts, and represent the soundscapes of various natural places and communities around the world. Of particular interest for our purposes was the BBC sound libraries' "Industrial Sounds" collection that preserves the sounds of rare historic machines and tools such as a wooden lathe, a double-handled wood saw, and even a butter churn. Each sound file was available as a 16/44.1kHz stereo audio sample, and multitrack editing of these samples into soundscape compositions facilitated the recreation of the Moravians' acoustic environments. The end result was a series of historically informed electronic compositions that provided a descriptive sense of historic acoustic environments.[68] 🌐 **See website Intro.5, Interactive sound map: "Moravian Soundscapes."**

For researchers interested in historic sound, soundscape compositions offer a model for reconstructing past soundscapes that simulate a coherent sense of an acoustic environment, even if formulated through the historical imagination of a composer. Both soundscape compositions and the performance or recording of past musical repertories stem from the desire to "sound" or to recreate historic aural experiences from an informed and academically rigorous perspective. Like other modes of historical performance, soundscape compositions are modern experiences and can only speculatively represent historical audible phenomena. Just as we can't know exactly how a particular musical tradition was articulated by practitioners in the past, we can still strive to create modern renderings that are informed and informative.

What can we learn from re-sounding past acoustic environments? What are the advantages or disadvantages of such historical recreations? What insights do we stand to gain from the spatial humanities and sound mapping? In the case of *Moravian Soundscapes*, GIS technologies and sound mapping have been invaluable research methodologies in creating both the book and the maps. They have allowed me to more accurately convey the inherent emphasis in Moravian communities on sound, and sound maps have allowed me to directly, rather than abstractly, represent the sounds of places such as Bethlehem. They have also helped me to articulate the often intangible and elusive qualities of the Moravians' sounded religious spaces and musical traditions. This has been especially important in my attempts to represent Moravians' spiritual understandings of sound. According to composer Isobel Anderson, sound maps are particularly useful for mapping the "in-between spaces" of culture and society—the imagined and invisible relationships that constitute human experience of sound in the past and present.[69] Moravian communities existed as much in sound as they did in space. This type of spiritual understanding of sound is certainly not unique to the Moravians. But as scholars studying religious traditions of music, we are often tasked with representing conceptions of sound and space that are imaginative, theoretical, and spiritual. In our attempts to document the soundways of religious communities, we often discover that ideas about sound are more important than the sounds themselves. Sound maps are just one way we might more deeply explore the sensory and imaginative aspects of religious traditions and communities, whether historical or contemporary.

Sensory data and knowledge can also be an important part of the research process itself. Since my background is in musical performance, my approach to historical research on music and sound has been greatly influenced by musicologist Elisabeth Le Guin's theory of applied musicology—the idea that there is a form of embodied knowledge gained through the physicality of playing an instrument or singing.[70] So I wondered if there could be a similarly embodied form of knowledge to be gained from "sounding" historical acoustic environments. Sound certainly has a direct relationship to emotion, memory, and instinct. Sensitivity to vibration is found in even the most primitive of life forms, and our brains have evolved over millennia to respond to acoustic signals and patterns.[71] If acoustic knowledge is fundamental to our experiences of the world, then research processes and methods of historical inquiry that take advantage of the acoustical properties of sound might impart embodied forms of knowledge that deepen our understanding of historical times and places.[72] In writing this book and composing the sound maps, I wanted to use my training as an academic *and* as a musician to imagine past soundscapes. I wanted to be a storyteller—a composer of both words and sounds. I have especially been inspired and drawn courage from Craig Womack's assertion that history should be dreamed and imagined, and that we are called as historians (and musicians) to create stories that are compelling and rich:

History means very little until we develop a relationship with it that in this cyberage we might call "interactive"... I am talking about more than developing a capacity to empathize with people from our pasts. This has to do with placing ourselves inside their stories, becoming participants in history, more specifically, turning ourselves into characters in a story. History must be dreamed. It has to be authored. It must be turned into a fiction before it can ever be true.... This is the responsibility of any human being who desires an ethical relationship to her past. History is a vision quest, the quintessential religious experience. How else, if not through vision, can we access these experiences from the past so we may also experience them? This is how we approach the paradox we are up against. How can we ever know what experience is in its original forms, apart from mediation, interpretations, our perceptions? We cannot. Reality may exist with or without us, but whatever we can know is affected by our thoughts, no matter how spiritual the message. But we can imagine the places where experiences originate.[73]

The book and the website are intended to be imaginative frameworks for Moravian places and experiences. They represent the journey that I have undertaken as author to understand how sound intersected with Moravian ideas of space, community, and spirituality. They also represent the journey that my collaborators and I undertook to understand Moravian places. And they represent the journey that you, as reader, might take. In digital and physical space, whether in person or by imagination, you might journey beyond the page, participating in these narratives out on the land in the places where these histories were shaped.[74] When you stand in these places, listen to the sounds around you. Imagine the layers of history underneath your feet, the traces and clues that previous generations have bequeathed us: names, stories, musical instruments, iron tools and wooden looms, buildings of stone and wood, and dug-out places in the earth. These are the sites of collective memory, of histories and soundscapes embedded in place. When we experience them in that way, we add our own stories, we add our own sounds.

Notes

1. Count Nikolaus Ludwig von Zinzendorf, "Lied bey den Liebesmahlen," *Herrnhuter Gesangbuch* (HG) Hymn 1340, verse 14. Hymn composed for the return voyage of a mission ship from St. Thomas in 1739.

2. Throughout this book, I have chosen to use "Mohican" rather than "Mahican." While most ethnohistorians and anthropologists prefer "Mahican" because it is close to the Dutch "Mahikander" (a term also used by the Moravians), the sole descendant community on the Stockbridge-Munsee reservation in Bowler, Wisconsin, uses "Mohican"

or "Muhheakunnuk" (https://www.mohican.com). Similarly, I also use "Delaware," rather than "Munsee" or "Unami," since it is the term currently used by many descendant communities. For more information, see the website of the Delaware Tribe of Indians, a descendant community in Oklahoma and Kansas (http://delawaretribe.org/services-and-programs/historic-preservation/removal-history-of-the-delaware-tribe/), and the website of the Delaware Nation at Moraviantown, a descendant community in Ontario, Canada (http://delawarenation.on.ca).

3. Obadiah Holmes's account of the Gnadenhütten massacre is recorded in T. Holmes, *The American Family of Rev. Obadiah Holmes* (Columbus, OH: 1915).

4. Earl P. Olmstead, *David Zeisberger: A Life among the Indians* (Kent, OH: Kent State University Press, 1997), 333.

5. Letter written at Bethlehem, April 5, 1782. MissInd 151.6.8a, MAB.

6. Olmstead, *David Zeisberger*, 333.

7. "Johann Heckewelder's Travel Diary of 1792," MissInd 213.7, MAB. Translated in Paul A. W. Wallace, *Thirty Thousand Miles with John Heckewelder, Or, Travels among the Indians of Pennsylvania, New York & Ohio in the 18th Century*. The Great Pennsylvania Frontier Series (Lewisburg, PA: Wennawoods Publishing, 1998), 261.

8. For detailed histories of violence during the American Revolution, including religiously motivated violence, see Holger Hoock, *Scars of Independence: America's Violent Birth* (Broadway Books, 2018); and John Corrigan, Lynn S. Neal, eds., *Religious Intolerance in America: A Documentary History* (Chapel Hill: University of North Carolina Press, 2010).

9. Olmstead, *David Zeisberger*, 333.

10. Quoted in Elizabeth Cobbs Hoffmann, Jon Gjerde, and Thomas G. Paterson, eds., *Major Problems in American History: Documents and Essays* I (Wadsworth: Cengage Learning, 2012), 205.

11. Gary Tomlinson, *The Singing of the New World: Indigenous Voice in the Era of European Contact*, New Perspectives in Music History and Criticism (New York: Cambridge University Press, 2009), 5.

12. For more information on the daily integration of hymns into Moravian life, see Sarah Justina Eyerly, "'Singing from the Heart': Memorization and Improvisation in an Eighteenth-Century Utopian Community" (PhD diss., University of California Davis, 2007).

13. Although hymnody has often been overlooked as a cultural and musical form, there has been a recent resurgence in scholarly interest in hymnody from disciplines as diverse as religious studies, anthropology, performance studies, Native American and Indigenous studies, and African American studies. Interdisciplinary methods of studying hymnody featured prominently in a recent roundtable at the 2019 meeting of the Society for Early Americanists, "The Hymn in Early America: A Roundtable," chaired by Chris Phillips (Lafayette College). The roundtable featured presentations on the revival hymn and the epic function in early America, poetry and hymnody in *Uncle Tom's Cabin*, Samson Occom's hymns and the articulation of Native space, and two different discussions of Moravian hymns.

14. This study is indebted to the work of Barry Truax and R. Murray Schafer in defining the key concepts of "acoustic ecology," "acoustic environment," and "soundscape." Truax's *Handbook for Acoustic Ecology* and *Acoustic Communication*, have been especially helpful in building a framework for study of the acoustic environments of eighteenth-century Moravian missions. Based on Truax's call to account for all environmental sounds within a given landscape, I aim to highlight the importance of studying the complex and interrelated

patterns of sound that surrounded Moravian Christians and how these soundscapes helped to construct personal, social, environmental, and religious identity. See Barry Truax, *Acoustic Communication, Handbook for Acoustic Ecology* (Burnaby, BC: Cambridge Street Records, 1999); Barry Truax, "Soundscape, Acoustic Communication and Environmental Sound Composition," in *A Poetry of Reality: Composing with Recorded Sound*, ed. Katherine Norman (Reading, UK: Harwood Academic Publishers, 1997); Barry Truax, "Paradigm Shifts and Electroacoustic Music: Some Personal Reflections," *Organised Sound* 20, no. 1 (April 2015): 105–110; R. Murray Schafer, *The Tuning of the World* (New York: Knopf, 1977); and R. Murray Schafer, "Soundscapes and Earwitnesses," in *Hearing History: A Reader*, ed. Mark M. Smith (Athens: University of Georgia Press, 2004), 417–431. I have also benefitted from pathbreaking work in the field of sound studies by Richard Leppert, Mark M. Smith, Bruce Smith, Douglas Kahn, Alain Corbin, David Samuels, and Steven Feld. See especially Raymond Leppert, *The Sight of Sound: Music, Representation, and the History of the Body* (Berkeley: University of California Press, 1993); Mark M. Smith, "Introduction: Onward to Audible Pasts," in *Hearing History*; Bruce Smith, *The Acoustic World of Early Modern England: Attending to the O-Factor* (Chicago: University of Chicago Press, 1999); Douglas Kahn, *Noise, Water, Meat: A History of Sound in the Arts* (Cambridge, MA: MIT Press, 1999); Alain Corbin, "Identity, Bells, and the Nineteenth-Century French Village," in *Hearing History*; Corbin, *Village Bells: Sound and Meaning in the Nineteenth-Century French Countryside* (New York: Columbia University Press, 1998); David W. Samuels, Louise Meintjes, Ana Maria Ochoa, and Thomas Porcello, "Soundscapes: Toward a Sounded Anthropology," *Annual Review of Anthropology* 39 (2010): 329–345; Steven Feld, *Sound and Sentiment: Birds, Weeping, Poetics, and Song in Kaluli Expression*, Publications of the American Folklore Society 5 (Philadelphia: University of Pennsylvania Press, 1982); and Steven Feld and Keith H. Basso, eds., *Senses of Place*, School of American Research Advanced Seminar Series (Santa Fe, NM: School of American Research Press, 1996).

15. Recent scholarship on Moravian mission work in general includes Stefan Hertrampf, *Unsere Indianer-Geschwister waren lichte und vegnügt: Die Herrnhuter Missionare bei den Indianern Pennsylvanias, 1745–1765* (Frankfurt am Main: Peter Lang, 1997); Carola Wessel, *Delaware-Indianer und Herrnhuter Missionare im Upper Ohio Valley* (Halle: Halle Verlag der Franckeschen Stiftungen im Niemeyer-Verlag, 1997); Jane T. Merritt, *At the Crossroads: Indians and Empires on a Mid-Atlantic Frontier, 1700–1763* (Chapel Hill: University of North Carolina Press, 2003); Merritt, "Dreaming of the Savior's Blood: Moravians and the Indian Great Awakening in Pennsylvania," *The William and Mary Quarterly* 54, no. 4 (1997): 723–746; Amy C. Schutt, *Peoples of the River Valleys: The Odyssey of the Delaware Indians*, Early American Studies (Philadelphia: University of Pennsylvania Press, 2007); Rachel Wheeler, *To Live upon Hope: Mohicans and Missionaries in the Eighteenth-Century Northeast* (Ithaca, NY: Cornell University Press, 2008); and A. G. Roeber, ed., *Ethnographies and Exchanges: Native Americans, Moravians, and Catholics in Early North America*, Max Kade German-American Research Institute Series (University Park, PA: Pennsylvania State University Press, 2008).

16. Walter Woodward, "'Incline Your Second Ear This Way': Song as a Cultural Mediator in Moravian Mission Towns," in *Ethnographies and Exchanges: Native Americans, Moravians, and Catholics in Early North America*, 125–142; Rachel Wheeler and Sarah Eyerly, "Songs of the Spirit: Hymnody in the Moravian Mohican Missions," *Journal of Moravian History* 17, no. 1 (2017): 1–26; Rachel Wheeler and Sarah Eyerly, "Singing Box 331: Re-Sounding Eighteenth-Century Mohican Hymns from the Moravian Archives," *The William and Mary Quarterly* 76, no. 4 (October, 2019): 649–696.

17. Glenda Goodman, "'But They Differ from Us in Sound': Indian Psalmody and the Soundscape of Colonialism, 1651–75," *The William and Mary Quarterly* 69, no. 4 (2012): 793–822; Kristin Dutcher Mann, *The Power of Song: Music and Dance in the Mission Communities of Northern New Spain, 1590–1810* (Palo Alto, CA: Stanford University Press; Academy of American Franciscan History, 2010); Geoffrey Baker, *Imposing Harmony: Music and Society in Colonial Cuzco* (Durham, NC: Duke University Press, 2008); and Geoffrey Baker, "Indigenous Musicians in the Urban 'Parroquias de Indios' of Colonial Cuzco, Peru," *Il Saggiatore Musicale* 9, no. 1/2 (2002): 39–79. For additional recent scholarship on transcultural musical exchanges involving vocal music and singing practices, see Linford Fisher, *The Indian Great Awakening: Religion and the Shaping of Native Cultures in Early America* (New York: Oxford University Press, 2012); Olivia Ashley Bloechl, *Native American Song at the Frontiers of Early Modern Music*, New Perspectives in Music History and Criticism (New York: Cambridge University Press, 2008); Patrick Erben, *A Harmony of the Spirits: Translation and the Language of Community in Early Pennsylvania* (Chapel Hill: University of North Carolina Press, 2012); Tomlinson, *The Singing of the New World: Indigenous Voice in the Era of European Contact*; Beverley Diamond, *Native American Music in Eastern North America: Experiencing Music, Expressing Culture*, Global Music Series (New York: Oxford University Press, 2008); Christine DeLucia, "The Sound of Violence: Music of King Philip's War and Memories of Settler Colonialism in the American Northeast," *Common-place: The Journal of Early American Life* 13, no. 2 (Winter 2013), www.common-place-archives.org/vol-13/no-02/delucia/; Joanna Brooks, "Six Hymns by Samson Occom," *Early American Literature* 38, no. 1 (2003): 67–87.

18. Mann, *The Power of Song*, 260; Daniel Vickers, ed., *A Companion to Colonial America*, Blackwell Companions to American History (Malden, MA: Blackwell Publishers, 2003), 120. This book is part of a growing body of scholarship that builds on more recent and nuanced studies of Native American communities pre- and post-contact, and historical narratives of the eighteenth century that focus on both Native peoples and colonial settlers. Although missionaries provided the sources at the heart of this project, and those sources must be read carefully and critically for how they may present information on Native Christians, they are still valuable as historical evidence. As Daniel Richter has argued, if we are not prepared to include missionary- and settler-authored sources, then we must assume that whole categories of people will be simply left out of histories of the eighteenth century. Daniel K. Richter, *The Ordeal of the Longhouse* (Chapel Hill: University of North Carolina Press, 1992), 4–5.

19. See Richard Cullen Rath, *How Early America Sounded* (Ithaca, NY: Cornell University Press, 2003); Peter Charles Hoffer, *Sensory Worlds of Early America* (Baltimore, MD: Johns Hopkins University Press, 2003); Geoffrey Baker, *Imposing Harmony: Music and Society in Colonial Cuzco*; and Sarah Keyes, "'Like a Roaring Lion': The Overland Trail as a Sonic Conquest," *The Journal of American History* 96, no. 1 (2009): 19–43.

20. Keyes, "The Overland Trail," 21–22.

21. Alexis de Tocqueville, *Democracy in America*, 322; quoted in Vine Deloria, Jr., "American Indians and the Wilderness," in *Religions and Environments: A Reader in Religion, Nature and Ecology*, ed. Richard Bohannon (London: Bloomsbury, 2014), 87.

22. Matthew Hunter Price, "Methodism and Social Capital on the Southern Frontier, 1760–1830" (PhD diss., Ohio State University, 2014).

23. George Tinker, *Missionary Conquest: The Gospel and Native American Cultural Genocide* (Minneapolis, MN: Fortress Press, 1993), 9–10.

24. See Sarah Rivett, *Unscripted America: Indigenous Languages and the Origins of a Literary Nation*, Oxford Studies in American Literary History (New York, NY: Oxford University Press, 2017).

25. Merritt, "Dreaming of the Savior's Blood," 736; quoted in Kyle Fisher, "After Gnadenhütten: The Moravian Indian Mission in the Old Northwest, 1782–1812," *Journal of Moravian History* 17, no. 1 (2017): 34.

26. Fisher, "After Gnadenhütten," 34. For recent works on Christian missions and Native communities, see Linford Fisher, *The Indian Great Awakening*; Joel Martin and Mark Nicholas, eds., *Native Americans, Christianity, and the Reshaping of the American Religious Landscape* (Chapel Hill: University of North Carolina Press, 2010); K. McCarthy, "Conversion, Identity, and the Indian Missionary," *Early American Literature* 38 (2001): 353–370; and Wheeler, *To Live Upon Hope*.

27. For information on the musical and compositional training of Native American Moravians, and the collaborative process of creating Native-language hymns, see Wheeler and Eyerly, "Songs of the Spirit," 1–26. See Wheeler, *To Live Upon Hope*, for a discussion of the indigenization of Christianity in Mohican communities.

28. See Luke E. Lassiter, Clyde Ellis, and Ralph Kotay, *The Jesus Road: Kiowas, Christianity, and Indian Hymns* (Lincoln, NE: University of Nebraska Press, 2002); Michael D. McNally, *Ojibwe Singers: Hymns, Grief, and a Native Culture in Motion*, Religion in America (New York, NY: Oxford University Press, 2000); Tom Gordon, "Found in Translation: The Inuit Voice in Moravian Music," *Newfoundland and Labrador Studies* 22, no. 1 (2007): 287–314; Tom Artiss, "Music and Change in Nain, Nunatsiavut: More White Does Not Always Mean Less Inuit," *Études/Inuit/Studies* 38, no. 1/2 (2014): 33–52; and Sarah Eyerly, "Mozart and the Moravians," *Early Music* 47, no. 2 (May, 2019): 161–182.

29. Lisa Brooks, "Digging at the Roots: Locating an Ethical, Native Criticism," in *Reasoning Together: The Native Critics Collective* ed. Craig S. Womack, Daniel Heath Justice, and Christopher B. Teuton (Norman: University of Oklahoma Press, 2008), 262, n. 30.

30. Craig S. Womack, "Theorizing American Indian Experience," in *Reasoning Together: The Native Critics Collective* ed. Craig S. Womack, Daniel Heath Justice, and Christopher B. Teuton (Norman: University of Oklahoma Press, 2008), 372.

31. Juanita Little, a Native Catholic nun, argues for an understanding of her own Catholic experience from a Native perspective: "No one asks can you be Irish and Catholic, or Peruvian and Catholic? What is so incongruous about being Indian and Catholic? . . . I want to tell my people. 'You can be Indian and you can be Catholic. They are both the same.' Except that in the Catholic Church, we are members, not just of the tribe, but of the world-wide family." Juanita Little, "The Story and Faith Journey of a Native Catechist," in *Native and Christian: Indigenous Voices on Religious Identity in the United States and Canada* ed. James Treat (Hoboken, NJ: Taylor and Francis, 2012), 218.

32. See Hoffer, *Sensory Worlds of Early America*, viii; and Merrell, "Indian History During the English Colonial Era," in *A Companion to Colonial America*, ed. Daniel Vickers, Blackwell Companions to American History (Malden, MA: Blackwell Publishers, 2003), 129.

33. Although this book focuses on the North American missions of the Moravian Church, there were Moravian mission settlements in Central and South America, the Caribbean islands, Greenland, Great Britain and Ireland, continental Europe, Africa, the Middle East, India, Tibet, Siberia, Sri Lanka, Australia, and the Andaman and Nicobar Islands. For more information on the geographic extent of the missions, see Annegrete Nippa, *Ethnographie und Herrnhuter Mission: Katalog Zur Ständingen Ausstellung im Völkerkundemuseum*

Herrnhut, Aussenstelle des Staatlichen Museums für Völkerkunde Dresden (Dresden, Germany: Staatliches Museums für Völkerkunde, 2003), especially the map and table of mission locations on pp. 12–13.

34. *Heiden Collegia*, MissInd 217.12b, MAB. The mission plan is also discussed in the *Bethlehem Diary* on December 24, 1742.

35. Zinzendorf was one of the first Europeans to travel north of the Kittatinny Mountains and into the river valleys of the Susquehanna and its western and northern branches.

36. The *Heiden Collegia* also included a plan for working with German communities in southeastern Pennsylvania by planting churches in Oley, Germantown, Philadelphia, Tulpehocken, and Fredericktown, and by establishing German schools in each area.

37. For more information on the transatlantic aspects of the Moravian Church, see Peter Vogt, "'Everywhere at Home': The Eighteenth-Century Moravian Movement as a Transatlantic Religious Community," *Journal of Moravian History* 1 (2006): 7–29.

38. From a letter of "Conrad Weiser to a Friend, 1746," quoted in *Memorials of the Moravian Church*, William Cornelius Reichel, ed. (Philadelphia: J. B. Lippincott, 1870), 89–90.

39. Shekomeko was the first mission established in an existing Native community. Bethlehem was the first fully Church-constructed Moravian mission town in North America. For more information about Rauch and the mission at Shekomeko, see Wheeler, *To Live Upon Hope*.

40. George Henry Loskiel, *History of the Mission of the United Brethren, vol. II* (London: printed for the Brethren's Society for the Furtherance of the Gospel: sold at No.10, Nevil's Court, Fetter Lane), 37.

41. Richard W. Pointer, *Encounters of the Spirit: Native Americans and European Colonial Religion*, Religion in North America (Bloomington, IN: Indiana University Press, 2007), 143. Wheeler, *To Live Upon Hope*.

42. David Zeisberger, "Foreword," in *A Collection of Hymns for the Use of the Delaware Christian Indians, of the Mission of the United Brethren in North America*, 2nd ed. (Bethlehem, PA: J. and W. Held, 1847).

43. Wheeler and Eyerly, "Songs of the Spirit," 1. Also see Pointer, *Encounters of the Spirit*, 144; and Woodward, "Incline Your Second Ear This Way."

44. See Wheeler, *To Live Upon Hope*, 95–104.

45. Paul Peucker, *A Time of Sifting: Mystical Marriage and the Crisis of Moravian Piety in the Eighteenth Century*, Pietist, Moravian, and Anabaptist Studies (University Park, Pennsylvania: The Pennsylvania State University Press, 2015), 19.

46. For an excellent overview of the use of lots in Moravian communities, see Gillian Lindt Gollin, *Moravians in Two Worlds: A Study of Changing Communities* (New York: Columbia University Press, 1967).

47. Eventually, the watchwords were standardized and chosen at the beginning of each new year by Church elders by drawing slips of paper randomly from a bowl to represent each day of the calendar year. This is a practice that continues in the Moravian Church to the present day.

48. Sarah Eyerly, "*Der Wille Gottes*: Musical Improvisation in Eighteenth-Century Moravian Communities," in *Self, Community, World, Moravian Education in a Transatlantic World*, ed. Heikki Lempa and Paul Peucker, Studies in Eighteenth-Century America and the Atlantic World (Bethlehem, PA: Lehigh University Press, 2010), 201–227; also see Katherine M. Faull, "Speaking and Truth-Telling: *Parrhesia* in the 18th-Century Moravian Church," in *Self, Community, World: Moravian Education in a Transatlantic World*, 204–230.

49. Wheeler, *To Live Upon Hope*, 6.

50. Joanne van der Woude, "Polyglot Harmony: Moravians among the Indians," in "Towards a Transatlantic Aesthetic: Immigration, Translation, and Mourning in the Seventeenth Century" (PhD diss., University of Virginia, 2007).

51. Peucker, *A Time of Sifting*, 26.

52. Wheeler, *To Live Upon Hope*, 63.

53. Hans Rollman, *Moravian Beginnings in Labrador: Papers from a Symposium Held in Makkovik and Hopedale* (St. John's, Newfoundland: Newfoundland and Labrador Studies, Faculty of Arts Publications, Memorial University, 2009), 135.

54. Peucker, *A Time of Sifting*, 4–5. The designation "Moravian" was a spiritual designation in the eighteenth century, and was not considered to be associated with race or ethnicity. The mission movement that began in the original community of Herrnhut may have started in Germany, but it has since that time evolved to become a worldwide church denomination. Currently, the African synods constitute more than half of the membership of the Moravian Church. For a discussion of the Moravian missions in Tanzania, see Anna Maria Busse Berger's article, "Spreading the Gospel of *Singbewegung*: An Ethnomusicologist Missionary in Tanganyika of the 1930s," *Journal of the American Musicological Society* 66, no. 2 (2013): 475–522.

55. English translation of "Catalogue of baptized Indians in North America," MissInd 3191.1, MAB.

56. This study aims to present what Anne Kelly Knowles and Amy Hillier have proposed as a new way of writing history that includes human experiences of space and place and affective geography. Scholars in historical GIS are increasingly interested in mapping frameworks that are capable of visualizing relations, networks, connections, emotions, and nonstandard patterns of movements. But it is my contention that spatial humanities approaches combined with audible histories have the potential to restore an almost multidimensional quality to the past. It is my hope to achieve a more holistic, sensory experience of Moravian mission history by combining the fields of geography, including aural and sound cartography and geographical information systems (GIS) with the fields of sonic and acoustic ecology, sound studies, and musicology. For recent scholarship on historical GIS, see Anne Kelly Knowles and Amy Hillier, eds., *Placing History: How Maps, Spatial Data, and GIS Are Changing Historical Scholarship* (Redlands, CA: ESRI Press, 2008); David J. Bodenhamer, John Corrigan, and Trevor M. Harris, eds., *The Spatial Humanities: GIS and the Future of Humanities Scholarship*, The Spatial Humanities (Bloomington, IN: Indiana University Press, 2010); and David J. Bodenhamer, John Corrigan, and Trevor M. Harris, eds., *Deep Maps and Spatial Narratives*, The Spatial Humanities (Bloomington: Indiana University Press, 2015); Ian Gregory, *A Place in History: A Guide to Using GIS in Historical Research*, History Data Service (Oakville, CT: David Brown, 2003); Ian Gregory and Alistair Geddes, eds. *Toward Spatial Humanities: Historical GIS and Spatial History*, The Spatial Humanities (Bloomington, IN: Indiana University Press, 2014); Stephen Daniels et al., *Envisioning Landscapes, Making Worlds* (London, England: Routledge, 2012); and John Lewis Gaddis, *The Landscape of History: How Historians Map the Past* (New York: Oxford University Press, 2002).

57. Since maps are best at representing particular points in time, the sound maps that form the *Moravian Soundscapes* project are sited in 1758. By 1758, most of Bethlehem's communal and industrial buildings had been completed, with the exception of the Widows' House and the final addition to the Single Sisters' House in 1768. Also, 1758 is the year best

represented by archival materials (maps, diaries, artistic representations) from the Moravian Archives in Bethlehem. The soundscape compositions embedded in the sound maps are more specifically representative of a typical mid-morning in the month of May 1758. This project is also a part of a new and interdisciplinary field—digital sound studies—that lies at the intersection of sound studies and the digital humanities. For recent works on digital sound studies, see Mary Caton Lingold, Darren Mueller, and Whitney Trettien, eds., *Digital Sound Studies* (Durham, NC: Duke University Press, 2018); and Rebecca Geoffrey-Schwinden, "Digital Approaches to Historical Acoustemologies: Replication and Reenactment," in *Digital Sound Studies*, 231–249. For an excellent discussion of the difficulties involved in writing and researching aural history, see Mark M. Smith's "Introduction: Onward to Audible Pasts," in *Hearing History*, 417–431.

58. Christine DeLucia has argued for the importance of "digging deep in small places over time" and paying attention to artifacts, rituals, and other gestures of human experience. DeLucia, *Memory Lands: King Philip's War and the Place of Violence in the Northeast*, Henry Roe Cloud Series on American Indians and Modernity (New Haven: Yale University Press, 2018), 3, 10. Like Karen Halttunen, she advocates for historians to eschew the privileging of bigger histories of early America over smaller histories that are attentive to local and regional place. Karen Halttunen, "Grounded Histories: Land and Landscape in Early America," *The William and Mary Quarterly* 68, no. 4 (2011): 513–532. The maps for this book are created with the idea of representing localized and intimate Moravian ideas of space and place. The terms "space" and "place" are used in this book in ways that are reflective of how eighteenth-century Moravians envisioned their communities. "Space" and "place" may represent specific locations in the physical (i.e., human and natural) world, as well as Moravians' conceptions of the environments they inhabited. However, it is also important to recognize a distinctly spiritual and intangible sense of location, which was an important concept in early Moravian mission communities. In this sense, the terms "space" and "place" are not tied to specific physical locations but represent instead an overlay of the spiritual world onto the physical geography of landscape. Like DeLucia, I hope that the recentering of place as a lens of analysis—rather than time, or the typical periodizations used in academic historical studies—can reveal alternative understandings of the past and geography. DeLucia, *Memory Lands*, 2–3. Also see Lisa Brooks, "The Primacy of the Present, the Primacy of Place: Navigating the Spiral of History in the Digital World," *PMLA* 127, no. 2 (2012): 308–316.

59. After the first mission at Gnadenhütten, Pennsylvania, was destroyed in 1755, it was eventually rebuilt in Ohio in 1772. Moravian scholars typically refer to these two communities as Gnadenhütten I and Gnadenhütten II. I have chosen to avoid those designations in this book.

60. I would like to thank Janet Rice for sharing her work and that of her collaborators in mapping the archaeological sites for Native communities in Pennsylvania. Barry C. Kent, Janet Rice, and Kakuko Ota, "A Map of 18th Century Indian Towns in Pennsylvania," *Pennsylvania Archaeologist* 51, no. 4 (1981): 1–18.

61. DeLucia, *Memory Lands*, xxv; Lisa Brooks, *Our Beloved Kin: A New History of King Philip's War*, Henry Roe Cloud Series on American Indians and Modernity (New Haven: Yale University Press, 2018), 13. Recent mapping projects by Lisa Brooks and Christina DeLucia of the indigenous geographies of New England have encouraged me to consider the wonderful benefits of "research road trips" and "place-visits." I have also been encouraged in my efforts to create sound maps of Moravian places by DeLucia's call to enliven historic places with alternate modes of seeing, touching, traveling, and mapping. She encourages historians of

early American history to consider new ways of writing and mapping that reflect different stories than the ones traditionally told in settler colonial contexts. DeLucia, *Memory Lands*, 21, 330.

62. I wish to thank Philip Trabel and Charlene Donchez-Mowers and the staff of the Historic Bethlehem Partnership and Burnside Plantation, for their assistance with this project.

63. For the purpose of this project, we were interested in general information about the spread of sound to elucidate how Moravians may have heard and understood their community. Thus, sound recordings were assigned a general weight in the calculations with an assumed range of human hearing set at 0 dB with a lower threshold at −9 dB. It is our hope that additional studies may take into account deeper and more nuanced views of the spread of sound in Bethlehem.

64. There are a growing number of artists and researchers using GIS technologies to inscribe meaning onto space through sound. Some important examples include the soundwalks created by Hildegard Westerkamp and Frauke Behrendt; the "Under Living Skies" project by Eric Powell that recreates the soundscapes of Saskatchewan, Canada; Isobel Anderson and Fionnuala Fagan's collaborative project entitled, "Stories Of The City: Sailortown," which explores the soundscapes of the old docks area of Belfast, Ireland; Janet Cardiff's soundwalks, such as *"Her Long Black Hair"* and *"A Large Slow River,"* that combine recorded voice with composed soundscapes in order to map a narrative onto a specific sound journey; and Jennifer Heuson's "Soundscapes of The Black Hills," which records various locations in the Black Hills of South Dakota.

65. There are not many researchers or composers experimenting with sound in time even though acoustic ecology is a growing field. In the field of archaeoacoustics, Miriam Kolar's project on the acoustic architecture of Chavín de Huántar, Perú, uses computer modeling to understand how this 3,000-year-old ceremonial center in the Incan Andes may have been acoustically designed. Miriam A. Kolar, "Sensing Sonically at Andean Formative Chavín de Huántar, Perú," *Time and Mind* 10, no. 1 (January 2, 2017): 39–59. In terms of soundscape compositions based upon historic sound recreation, Maile Colbert's sound projects, *Passageira em Casa* and *Passageira australis*, explore various sounds or locations in Portugal and Australia as heard and experienced through time. Several university-based research groups have published websites dedicated to sounding historical places, including the University of Cambridge's "Seventeenth-Century Parisian Soundscapes Project," and the "Sound of Paris in the Eighteenth Century" project at the University of Lyon. Some cities and cultural regions have also funded similar projects, including the "Vancouver Soundscape Project," and the *"Paisajes sonoros históricos de Andalucía* (c. 1200–c. 1800) [Historic soundscapes of Andalucía project (c. 1200–c. 1800)]."

66. Barry Truax, "Sound, Listening and Place: The Aesthetic Dilemma," *Organised Sound* 17, no. 3 (December 2012): 193, 196–197.

67. Barry Truax, "Genres and Techniques of Soundscape Composition as Developed at Simon Fraser University," *Organised Sound* 7, no. 1 (April 2002): 12. Soundscape compositions are a style of composition pioneered at Simon Fraser University by the World Soundscape Project. Some early examples include *The Vancouver Soundscape* (1973), and *Soundscape Vancouver* (1996). Hildegard Westerkamp defines soundscape composition as electronic compositions that are created with recorded environmental sounds (Westerkamp, "Linking Soundscape Composition and Acoustic Ecology," *Organised Sound* 7, no. 1 (April 2002): 51). Soundscape compositions might explore structures and perspectives that mirror real-

world experiences, such as listening from a fixed spatial perspective or moving through a connected series of acoustic spaces. They are also meant to convey a sense of real sound environments. In this sense, they are similar to the soundtracks of wildlife films, which are typically a combination of sounds recorded in the wild during the filming or previously, as well as sounds that must be recreated in a studio. These soundtracks are meant to provide insight into animal behavior, and to create a sense of a wild place, as well as to heighten the emotion and drama of the film. As a result, wildlife filmmakers often turn to sound designers to recreate something that simulates the sounds of wild places—a soundtrack that is in its essence true to nature, yet recreated from sound samples that may not have been recorded along with the film itself. This is similar to the comprehension of visual materials advocated for by photographer Dorothea Lange. For Lange, the camera was to be a tool to learn to see without a camera. In this way, recordings can also be envisioned as tools to hear places we have not experienced or cannot experience without the aid of recordings. See Bernard Krause, *The Great Animal Orchestra: Finding the Origins of Music in the World's Wild Places* (Boston: Back Bay Books Little Brown, 2013), 16.

68. Barry Truax makes a distinction between compositions based upon "sound effects libraries" and "soundscape documentation projects" (Truax, "Paradigm Shifts and Electroacoustic Music," 109). The soundscape compositions in *Moravian Soundscapes* are not documentation projects, but historically informed recreations.

69. See Isobel Anderson, "Soundmapping Beyond The Grid: Alternative Cartographies of Sound," *Journal of Sonic Studies* 11 (January 14, 2016); and S. Caquard, G. Brauen, B. Wright, and P. Jasen, "Designing Sound in Cybercartography: From Structured Cinematic Narratives to Unpredictable Sound/Image Interactions," *International Journal of Geographical Information Science* 22, nos. 11–12 (2008): 1220.

70. See Elisabeth Le Guin, *Boccherini's Body: An Essay in Carnal Musicology* (Berkeley, CA: University of California Press, 2006).

71. Seth S. Horowitz, *The Universal Sense: How Hearing Shapes the Mind* (New York: Bloomsbury, 2013), 32.

72. Mark M. Smith, Mitchell Snay, and Bruce R. Smith, "Coda: Talking Sound History," in *Hearing History: A Reader*, ed. Mark M. Smith (Athens: University of Georgia Press, 2004), 365–366. Also see Truax, *Handbook for Acoustic Ecology*, 126.

73. Womack, "Theorizing American Indian Experience," 372–374.

74. See Brooks, *Our Beloved Kin*, 124.

PEALE

ON A FALL AFTERNOON IN 1989, PAST THE season of rattlesnakes, I hiked into the forest near my home to a mining community that had flourished and vanished in the early twentieth century. Peale, Pennsylvania, was once home to a bituminous tunnel mine and several thousand people. It was a hub of activity and commerce, resounding day and night with voices, the groans and creaks of mine machinery, and the keynote sounds of shops, homes, and businesses. Now it lingered in the local imagination as a sketchy, silent "ghost town" lost in the woods. Few people knew how to find it. Its buildings, roads, and railroad tracks had long been subsumed under the plants and trees that quickly reclaimed the landscape. Its human soundscapes were replaced with those of the forest: birds, rustling leaves, white-tailed deer, and chipmunks. But if you knew where to look, you could still see the town's graveyard, and the muted remnants of cherished hymn and Bible verses scrawled on tombstones.

I had been told that to locate Peale I should walk along the banks of the Moshannon Creek to where it joined a smaller run coming out of the hillside near the town of Grassflat. Like the verses that lingered on tombstones, the name of this small waterway persisted long after anyone remembered its significance: Moravian Run. I wondered about the name. Perhaps it was simply named after immigrants from the region of Moravia, as were many mining communities and places in the area that retained traces of their ethnically segregated roots. But I puzzled over its parallel with the name of the Moravian Church, the German church community that had founded Bethlehem, Nazareth, and Lititz, Pennsylvania, in the eighteenth century. There were no Moravian churches around my home, so the connection seemed highly unlikely. Still, it was a mystery that continued to intrigue me.

Less mysterious perhaps was a beautiful, secluded island in the Moshannon Creek about one mile further east of Peale. My mother called it Post's Island, and often recounted stories of my Swedish grandfather, Morton Erickson, who loved to spear fish from the island and hunt in the forests along the Moshannon. In the late nineteenth century, when the first Swedish settlers came into the area, the waters around Post's Island were exceedingly clear and teeming with fish, otters, water striders, and mink. White-tailed deer and elk roamed the river's edge, and bears haunted the deeper parts of the forest. The acoustic ecology of the Moshannon forests was vibrant with animals, plants, insects, and the geophony of weather, water, soil, and rock. My grandfather was a good singer. I imagined him singing the latest gospel hymns while sitting on the banks of Post's Island hoping to lure a fish. That was before the days of Peale and other mining communities that caused the Moshannon to run red with sulphur.

I never wondered about the naming of Post's Island. It seemed unremarkable and rather ordinary. But many years later, I would learn about the man Post and the journey that had taken him, like me and my grandfather, along the Moshannon. I would learn of another grave that lay under the remains of Peale, never marked by a stone, which also carried the history of human settlement and passage in that place at the confluence of the Moravian and the Moshannon. It was a time when the island and the creek had received their names and when the riparian forests had echoed with the songs of even more ancient ancestors. But these were mysteries that would take many years to solve.

1

PENN'S WOODS

Ein GOttes närrgen	God's little fool
ist schon so imaginatif	Is so beautifully imaginative
ins Lämmleins seine Pleura tief:	Inside the little Lamb's side-hole deep:
kein fischgen schwimmt,	No little fishes swim,
kein vöglein singt,	no little birds sing,
kein bäumgen blüht,	No little trees bloom,
kein hirschgen springt	no little deer spring
so applicirts das selgen	So applies the soul itself to thee,
auf sich unds wunden-höhlgen.[1]	And to your little wounds-hole.

THE FORESTS, MOUNTAINS, AND RIVER VALLEYS OF PENNSYLVANIA had been named and mapped by Native American cultures for thousands of years before the colonial era.[2] Long before Pennsylvania became a territory ceded by Charles II of England to Admiral William Penn to discharge his debts, this was a landscape that had been traversed and interpreted, worshipped, storied, and sung by the people who lived there. Native places on the land and water were often endowed with names that carried mnemonic, descriptive qualities: Ahkokwesink (The Place of Mushrooms) or Ahsenesink (The Place of Rocks). The acoustic environments of forests, fields, and streams were also remembered with names that spoke of their sounds: Chekhonesink (The Place Where There Is a Gentle Sound) or Oniska (The Ringing Rocks). These place names were sounded metaphors that embedded generations of memories of animals and birds, the natural topography, or the sounds of falling water and lithophonic rocks. They were charted and mapped in rock cairns and painted trees, bark scrolls and songs. Places became intertwined with their names. To sound them was to honor and remember the ancestors who had once claimed that ancient landscape with words.

In the eighteenth century, this familiar landscape became a liminal space, wedged between the competing land claims of France, Great Britain, and the Six Nations. It would be claimed under a new name, Penn's Woods, or *"muni khikhakan eheluwensink Pennsylvania* (This State Which Is Called Pennsylvania]." In this renamed landscape, both settlers and Native peoples sought to construct new and changing identities in response to each other and to rapid changes in their natural, political, and cultural environments. Along with an influx of immigrants from Europe, Africa, and the Caribbean, seeking new opportunities in Penn's Colony, the first Moravian missionaries arrived in Pennsylvania in 1740. Within two years, they had established communities in the region around the Lechewuekink (Lehigh) and Lenapei Sipu (Delaware) Rivers. ⊕ **See website chap1.1, Static map: "Early Moravian Missions in Pennsylvania and Ohio."** Moving outward from these mission centers, Moravian missionaries traveled frequently to Native American villages and settlements throughout eastern Pennsylvania, New York, and Connecticut, via water or the complex network of forest trails that had been used by Indigenous communities for centuries before the arrival of Europeans. On these forest journeys through the Mahantango and Kittatinny Mountains, and the river valleys of the Lehigh, Juniata, and Susquehanna, missionaries renamed indigenous places. Mountain ridges, rivers, valleys, and springs were memorialized with new Moravian perspectives on the landscapes of Penn's Woods: Ludwig's Fountain, Erdmuth's Spring, Ludwig's Rest, Anna's Valley, Benigna's Creek, Jacob's Heights. Even a Native hunting cabin in the Tiadaghton Forest came to be christened the "Coffee House" in remembrance of European places left behind.

Confronted with new and unfamiliar landscapes, European settlers, including Moravian missionaries, often fell back on familiar patterns of naming and claiming space that would transform Native country into a European-inflected landscape.[3] ⊕ **See website chap1.2, Interactive map: "The Pennsylvania Frontier."** For Moravians, this process of claiming Pennsylvania as a Christian, and more specifically a Moravian Christian, space began immediately upon their arrival in Pennsylvania. In 1742, Count Nikolaus Ludwig von Zinzendorf became one of the first Europeans to travel north and west of the Kittatinny Mountains into the dense forests that covered much of eastern and northern Pennsylvania. As he journeyed, he named and interpreted the places and people he encountered, committing them into an ongoing Christian narrative told through maps, diaries, names, and hymns that came to symbolize the essential features of the American colonies for a pan-Moravian audience who had never seen or heard Pennsylvania's forests.

Hearing the Forest

The world was a library and its books were the stones, leaves, grass, brooks, and the birds and animals.

Luther Standing Bear, Land of the Spotted Eagle

Our interactions with the natural environment are framed by the maps we draw, the stories we tell, and the songs we sing. As historian Simon Schama has argued, "landscapes are culture before they are nature; constructs of the imagination projected onto wood and water and rock."[4] Whether those landscapes are forest, grassland, mountain, sea coast, or other setting, the natural environment is the most fundamental place that we inhabit. It is in these environments that we develop our understandings of the world: attachments, connections, meanings, experiences, belongings, and exclusions.[5] These simultaneously imagined and physical landscapes constitute the reality of our human experience on a daily level. But landscapes are not merely physical topographies: they also exist in sound.[6] Our sense of spatiality is not grounded in only sight but in sound; we listen to perceive distance and space. Our interactions with the world fully engage the senses, and our ears are constantly attuned to a wide range of sounds: language, music, rain, even birdcalls.[7] It is from these sounds, and other sensory data, that we form Schama's "constructs of the imagination." These are the landscapes of songs, maps, names, stories, rituals, and histories.

Long before Count Zinzendorf's mid-eighteenth-century journey through Pennsylvania's forests, this was a landscape that had been sung, storied, and mapped by Indigenous communities. According to current archaeological data, as glaciers receded from Pennsylvania around the end of the Ice Age, people migrated into the Ohio, Susquehanna, and Delaware river valleys from the more populous interior regions of the continent.[8] Pennsylvania's first residents would have encountered an ecological patchwork of environments in the lands south of the glacial ice. Thick forests of spruce, fir, birch, pine, and alder dominated the lower slopes of the Appalachian Mountains and the ridge and valley systems that stretched to the east and south. Over time, as the climate warmed, hardwoods such as oak, chestnut, hickory, and beech began to populate the forests along the Appalachian Plateau. These new forests supported a rich understory of edible and medicinal plants: mushrooms, berries, ginseng, chestnuts, walnuts, and hazelnuts.[9] In the middle canopy, dogwood, ironwood, viburnum, spicebush, witch hazel, and honeysuckle vied for sunlight and sustenance.[10] On the alluvial plains along the Susquehanna and Delaware Watersheds grew carpets of wild strawberries, so notable a feature of the riparian landscape for hundreds and perhaps thousands of years that early European travelers would eventually write of "whole plains covered with them as with a fine scarlet cloth."[11] Even during the sixteenth century, and the first recorded journeys of Europeans along these eastern river valleys, more than 90 percent of Pennsylvania's landscape was covered in densely packed forests.

Pennsylvania's diverse geographic regions and forests supported a great variety of animals. The grasslands of the higher elevations and pools of salt and brackish water along creeks and streams were home to deer, elk, moose, and buffalo.[12] These salt holes attracted predators such as wolves, panthers, lynx, cougars, and foxes. In the denser parts of the forest, thickets of mountain laurel, witch hazel, and chokeberry harbored bears' dens and crouching panthers, wild cats, mountain lions, and boars hiding in the underbrush. The dense carpet of leaves on the forest floor teemed with field mice, moles, chipmunks and squirrels, as well as ticks, fleas, and beetles that carved the bark of trees or fed on the blood of passing animals. Minks, otters, muskrats, and beavers flourished along the many creeks and streams that flowed off of the Appalachian Front.

The acoustic ecologies of Pennsylvania's forests were dynamic. Depending on the particular place, time, and season, the sounds of wind, water, fire, rustling plants and trees, and falling rocks carried quickly over dry terrain or were muffled in the humidity of a rainy day. Even the dynamic sounds of water fluctuated from ice to snow to rain, or from stream, to creek, to river. The quiet sounds of winds moving through the dense hardwood stands had their counterpoint in the vigorous blowing of salt breezes on the riparian plains of the Susquehanna. The branches of trees that remained silent and still in the heat of summer crackled in the brittle cold of winter. These natural sounds were augmented by birds, insects, and animals who responded in their calls and communications to patterns of light and dark, fluctuating seasons and climates. The dense heat of a summer day could suddenly transform into a cacophony of birds, insects, and frogs after an afternoon thundershower. Spring evenings resounded with the dense soundscapes of insects and amphibians that resonated over wetlands and along the margins of ponds.[13] Common horseflies, mosquitos, grasshoppers, yellow jackets, wasps, and locusts clicked and scraped in densely layered soundscapes in the upper canopies of forests and along the grassy edges of meadows. The calls and songs of forest and meadow birds—pigeons, turkeys, turtle doves, woodpeckers, bald eagles, owls, wrens, bluebirds, hummingbirds, and thrushes—resounded through the skies.[14]

Within this densely layered landscape and soundscape of plants, animals, insects, and birds, Native American settlements clustered around Pennsylvania's distinct geological regions and river systems. A majority of travel and commerce centered on the three major watersheds of the Delaware, Susquehanna, and the confluence of the Allegheny, Monongahela, and Ohio Rivers. An intricate network of trails and pathways linked different settlements in the river valleys, although only a few pathways, such as the Great Shamokin Path and Kittanning Path, traversed the high mountains of the Appalachian Front.[15] 🌐 **See website chap1.3, Interactive sound map: "The Great Shamokin Path."** Rivers, streams, and springs were also crucial hubs for the spatial distribution of settlements and territorial boundaries, creating zones of human activity interspersed with forested borderlands.

For Native Americans, these borderlands of the forest were filled with the sounds and voices of stones, dirt, animals, plants, wind and air, water and fire, trees, insects. The patterns of these geophonic and biophonic soundscapes articulated distinct sonic languages that characterized particular geographic areas.[16] Careful attention to and meaningful interpretation of aural cues from animals, insects, and birds, as well as wind, water, and storms were important in a typically dense, forested environment where distance could not be adequately judged or easily remembered by sight.[17] The paths between well-established and populous towns such as Shamokin, Kittanning, and Onondaga were carefully charted through trail markings, painted trees, and mental maps that preserved the spatial relationships of the structured world of villages and agricultural land, to lighter forest thickets and dark "swamps," places where the trees grew so close and so high that they blocked the sun.

Human journeys into the forest were often accompanied by a complex system of songs and offerings that could be sung to appease or beguile the spirits who resided there. These symbolic methods of naming, remembering, and sounding the landscape were especially important in navigating the miles of forest lands that lay between villages and towns. These forested spaces were crowded with spirits, who sometimes helped or hindered the people who encountered them. In the darkness of the liminal under-canopy, accidents could easily happen: broken legs, starvation, mental illness, and stripping of the powers of sight and hearing. People listened carefully for the dynamics and counterpoint of the natural environment, observing climate, season, weather, and time of day through the soundscapes of frogs, trumpeter swans, or wild geese. Shades of darkness were measured by the calls of nocturnal birds: spring and summer nights resonated with the songs of the whip-poor-will and the noisy calls of owls. But in the liminal spaces of the forest, these sounds could also disrupt human activities. Upon their return to villages and human spaces, travelers were immersed in complex rituals designed to counter the ill effects of the woods. For eastern Woodlands cultures, "Edge of the Woods" ceremonies were critical pathways to healing that cleared the eyes and unstopped the ears of those who had ventured "thro' dangerous places, where evil Spirits reign."[18] The particular ability of owls to imitate the human voice could create misunderstandings or cause messages to go astray.[19] According to a Delaware legend, screech owls were particular bearers of misfortune: "*Enta wa chululhuwe pèchi lihëlak hìtkunk tali kochëmink ènta awèn wikit luweyok hùnt, 'O, mata wëlëtu.' Alëmi wishas'hatuwàk, wëluseme̊neyo në sikhay òk patamaok. Elaihòsihtit hùnt lòmwe Lënapeyunkahke lòmëwe.*" (When a screech owl comes to your home and lands outside in a tree where a person lives they say, "Oh, that is not good!" They began to be afraid and they burned some salt and they prayed. That is the way the old Delawares did long ago.)[20]

The perching of a nighthawk on the roof of a house was also to be avoided at all costs. Its "singing with a mournful note" portended impending disaster to those who heard it. The cooing of turtle doves was even worse—a harbinger of death.[21] The bald eagle could cause thunder if angered in spirit.[22] But rather than bringing ill luck, some animals were simply noisy nuisances. Meadows and agricultural plantations were frequently inundated with flocks of wild pigeons so loud that they prevented people from hearing each other. These birds could appear suddenly in large groups and descend like a cloud, forming "a ceiling between earth and sky."[23]

But in the woods, where sight was often limited to the next stand of trees, recognition of animal and bird sounds was crucial to survival.[24] Birds and animals possessed a keenness of hearing that rendered them particularly dangerous. Packs of wolves roamed the woods at night, listening and smelling for prey. Those humans and animals who traveled or slept out at night took care to be silent. The howling of wolves was an omnipresent sound of the nocturnal forest—their ostinatos called out the hunt or served as directional locators for the pack in the darkness.[25] Animals also listened and responded to human sounds. An early Moravian Church history, George Loskiel's *History of the Mission of the United Brethren*, relates the story of a young Native man named Joshua. One summer's day, as Joshua was traveling in the woods near his community, he surprised a mother bear and cub near their den. With a roar, the mother rushed at him. The terrified Joshua screamed so loudly that the bear was unnerved and "suffered him to escape."[26]

Sounded communications between humans and animals were especially critical for those who entered the woods to hunt. Hunters needed to possess inward spiritual knowledge to sing songs and make sounds that would summon prey or appease the spirits of animals killed for food. Often, hunting songs came through dreams produced after fasting or ingesting spiritual medicines called *besons*. Hunting besons were typically prepared by older men, who may have been too weak to join in the hunt, and consisted of roots, herbs, and seeds. Some were emetic and produced vomiting; but ingested in small doses they could ensure success in hunting. Besons also had the power to yield potent dreams.[27] In beson-induced dreams, hunters could learn of the locations of animals or the best methods to appease the wrath of evil spirits. Dreamers especially hoped to encounter the dead and to hear them speak. With the right prayers and sacrifices, an ancestor might guide the hunter to game.[28] Young boys were encouraged to solicit dreams of communication by envisioning animals or animal spirits.[29] According to Loskiel's *History*, dreams of predatory birds and vultures could signal either the success or failure of an upcoming hunt, depending on the nature of the dream. Hunters could also offer a preparatory sacrifice of a deer, divided into many small pieces, so that they could observe the pattern created by carrion birds as they ate the pieces of meat. These patterns possessed symbolic meaning that alerted hunters to the nearby presence of game, or an impending injury or disaster that may occur if they chose to hunt at that particular time. Nocturnal birds such as owls could also serve as an aid to hunters. Those who heard their calls in the woods at night could offer a sacrifice of tobacco to the campfire and expect to receive a blessing in the next day's hunt.

Hunters also learned to communicate through sound with animals and their spirits by "calling the game"—vocally imitating animal noises, or using natural objects to produce sounds attractive to particular animals. These calls appealed to an animal's sense of hearing and their instincts.

Archaeological finds in various locations throughout Pennsylvania have verified that turkey-wing calls have been used in Pennsylvania for at least the past four thousand years. These small, delicate wing bones could be carefully joined together, and used as a type of mouth organ that produced sounds remarkably similar to a turkey's call. But mastering the art of communication with animals also required practice and an understanding of the syntax of animal sounds: greeting calls, hailing calls, feeding calls, lonesome calls, and mating calls. Clucks, yelps, quacks, purrs: each animal had a range of sounds embedded within a complex sonic language. Even with the aid of a wing call, learning these animal languages required years of practice.

Boys began to learn the art of hunting in their infancy. Toddlers were encouraged to begin by climbing trees in search of birds. This strengthened their connection with the spirits of birds and honed their eyesight for even the smallest movements in the bushes or trees. Woodlore and hunting lore were skills honed by weeks or months spent in the woods. Many years of a man's life could be spent in the forest, beginning with a rite of initiation around the age of ten:

> At the age of about ten, a boy underwent a test of endurance. He was sent into the woods with bow and arrows and told not to return until he had shot something to eat. Before he set out, his face was blackened with charcoal, a sign to all whom he met that he was on his test and was not to be helped. Little Wildcat Alford, when undergoing this ordeal, was two days alone in the woods without food. He became too weak to shoot straight; but he managed somehow to kill a quail and returned to his family, a man.[30]

Another important goal of these solitary experiences was to establish lifelong relationships with natural spirits. The forest could be a place of revelatory experiences and visions, an in-between space where one encountered the divine. Listening was central to understanding and mapping these sacred spaces within the forest. The cawing of a raven often preceded a period of religious revelation.[31] In pan-Native American lore, even inanimate objects had the potential to speak. According to a Seneca tale, "a young boy, tired from hunt, rested his head against a great stone. The stone began to talk. Tired and hungry as he was, he listened, for every hunter knows that spirits dwell in objects. From the stone he learned the ways of the various game animals."[32]

Since the power of spirits could be revealed in the rushing of mountain streams or the roar of waterfalls, the murmur of springs or the groaning of trees as they swayed in the wind, naming these acoustic environments was one important way of claiming their potential for human use.[33] For eastern Woodlands cultures, geographic information, and therefore a sense of distance and place, was conveyed through place names that reflected the natural features of the environment: Kighalampegha (Standing Pond), Moghwheston (Worm Town), Sughchaung (Salt Lick), Oghkitawmikaw (White Corn) or Wapwallopen (Where the White Hemp Grows). Each of these place-names represented the plants, insects, animals of the local ecology. The town of Wyoming (Large Plains) was named after a stretch of riparian meadow on the north branch of the Susquehanna River. The Delaware River, whose western branch flowed by the future site of Bethlehem, Pennsylvania, was sometimes referred to with a name that represented the speed of rushing water: Lenape-whituttuck (Rapid Stream).[34] And the people who lived around the river took their name from that place: the Lenape (Delaware). Place-names could sometimes signify two different meanings. The Delaware settlement of Tulpehákink (Turtle Land Place) was named for the Tulpehocken Creek, which once abounded with water turtles, but it also indicated the area's human inhabitants, members of the Delaware Turtle phratry.[35]

Areas with notable soundscapes were also named. A place along the banks of the Ohio River populated by toads so loud that travelers reported being unable to sleep received the fitting name Tsquallutene (Town of Toads).[36] The name of Sheshequanink (At the Place of the Gourd Rattle), on the north branch of the Susquehanna River reflected the sound of the gourd rattles that accompanied religious ceremonies often conducted in the area. Paupaunoming (Cave of the Winds) in the Kittatinny Mountains resounded with whistling gusts of air from small holes in the cave walls.

Once names were given to places, it was possible to link them into a mental framework for quick reference and travel. This type of mental mapping could be sounded through song or speech for personal or communal use.[37] Mental maps also incorporated an extensive network of trails and pathways that linked named places across Pennsylvania and extended beyond to the entire North American continent and the many hundreds of cultures and civilizations of Native America.[38] Pennsylvania's Allegheny Path likely once joined other trails that connected to places as far away as Cahokia (modern-day St. Louis, Missouri). From Pennsylvania, it was also possible to connect southward to the Shenandoah Valley through the Appalachian Mountains to the Cumberland Gap—a distance of nearly eight hundred miles.³⁹ Although these paths could shift with seasons and weather, they still served as important intersections of cultural communication and trade. Understanding how to traverse them was a skill that depended on long practice and knowledge of physical and sonic geography.⁴⁰

Places could be linked and remembered through maps drawn on tree bark, painted trees, and rock cairns. Eastern Woodlands people envisioned these complex symbolic systems to be visual as well as aural in nature. Hieroglyphics, painted trees, and symbols such as trail markers and maps created from stones, imparted visual cues, but they could also contain spirits with the power to speak.⁴¹ Rather than a strictly visual system, these methods of communication intersected with oral traditions of speech and song, and religious understandings of the spiritual world and its soundscapes. Early European travelers in Pennsylvania often encountered these systems of visual and aural communication, although they seldom had the cultural knowledge to understand their significance. When the Moravian missionary Johannes Ettwein observed a rock cairn on a mountain near the community of Wajomick (Wyoming) as he journeyed from Bethlehem to the Wyoming Valley of Pennsylvania, he relied on a Native Moravian traveling with him to interpret its meaning. Ettwein was told that that the cairn "indicate[d] the number of Indians who had already climbed the mountain; it being a custom for each one to add one to the heap on passing that way."⁴²

Painted trees might also be used to mark the junctions of important pathways or notable trail areas. According to Loskiel's *History*, "Their hieroglyphics are characteristic figures, which are more frequently painted upon trees than cut in stone. They are intended, either to caution against danger, to mark a place of safety, to direct the wanderer into the right path, [or] to record a remarkable transaction."⁴³ Loskiel's *History* also makes note of the custom of creating "warrior's trees" that recorded important achievements in battle: "They [warrior's] generally chuse a tall well-grown tree, standing upon an eminence, and peeling the bark on one side, scrape the wood till it becomes white and clean. They then draw with ruddle, the figure of the hero whose exploits they wish to celebrate, clad in his armor, and at his feet as many men without heads or arms as fell by his own hand."⁴⁴ Members of hunting parties also sometimes recorded their camping locations or the number of bears or deer killed in addition to their tribal affiliation. Some trails were bordered with so many painted trees that they became legendary. A section of the Towanda Path was at one time known as "the Painted Line" because of the many peeled and painted trees along its course. These tree paintings could remain visible for as long as fifty years, and were important methods of communication that superseded language or tribal affiliation.[45]

After the arrival of European settlers in Pennsylvania, these ways of mapping, naming, and remembering were often adapted to include new modes of communication and understanding. In the vicinity of some Moravian missions, Native Moravians began to paint scripture passages and hymn texts on trees as a symbol of their affiliation with the Moravian Church, referencing both the visual and aural nature of hymnody.[46] Despite changing cultures and ecologies, many Native names, maps, and trails survived colonization to endow the natural environment of Pennsylvania and its places and soundscapes with multiple forms of visual and sonic meaning.

Into the Woods

In the eighteenth century, the complex ecological and geographical regions of Pennsylvania—its forests, rivers, valleys, and mountains—became geographic crossroads that connected both settler and Native communities. But there were also fundamental differences in how settler and Native groups understood Pennsylvania's natural environment and their place within it. Like Native Americans, Europeans who settled in Pennsylvania also encountered its physical and acoustic environments. But in dense thickets where Native people had heard the voices of particular ecologies, European travelers and settlers most often heard danger and evil, emanating from oceans of woods "without boundary, markers, or end."[47] Virgin forests of exceeding darkness were labeled by early settlers as the "Shades of Death."[48] In these lightless forests, where hemlock and pine grew so thick that their evergreen foliage shut out the light of day, travelers could journey for days without sunlight.[49] Well into the eighteenth century, when the Moravian elder August Spangenberg traveled on the Great Warrior's Path to Onondaga, his most pressing concern was the darkness of the forest: "This is a wilderness where one does not see the sun all day long. The woods are so thickly grown that sometimes one can hardly see twenty paces ahead."[50] The Quaker botanist and explorer John Bartram also recorded in his diary a desperate journey through stands of white pine so high that "the Sun could hardly shine through. . . . The tops of the trees [are] so close to one another for many miles together, that there is no seeing which way the clouds drive, nor which way the wind sets: and it seems almost as if the sun had never shone on the ground, since the creation."[51]

Early written accounts and maps by Europeans all indicate the omnipresence of trees—blocking the sun, blocking the path, stretching on for endless and wearying miles (fig. 1.1). When traveling on the Sheshequin Path from Otstonwakin, Johannes Ettwein noted that the pathway was "full of wood and trees which the wind has piled up sometimes three to four logs upon one another that often one does not know how one may get through."[52] According to Ettwein, the undergrowth of the forest was often so dense "that ofttimes it was impossible to see one another at the distance of six feet. . . . It was a daily matter of astonishment to me, that any man should presume to traverse this swamp, and follow what is called a path."[53] He complained vociferously at the vagaries of traveling along a path that forced him to cross the Muncy Creek no less than thirty-six times, frustrating his attempts to hurry through the woods.

But not all European travelers or settlers saw darkness and danger in Pennsylvania's forests. For some, these new landscapes offered opportunities for study and inspiration. When Englishman Thomas Pownall visited Pennsylvania in 1754 in the midst of a two-year tour of the American colonies, he took detailed notes and observations of the natural environment, which he eventually published as *A Topographical Description of the Dominions of the United States of America*, accompanied by a "new and accurate" map produced in Philadelphia by the Welsh cartographer Lewis Evans.[54] Pownall was fascinated by the landscape of Pennsylvania, and attempted to portray the density and prevalence of American forests to audiences across the Atlantic: "Wherever the Waters do not prevail, it [America] is covered with Woods, so that viewing this great Continent America (as yet a new World to the Land-workers of Europe) we see it a Country of Woods. . . . I know of but one Place which is totally without Trees, and that is a Tract of Land upon Long Island, in New York State called Jamaica or Hampstead Plain, on which a shrubby Kind of Heath only grows."[55]

He described Pennsylvania's ridge and valley topography as a great ocean of woods, swelling with trees like the waves on the ocean's shore: "The Vales between the Ridges of these Mountains have all one and the same Appearance, that of an Amphitheatre enclosing, as it were, an Ocean of Woods swelled and depressed with a waving Surface like that of the great Ocean itself."[56] For Pownall, the untamed nature of the American landscape connected with his innermost instincts as a "*mere Sylvan Animal* of the Woods," instincts that had lain dormant in the treeless expanses of the

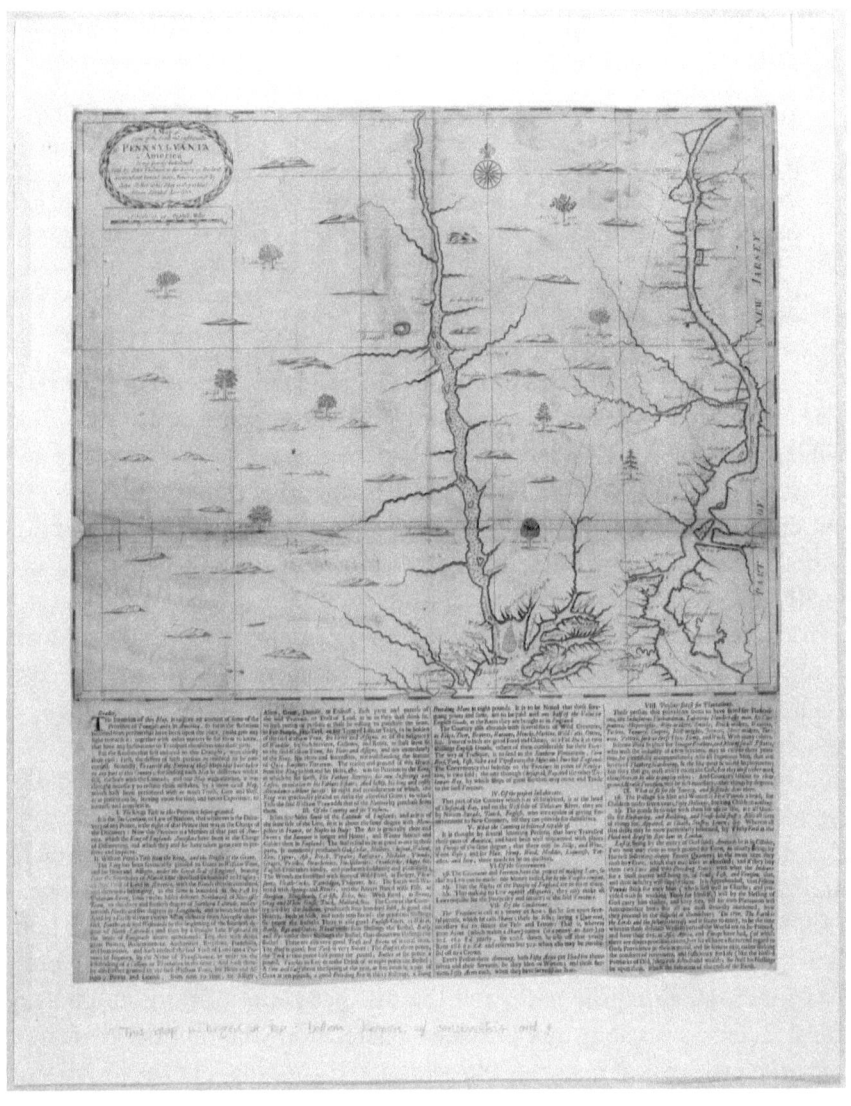

fig. 1.1 "A Map of Some of the South and East Bounds of Pennsylvania in America Being Partly Inhabited, Sold by John Thornton and by John Seller, London," 1681. One of the first known maps of Pennsylvania. Depicts settlements and forests north of the 40th parallel north, including the Delaware Bay, the Chesapeake Bay, and southern stretches of the Delaware and Susquehanna Rivers. Reprinted with permission of the Historical Society of Pennsylvania.

English countryside. In this new and wild landscape, there was enormous potential for the imagination:

> The general Face of the Country [America], when one travels it along the Rivers through Parts not yet settled, exhibits the most picturesque Landscapes that Imagination can conceive, in a Variety of the noblest, richest Groupes of Wood, Water, and Mountains. As the Eye is lead on from Reach to Reach, at each Turning of the Courses, the Imagination is in a perpetual Alternative of curious Suspense and new Delight, not knowing at any Point, and not being able to discover where the Way is to open next, until it does open and captivates like Enchantment.[57]

Europeans' descriptions of the acoustic ecologies of American forests also evidenced a wide variety of perspectives. While some heard "a howling wilderness," filled with screeching and noise, others sensed interesting and unusual sounds.[58] Whether terrifying or enchanting, sound helped to orient European travelers, as it did for Native Americans, to distance and direction. Hearing and understanding the sounds of rivers, wind, birds, and wildfires were important to survival in an environment where sight was not as useful as sound. Many travel accounts of journeys in the woods featured sound prominently in the foreground of sensory perceptions. Johannes Ettwein did not just see trees, he also listened carefully to the biophonic and geophonic soundscapes of the forest. It was the sounds of a sudden and violent summer storm that gained the foremost place in his narrative. These were sounds of "thunder [that] rolled like siege artillery," "rain [that] came down with the sound of many waters, or the rushing of a mighty cataract," and tall oaks that bent in "the grasp of the roaring wind."[59] Other travelers also listened carefully to the natural environment. The diary of John Bartram details the alternatively soft and hard sounds of rain in the forest and meadows along the path, and the distinct patter of rain on a hastily constructed bark cabin that provided shelter from the thunder.[60] He also remarked with joy at hearing "the great green grass-hopper [begin] to sing (*Catedidist*)," or the enchanting yet musical howling of a wolf.[61] For other travelers such as the Moravian missionary David Zeisberger, the noise of wolves was less welcome: "The wolves made a terrific noise around us during the night."[62] Their howling could signal potential danger and even terror: "[We] lost several nights rest, by the dreadful howlings, and even bold attacks of the wolves."[63]

Sound framed not just settlers' and travelers' interactions with the natural environment, but also cross-cultural encounters. Music and language were important modes of cultural contact between European settlers and travelers and Native communities. When John Bartram attempted to describe the different tribal groups that comprised the Six Nations in his diary, he chose metaphors of sound and music. Music and the soundscapes of village life had characterized his experience of Native communities throughout his travels north of the Wyoming Valley in Tioga and Onondaga. So, in attempting to describe the Haudenosaunee, Bartram characterized them as "grave, solid, and still in their recreations." He used this perceived sonic stillness to contrast the soundscapes and cultural ways of the Haudenosaunee with Delaware and Susquehannock communities, which he described as "noisy in their recreations, loud in discourse."[64]

Bartram also had numerous occasions to observe music and musical practices in his travels. In writing about musical encounters, he struggled to place the sounds he heard within a familiar framework of musical, social, and spiritual meaning, seeking help from Native interpreters to understand the meanings of songs: "We had a fine warm night, and one of the *Indians* that had so generously feasted us, sung in a solemn manner, for seven or eight minutes, very different from the common *Indian* tune, from whence I conjectured it to be a hymn to the great spirit as they express it. In the morning I asked the Interpreter what the *Indian* meant by it but he did not hear him, and indeed I believe none of the company heard him but myself, who wake with a little noise, rarely sleeping sound abroad."[65]

Although these inquiries sometimes yielded little information that was meaningful to Bartram, he continued to write detailed descriptions of music and performances that reflected the multicultural and multilingual worlds of mid-eighteenth-century Native villages. One night, as Bartram was resting by the fire with Conrad Weiser and Shikellamy, he was "entertained by a comical fellow: . . . he carried in one hand a long staff, in the other a calabash with small stones in it, for a rattle, and this he rubbed up and down his staff; he would sometimes hold up his head and make a hideous noise; he came in at the further end, and made this noise at first, whether it was because he would not surprise us too suddenly I can't say." Bartram was interested in the performance, and asked what those vocal sounds represented. Shikellamy answered in English, "Lye still John." Bartram was shocked to hear his own language spoken: "I never heard him speak so much plain English before."[66]

Bartram's diary entries provide one traveler's sense of the musical and linguistic soundscapes of mid-eighteenth-century Native and settler towns throughout Pennsylvania and New York. Certainly, Bartram did not fail to note that these communities were shared spaces that resounded with a veritable babel of voices, accents, dialects, and languages. Travelers like Bartram likely experienced these communities as intensely multilingual places where every person seemed to be speaking a different language or dialect—sometimes Algonquin dialects, sometimes Haudenosaunee, sometimes German, sometimes various English regional patois. Even Bartram's traveling companions were diverse in their linguistic backgrounds—Shikellamy was Haudenosaunee, although he lived in the Delaware settlement of Shamokin, and Weiser was originally a Palatine German.[67] Bartram himself was a third-generation English-speaking Pennsylvania Quaker.

The musical soundscapes of mid-eighteenth-century village life were also diverse and similarly served as important conduits of cross-cultural encounters. European settlers sometimes adopted Native cultural ways, dress, instruments, and musical traditions, and Native people also adopted features of European culture that were deemed useful. This included the playing of European musical instruments. When Bartram entered the town of Tohicon, Pennsylvania, he observed that "the *Indians* welcomed us by beating their drum, as soon as they saw us over the branch, and continued beating after the *English* manner as we rode to the house."[68] He was also received in Onondaga with "the music of violins, flutes, and a drum."[69] During the visit of Moravian missionary Bernhard Grube to the home of the Shawnee chief Paxinosa, Grube sang several hymns in Munsee upon request and also demonstrated how a German would play the violin: "Paxinosa desired to have a meeting to-night, because he would like to hear about the Saviour. About 30 Indians and the whole family of Paxinosa assembled. The men sat at one end of the hut and the women at the other, while we were in the middle. Then I preached the Gospel to them. Both before and after the address we sang a few Delaware verses. The youngest son of Paxinosa and another Shawanese came to us with two violins, and desired to hear our melodies. We played a little, at which they and our Brethren and Sisters were well pleased."[70] Clearly, the men and women of Paxinosa's village were interested to learn about new performance practices and musical traditions from European visitors. But even when Europeans were not present, Native ceremonies in the mid-eighteenth century may have sometimes incorporated European instruments such as violins and flutes in addition to traditional Native instruments.[71]

As Katherine Faull and James Merrell have written, Pennsylvania in the mid-eighteenth-century was a diverse place, both culturally and linguistically, and in terms of its broader connections with the colonial Atlantic World and Native America.[72] Despite incidents of violence and misunderstanding, curiosity about cultural ways including music was an important source of common ground in the years before the Seven Years' War and later violent conflicts between and among settler and Native communities that would render such attempts at meaning-making and understanding virtually impossible. Through trade of cultural knowledge, including singing and playing instruments, both settler and Native communities learned and began to incorporate new cultural traditions.[73] Moravian settlers were no different from other European immigrant groups in this respect. Moravian hymn texts from the early period of Moravian settlement in the 1740s, as well as early maps and travel writings from Bethlehem and other Moravian mission communities in the Susquehanna, Delaware, and Lehigh Valleys reveal the desire of Moravians to adopt new and hybridized traditions of environmental experience that capitalized on the traditions of both Native and European Moravians. But they also reveal fundamental differences in the interactions of both groups with the natural environment of Pennsylvania. While Native Moravians may have sung traditional sacred songs or newer Christian hymns to bridge the spaces between natural and human environments, early European Moravians such as Zinzendorf and Ettwein sang hymns to claim and sanctify a Christian space within what they heard as a vast wilderness.

Singing away the Wilderness

For Moravians, Pennsylvania's forests were a liminal place where the divine could easily be encountered. Natural events or disasters such as floods, earthquakes, lightning strikes, or even illness or injury were interpreted through a sacred ecological cosmology that projected the "hand of God" onto the natural world. Prayers, songs, rituals, and copies of the Bible or other spiritual texts were not just religious paraphernalia, they were wards against unseen evils. Practicing the rituals of Christianity in "the wilderness" was an essential act of survival in what Moravians often viewed as a foreign and hostile landscape. On August 2, 1747, as missionary Johannes Hagen held the first communion and *pedilavium* (foot-washing) service in the town of Shamokin, he experienced an enormous sense of relief at casting the ritual boundaries of this familiar Christian service upon the unfamiliar space of Shamokin: "How we felt at this communion with our Lord I cannot describe. It was also important to us because it is the first one here in the wilderness, perhaps as long as the forest has stood."[74]

For Moravian missionaries and travelers, the singing of hymns and the performing of sung rituals such as the Singstunde cast an almost magical spell on Pennsylvania's wild places—encompassing all within earshot in a familiar and protective boundary. Hymns were important to all aspects of Moravian life, but perhaps nowhere more essential than on journeys through the forests of Pennsylvania. Countless diaries and travel journals of mid-eighteenth-century Moravians are filled with references to singing hymns "in the midst of the forest." Whether these hymns were sung for protection, spiritual connection, or simply for the joy of having caught eight trout after several days of near starvation, they accompanied almost every Moravian journey. The singing of hymns and entire Singstunden framed (Johann) Martin Mack's journey to Onondaga in 1752. Mack and his companions sang hymns every morning, throughout the day, and in the evening. On the evening of Friday, August 25, while camped along a small stream, they "kept a happy singing hour." The following morning, Mack's diary records that "having rested well, we arose early and sang some verses. After passing through Anajot [a town on the way to Onondaga] we came to a hill, about a quarter of a mile beyond where we rested. . . . We sang some verses."[75] Even after Mack reached his destination at Onondaga, he continued to sing and to write about singing in his diary: "In the evening we sat by the fire and sang hymns, having a strong sense of the Lord's watchful presence."[76] When the Moravian missionaries David Zeisberger and Andrew Frey returned to Onondaga the next year, they sat in the same place as Mack and also "sang hymns together around [the] camp fire."[77]

Hymns could serve as protective charms, reinforcing the saving power of God, as recounted in an earlier diary kept by Mack. Eight years before his Onondaga journey, Mack had confronted the terrible heat of a forest fire on April 19, 1744. It was only through the reassurance of a hymn verse spoken by his companion Christian Fröhlich that he was able to pass safely through the flames:

> The woods were on fire all around us, so that in many places it looked very terrible, and many times we scarce knew how to get through. The burning trees

fell down all about. We could not easily get out of the way, because there are such high mountains on each side. After dinner we came between two great mountains, and the fire burnt all around us, and made a prodigious crackling. Before us there was sent such a great flame that we were a little afraid to go through it, and we could find no other way to escape it. Brother Christian went first through. The flame went quite over his head; it looked a little dismal. He got through but I did not know it, because I could not see him for the smoke. I called to him; he answered me immediately. I thought I would wait a little longer till it was burnt away a little more, but the fire grew still fiercer. He called again and prayed me to come through, saying our dear Savior had promised "When thou walkest through the fire thou shalt not be burnt, neither shall the flame kindle upon thee." I ventured and went cheerfully into the flame, and got safe through. We thanked the Lamb for it, that he had preserved us so in the fire.... When we came to Bethlehem we found that the watchword for that day had been: "When thou passest through the waters I will be with thee; and through the rivers they shall not overflow thee; when thou walkest through the fire thou shalt not be burned, neither shall the flame kindle upon thee. Fire, hail, snow, vapor and stormy wind are servants of His will."[78]

Safety for Moravians like Mack and Fröhlich was assured by a God who was a master of nature. This was a God who subdued the forces of fire, hail, snow, vapor, and stormy wind. The desire for divine deliverance from hostile natural environments was an important part of Moravian cosmology, but its roots lay in a broader pan-Christian worldview on the natural environment.[79] In the long history of Christian and Jewish writings on wild places, the natural environment was viewed as a possible site of intense religious experience—of stark need for food and water, of isolation, of danger and divine deliverance, and of renewal and encounters with the divine. For Moravian missionaries, as for other historic and modern Christian groups, the forest therefore represented the potential benefits of a divine encounter, coupled with inherent personal and spiritual danger.

Seeking and longing for divine deliverance, especially in the midst of the forest, is an important feature of many Moravian travel diaries. When Johannes Ettwein prayed for guidance to cross the rain-swollen Susquehanna, he ascribed a sudden meeting with the Mohican Moravian Tassawachamen (Joshua Sr.) along the river bank as an answer to his prayer. After safely crossing the river in Joshua's canoe, Ettwein recorded in his diary that the meeting with Joshua was "Providential deliverance."[80] And, when European Moravians scaled the summits of the Kittatinny and Mahantango Mountains, mountains so impassable that their crossing could only be accomplished by "the Saviour's children," they saw the hand of God at work.[81] Just after Martin Mack was saved from a forest fire by a hymn verse, he was delivered from being crushed to death by a falling tree limb: "Saturday, August 19, 1752. We went on a little farther and lodged in a cold and dark wood. Just as we were seating ourselves around a fire which we had made, there began such a cracking and rattling over our heads, that we did not know in what direction to run; and there fell a huge tree close by our fire. We thanked our Savior for His protection over us. Before going to sleep, we had a 'singing hour' together."[82] Mack sang an entire Singstunde in response to his salvation from the double perils of fire and forest.

For Native people who affiliated with the Moravians, the woods had always been a place of sanctuary and solace, a borderland for encountering the spiritual world. But as many Native Moravians began to incorporate Christian concepts of divinity into their spiritual practices, the woods now became a place where encounters with a specifically Christian-inflected version of the divine were possible:

> Another [Native] visitor, who had formerly heard the Gospel in Gnadenhütten, but then resisted convictions, related, that soon after his return, his child was taken dangerously ill.... He ran into the woods, and cried to God, in the anguish of his soul, that he would in mercy restore its health.... After giving vent to his tears, his heart was comforted, and on his return he found the child better; he therefore came now to Gnadenhütten, to request the Brethren, to take him and his family under their protection. Tears flowed while he spoke; he obtained permission to live in the place, and was baptized with his whole family.[83]

In this particular story from Loskiel's *History*, as well as many passages in the communal diaries of the Nain, Shekomeko, Pachgatgoch, and Gnadenhütten communities, the desire of Native Moravians to incorporate differing religious traditions often appears to be in response to experiences of pain, loss, and suffering. These experiences likely reflected the psychic toll of colonialism and the fundamental changes in cultural, social, and economic structures that took place in many Native communities following European colonization. Loskiel's *History* relates the account of an anonymous Mohican visitor to Gnadenhütten who had tried traditional remedies to heal his extreme sorrow after the death of his first child, with little success, and later came to the Moravians seeking a further solution to his grief:

> When my wife was going to lie in with her first-born, I was impatient to see the child. When I saw it, I thought: This child God has made; and I loved it so

much, that I could not forbear looking at it continually. Soon after the child died, and I mourned to that degree, that nothing would comfort me. I had no rest, day nor night, and my child was always in my thoughts; for my very heart cleaved to it. At last I could bear the house no longer, but ran into the woods, and almost lost my senses. The Indians then advised me to take an emetic to get rid of my sorrow. I complied, but the love for my child, and my sorrow for its loss, were not removed, and I returned to the woods. There I beheld the trees and the birds, and considered, that the same God created them who made my child. I then said: "Thou, O God! who mad'st all things, I know not where thou art, but I have heard that thou dwellest in heaven. Thou hast taken my child, take my sorrow and grief likewise from me." This was done, and I then could forget my child.[84]

Native Moravians began to carry Christian rituals and hymns into the forest as tokens of spiritual and divine presence and favor. Moravian-affiliated hunters often turned to Jesus as a provider, or at least arranger, of a successful hunt, one who could be appealed to through song in the manner of *manitou*. On one occasion, when the missionary Johann Schmick visited Mohican hunters in their lodges in the forest, he reported that they had been holding evening Singstunden regularly, at which they discussed the "astonishing love of the Savior," and that both morning and evening, the hunters sang hymns. Schmick subsequently reported that "in the hunt, they are especially successful."[85] Native Moravians also sought communication with the divine through hymn singing in the forest. According to the Gnadenhütten diary, Nicodemus, a resident of Gnadenhütten, was reportedly "happy and loves the Savior and his wounds. He often walks alone in the woods and sings verses."[86] Nicodemus also experienced a vivid dream in which he saw "Jesus sitting in a tree, and he had gone to kiss the Savior's side wound."[87] Jephta, another resident, was also apparently fond of walking in the woods and singing hymn verses to the wounds of Christ.[88] Some of the earliest Mohican Moravians from Shekomeko were encouraged by the missionaries to never forget the suffering of Christ and to "not only to think on it in Shekomeko, but in the woods, and when out a-hunting."[89]

The religious identity of both European and Native Moravians was therefore intertwined with their spiritual experiences of the natural environment. But, for many European Moravians, the potential danger represented by the Pennsylvania forests impelled them to fall back upon familiar patterns of meaning and representation. European Moravian-authored stories, maps, and songs sought to tame this wilderness of trees, animals, fires, and rocks. They dampened the biophonic, geophonic, and anthrophonic soundscapes of Pennsylvania and transferred them into more pleasing sounds: the regular meter and predictable chorale tunes of Moravian hymns. Throughout Zinzendorf's first journeys in 1742 into Native places beyond colonial settlements, he interpreted in song the people, places, and environments he encountered. Once these hymns became a part of the regular Moravian corpus of song, they were sent via ship across the Moravians' geographic network in the Atlantic World and became a sounded part of Moravian congregational life in places as far away as Herrnhut, Germany, and Paramaribo, Suriname. Zinzendorf's hymns sonically claimed an imagined Moravian space for a worldwide church body who had never seen or heard the Pennsylvania forests.

Zinzendorf's Pennsylvania Hymns

On December 2, 1741, Zinzendorf landed for the first time in North America. From the port of New York, he began his overland journey to Philadelphia and the newly founded Moravian settlement of Bethlehem. During the winter, spring, and early summer of 1742, he made various preaching trips to German settlements in southeastern Pennsylvania, traveling as far west as Conestoga. 🌐 **See website chap1.4, Timeline: "Zinzendorf's Pennsylvania Journey."**

In late July, he embarked on his first trip outside of colonial territory and into Native country. From July 24 to August 7, Zinzendorf, his seventeen-year-old daughter Benigna, and eleven other Moravians traveled to the Delaware town of Meniolagomeka and other Native communities around the region of the Delaware River and the Kittatinny Mountains. The first part of their journey took them to Bushkill Creek near the agricultural plantations of the Native Christian Moses Tatemy. After resting at Tatemy's home, they continued northward through the mountains to the village of Clistowackin (Fine Land) on Martin's Creek, a small village of about ten houses. That night it rained steadily, and the Moravian party spent the night with the chief of a neighboring village. In the morning, they crossed the Kittatinny Mountains. It was July 27. Sometime that day, Zinzendorf sent a written message by messenger back to the congregation in Bethlehem: "I kiss you and take each one's arm and lead him with me to *that neck*, you well know Whose I mean. He sanctifies everything. Your faithful and happy little foster son, Ludwig, Johanan."[90]

By the next evening, the Bethlehem congregation had received the letter, and the *Bethlehem Diary* records that it was read at the nightly *Liebesmahl* (Lovefeast). During the service, the congregation sang a litany as a protective prayer for Zinzendorf and his traveling companions. They invoked God's protection for the travelers as they journeyed through the Pennsylvania wilderness, and also sang several hymns the next day for the same purpose. As the congregation meditated on Zinzendorf's message from the Kittatinny Mountains, "*that neck*, you well know Whose I mean," they sang a hymn about the neck of Christ and performed a prostration, falling down with their faces on the wooden floor of the worship hall to sing the *Te Christum laudamus* litany.

While Moravians in Bethlehem were offering sung prayers for the travelers, Zinzendorf and his companions had lodged for the night in a tent pitched near the house of the medicine man in the village of Captain Harris, father of the Delaware leader Teedyuscung. Although Captain Harris could speak English, Zinzendorf was not adept at speaking English, so two of his fellow travelers, J. William Zander and his wife Johanna Magdalena, served as interpreters. Zinzendorf attempted to share with Captain Harris and local residents the good news of the "Creator and Redeemer and his plan for the salvation of all men from sin and eternal death."[91] Zinzendorf recorded in his travel diary that he was concerned about whether these Christian messages had resonated with Captain Harris, but he speculated that perhaps it was too soon to tell. The next day, the travelers moved on and reached their destination at Aquashicola Creek and the town of Meniolagomeka.

After visiting Meniolagomeka, Zinzendorf intended to return to Bethlehem, but he experienced a spiritual presentiment telling him that he was urgently needed at Conrad Weiser's house in Tulpehocken, a Delaware community south of the mountains now also inhabited by Palatine German settlers. According to his diary, Zinzendorf was "drawn thither by an irresistible power, and in strong faith I obeyed the call, although knowing neither why nor wherefore."[92] Zinzendorf set out to the southwest, fording the Lehigh River and passing through the Allemängel (Destitute Place) and the Ontelaunee Valley to the bank of the Schuylkill River that flowed southward toward Philadelphia. There he and his companions camped for the night. It was August 1.[93]

That evening, while his companions slept, Zinzendorf recorded in his diary that he lay awake. As he gazed out over the Schuylkill from its western bank, a region known as Sikihillehocken, he began to improvise a hymn to record his journey and the landmarks and waymarks of the American landscape and its people.⁹⁴ 🌐 **See website chap1.5, Interactive sound map: "Zinzendorf's Journey to Shamokin and Wyoming."** For a tune, he chose one of his favorite chorales, "In dulci jubilo," originally a German and Latin macaronic carol supposedly gifted by angels to German mystic Heinrich Seuse in 1328.⁹⁵ Onto this ancient tune, Zinzendorf mapped new landscapes and places:

Hier schrieb ich einen Brief,	I here a letter write
Als alles um mich schlief	While 'round me all tonight
In der finstern Wüsten	Sleep within this lonely
Sickihillehoken,	Sickihillehoken
Wo wenig Vöglein nisten;	Where tarry few birds only.
Wird ich doch kaum inn'	Scarcely I'm aware
Dasz die Schuylkill *rinn*	Of the Schuylkill fair
Ueber [sic] *Nachbar* Green.	In Green's farmland there.⁹⁶

As he began to sing and to improvise the eighteen verses of the hymn, he may have envisioned the worldwide Moravian Church—linked across oceans and frontiers. People, landscapes, places, even the trifold manifestations of the Holy Trinity that guarded and guided Moravian Christians regardless of place. All of these images were committed into the song's text. Also into the song's text he recorded descriptions of the Pennsylvania wilderness, "*Penns Wüste* (Penn's Wastes)," and the places he had traveled that day: the German settlements at the Allemängel and the Schuylkill River itself. Even the tent where he sat that night was referenced in the song: "My night in part shall be / devoted here to Thee, / 'Neath the open tenting / In Indian River's lee." He sang about the German congregations of Herrnhut and Marienborn "there and beyond the ocean." And, the "warrior's gate" of the Americas, Bethlehem. He would have certainly known that in Bethlehem at that very moment, while most of the settlement was sleeping, a Moravian intercessor (a person assigned to pray and sing hymns at every hour of the night) would be awake, keeping a sounded, wakeful vigil along with him: "At the warrior's gate [Bethlehem] / They who supplicate [the hourly intercessors] / Vain through no night wait."

He also considered the residents of Tulpehocken, whom he would encounter in the morning: "In Tulpehocken, flames! [camp fires] / A glow that darkness tames! / Friendly Indian nations / Which have made known their aims / Return now to their stations. / By their trail I'll know / Whither I must go / Them the Cross to show." He also sang of the first Mohicans to request baptism at Shekomeko, "In songs, Him glory giving, / *Abrah'm, Israel* [Jacob], *Isaac, Hans* [Johannes], as well, / In His side-wound dwell." In Zinzendorf's grand mission plan these Mohican men would be just the first of many Native Christians, for "Where *ten* were found to go / Preaching Christ's salvation / A *hundred* now do show / The blessed Revelation / Soon a *thousand* there / Shall our mission share / And His name declare." In the morning, Zinzendorf immediately dispatched a manuscript containing his improvised hymn to Bethlehem, where it was received on August 4 and sung at the Liebesmahl at two o'clock that afternoon.

From the banks of the Schuylkill to the banks of the Lehigh, a single hymn now became a nexus of spiritual and communal meaning that linked people across geography and space. Not only did this newly composed hymn form a sonic link between Zinzendorf and the congregation in Bethlehem, but the Liebesmahl at which it was performed was also conducted in synchronized time with worship services held by Moravian congregations in Europe. According to the *Bethlehem Diary*: "At 2 o'clock this afternoon the congregation held a lovefeast to show its fellowship with our congregations in Europe, which today partake of the Lord's Supper at this time." At three o'clock, after the Liebesmahl, Henrich Müller set out with the hymn manuscript and other written communications for Philadelphia and New York before embarking across the ocean to England and Germany. Thus, a hymn improvised in Pennsylvania's forests made its way swiftly across the Atlantic.

The day after composing the hymn, Zinzendorf entered the Tulpehocken Valley, impelled as he claimed by the Holy Spirit to visit Conrad Weiser. It was indeed a fortunate day, for one of the most important events of his stay in Pennsylvania occurred at Weiser's house. Lodging with Weiser were several leaders of the Six Nations who were traveling northward from Philadelphia toward the Shamokin Path that led from the Tulpehocken Valley to Shamokin.[97] This fortunate encounter allowed Zinzendorf to meet Shikellamy, the Haudenosaunee official stationed at Shamokin, and to request permission for Moravians to travel in the regions between the Delaware River and the lands of the Haudenosaunee Confederacy. Shikellamy offered his consent and invited the Moravians to visit him at Shamokin and to stay with him as "friends, not strangers." To seal this pact, Shikellamy gifted

Zinzendorf a belt with 186 pieces of wampum.⁹⁸ Thus began the next phase of the Moravian mission plan, commemorated by an oil painting created in London of Zinzendorf, Weiser, Shikellamy, and other leaders (fig. 1.2). Like Zinzendorf's Sikihillehocken hymn, the painting became a symbolic representation of Pennsylvania and its Native residents for an eager audience of Moravian Christians across the Atlantic.

After returning from the meeting at Weiser's house, Zinzendorf again left Bethlehem from August 10–31, 1742, to visit the community of Mohican Moravians at Shekomeko. Zinzendorf's companions on this journey were his daughter, as well as the Moravian eldress Anna Nitschmann, and Anton Seiffert. Conrad Weiser served as their guide. During the visit to Shekomeko, Zinzendorf officially organized a Christian congregation with the first ten Mohican Moravians on August 22. He also communicated in song with the congregation in Bethlehem, just as he had done on his visits to Meniolagomeka and Tulpehocken. At the evening Singstunde on Monday, August 13, the fifteenth anniversary of the first spiritual awakening of the Renewed Moravian Church in Herrnhut, the congregation in Bethlehem sang a hymn that Zinzendorf had composed in the Kittatinny Mountains on his way to Shekomeko.⁹⁹ These hymns quickly became a part of the sonic life of the community. One month later, on Sunday, September 29, Zinzendorf's Shekomeko hymn was again sung at a Liebesmahl. The *Bethlehem Diary* records that the hymn verse, "Call to mind with longing, Cherish all belonging, To Christ's witness folk, [etc.]" was especially meaningful to the worshippers that day.¹⁰⁰ The hymn also elicited "joint remembrance of them [the Moravians in Shekomeko] as we sang."¹⁰¹ When Zinzendorf returned to Bethlehem, he also sang the hymn, in addition to others he had composed on his journey: "In the lovefeast the beautiful hymn was sung which Bro. Ludwig had composed on his journey to the heathen, first noted [above]: *How decidedly evil are we*, etc. . . . This afternoon we held our regular Sabbath lovefeast, attended by probably one hundred non-resident friends. In it the two hymns were sung which Bro. Ludwig Johanan had composed while on his visit to the heathen, namely, *At Christ's manger, Freed from danger, Clan of sinners*, etc. and *In truth He stays The same always*, etc."¹⁰²

After his return from Shekomeko, Zinzendorf did not remain long in Bethlehem. From September 21 to November 9, Zinzendorf, Martin and Jannetje Mack (Rau), Peter Böhler, Anna Nitschmann, Conrad Weiser, and two Mohican Moravians, Joshua and David, undertook their longest

Fig. 1.2 Zinzendorf mit Konrad Weiser und die Häuptlinge der 5 Indianernationen am 14.08.1742 [Zinzendorf with Konrad Weiser and the Leaders of the Five Nations on August 14, 1742]. Anna Arndt, after Johann Valentin Haidt, oil on canvas. GS.389, UA Herrnhut. The original painting was housed in the Provincial Archives in London and was destroyed during World War II.

trip into Native country. 🌐 **See website chap1.5, Interactive sound map: "Zinzendorf's Journey to Shamokin and Wyoming."** Two small maps carefully inscribed by Zinzendorf into his travel diary trace his travels from Bethlehem. The first part of his journey took him to Weiser's home in Tulpehocken, and from there over the mountains to Shamokin. From Shamokin, he would cross the western branch of the Susquehanna and travel as far west as Otstonwakin, and then return east to Skehantowa, or the Wyoming Valley. In his travel diary, Zinzendorf described the trip with great detail, including the people that he met and the natural features of the landscape. A second small map penned into the diary also records that the entire journey was undertaken through the favor of "*Jesu Nahmen* [Jesus's name]" and under the protection of divine grace (fig. 1.3a–b).

Zinzendorf's travel diary opens with numerous passages on the sheer difficulty of travel in the Pennsylvania forests, and his joy at finally reaching the home of a German family near the settler community of Oley on his first night away from Bethlehem: "We set out [from Bethlehem] and took the road to Tulpehocken, keeping between Long Swamp and the Oley Hills. We rode on until late at night. Before we reached our place of destination it grew dark as pitch, and riding became very difficult. I was struck on the cheek and on the left eye by the limb of a tree, and several of the Sisters fell from their horses. No one, however, was seriously injured. At last we entered the border of Oley, and reached Brother Bürstler's house."[103] Since it was late at night, and the Bürstler family was already asleep, Zinzendorf and his companions sang a hymn outside of their window to wake them up. 🌐 **See website chap1.5, Interactive sound map: "Zinzendorf's Journey to Shamokin and Wyoming."**

Herr Zebaoth!	Lord God!
Du wahrer Gott der Kreatur!	Truest God of all creatures!
Gott, Schöpfer der Natur!	God, creator of nature!
Gott, der die ganze Welt erhält!	God, who receives the whole world!
Und was verdarb	He who has out of ruin
Mit Blut erwarb	With blood
Und heiliget,—	Purchased and sanctified us,—
Sey von uns angebet![104]	Receive from us an offering!

The Bürstler family immediately recognized the travelers by their song and welcomed them. As they traveled on, passing north from the Bürstler home along the Shamokin Path, Zinzendorf noted in his diary the "exceedingly high mountains," littered with rocks and sharp stones and towering

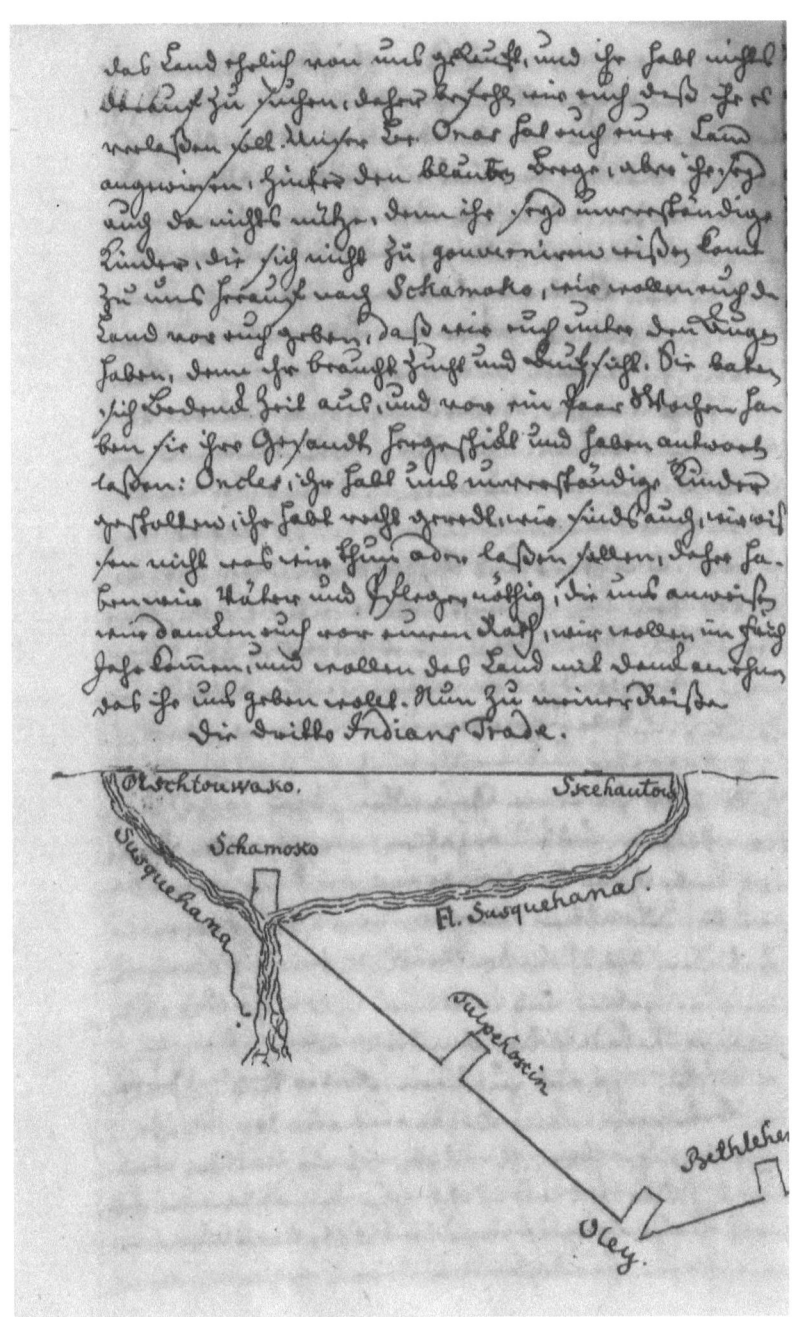

Fig. 1.3a–b Sketch maps as they appear in Nikolaus Ludwig von Zinzendorf's travel diary from his second and third visits to Native American communities in Pennsylvania ("Ludwigs Reise Diarium an der Susquehannah"), August–September 1742. MissInd 121.1, MAB.

wie die Delawar Indians kamen, die mußten antworten. Sie wüßten nicht viel zu sagen, als die Engländer wären Betrüger. Wenn Sie sich auch bei sie würden um ihr Land betrogen haben, so würden Sie so eben practiciren mit ihnen, als wir es jez ihnen den Delawar Indians gewiesen. Darauf sagten die 5 nationes Cousins. Ihr seyd ein zänkisches Volk, die Engländer haben das Land ehrlich von uns gekauft, und ihr habt nichts darauf zu suchen, daher befehlen wir euch, daß ihrs verlaßen sollt, unser Bruder Onas hat euch zwar Land angewiesen hinter den blauen Bergen aber ihr seyd euch da nichts nütze, denn ihr seyd unverständige Kinder, die sich nicht zu gouverniren wißen. Kommt herauf zu uns nach Schoanoko oder wir wollen euch da Land von hier geben, daß wir euch unter den Augen haben, denn ihr braucht Zucht und Aufsicht. Sie baten sich Bedenk-Zeit aus, und nach ein paar Wochen haben Sie ihre Gesandten hergeschickt, und haben antworten laßen; Oncles' Ihr habt uns un- verständige Kinder gescholten, ihr habt recht ge- redt, wir finden uns, wir wißen nicht, was wir thun sollen oder laßen, daher haben wir Väter und Pfleger nöthig, die uns anweisen, wir danken euch vor euren Rath. Wir wollen aufs Früh-Jahr kom̄en und wollen das Land mit dank annehmen, das ihr uns geben wollt. Nun zu meiner Reise.

Der dritte Indians Trade.

Otschonwako
Sus que ha onna   Skehantowa
        Schomoko   Susquehanna
        Tulpehokin
        Oley   Bethlehem

Von der gewißheit des Lichts in Jesu Namen, war am

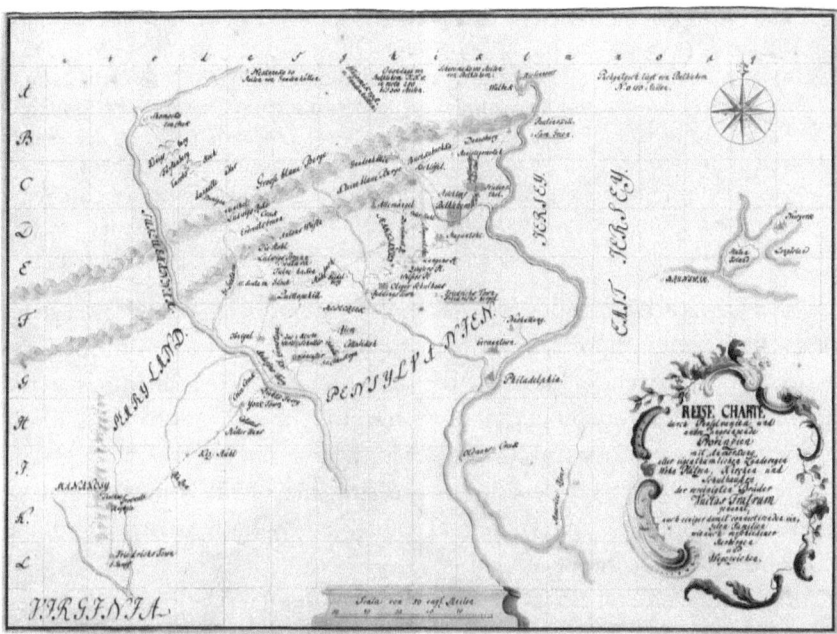

Fig. 1.4 Moravian travel map of Pennsylvania, northern Maryland, and New Jersey, "Reise Charte durch Pensylvanien und andre angränzende Provinzien..." c. 1752, DP f.037.6, MAB.

hemlocks, "symbols of a very wild region of country." Zinzendorf's response to this wild landscape was to name any notable features he encountered along the Shamokin Path. These names were later preserved on a 1752 map produced by the Moravian Mattheus Hehl. Hehl's map records the places that Zinzendorf had visited and named ten years earlier on his trip to Shamokin and Wyoming: Ludwig's Fountain, The Hole, Erdmuth's Spring, Ludwig's Rest, Anna's Valley, Benigna's Creek, Jacob's Heights, Fürstenberg, Königsberg (fig. 1.4).[105] Other travelers might have known these mountains, valleys, resting places, and creeks by various other names—the Second, Third, Peter's, and Berry's Mountains, or the Mahantango, Wicomsco, Mahonoy, and Shamokin Creeks—but for Moravian travelers, Zinzendorf's new names carved a persistent sonic pathway of meaning and home all the way from Bethlehem and Tulpehocken to Shamokin. 🌐 **See website chap1.5, Interactive sound map: "Zinzendorf's Journey to Shamokin and Wyoming."** Three years after Zinzendorf's journey, when August Spangenberg traveled the same pathway, the names claimed by Zinzendorf on his first journey to Shamokin served as important waymarkers for Spangenberg:

May 31, 1745—Set out from Tulpehocken, crossed the *Great Swatara*, and climbed the steep and rocky *Thürnstein*. On its summit drank of *Erdmuth's Spring*, descended the mountain, and nooned at *Ludwig's Rest*. Next came to *Anna's Valley*, and encamped on *Benigna's Creek*, near "The Double Eagle." June 1st. — Crossed *Leimbach's Creek*, ascended *Jacob's Heights*, and at noon struck the Susquehanna, fifteen miles south of *Shamokin*. Now passed through *Joseph's Valley*. Having rested at *Marienborn*, we climbed the steep *Spangenberg*, crossed *Eve's Creek* (the Shamokin), and arrived at Shamokin."[106]

Hehl's map itself, built upon Zinzendorf's renamed Pennsylvania geography, became an important way of marking the landscape for future Moravians.[107] But Zinzendorf was not the only Moravian traveler to name and map the places he encountered, or to commit those places to song. Many of the first exploratory travels of the Moravians into Native country were coupled with attempts to describe, name, sing, map, and claim these spaces. These were journeys undertaken, as expressed by both Spangenberg and Zinzendorf, "in the name of God."[108]

As Amy DeRogatis has argued, there was often a relationship between the establishment of Christian missions in North America and colonial efforts to map these new territories. Colonial governments and surveyors imposed physical order on the landscape, just as missionaries attempted to impose their version of moral order on Native communities.[109] But missionaries such as Hehl and Zinzendorf also created maps to shape and rename the physical landscape, and in so doing participated in larger colonial structures that enabled rapid European colonization of Pennsylvania. According to Judith Ridner, naming or renaming Pennsylvania's towns and landscapes was a mechanism used by the colonial government in Philadelphia to control and impose order on both settler and Native communities. For instance, in the settler community of Carlisle, Pennsylvania, established by Thomas Penn, only English names could be used for streets. In other communities, only street names that already existed in London were approved by the Penn brothers. In many cases, Native, Dutch, or Irish names for towns were not permitted. Thus, Carlisle and other early colonial settlements were kept squarely within the matrix of a British Atlantic World.[110]

Despite the fraught nature of naming, some attempts by Moravians to describe Pennsylvania's landscape may have been simply humorous or lighthearted ways to make difficult journeys in the woods more bearable. On his way to Onondaga, Spangenberg playfully named three islands in the north branch of the Susquehanna for the three Penn brothers, John, Thomas, and Richard.[111] After a particularly arduous day of traveling through the Tiadaghton Forest, he also christened the Native hunting hut where they rested for the night the "Coffee House."[112] Other travelers such as Friedrich Cammerhof and David Zeisberger simply enjoyed naming places they encountered as a way of remembering the events of their journey. A hill could become "Snake Mountain, because we saw snakes in great numbers, lying on the stones and rocks near the shore, basking in the sun." And good fishing days could be commemorated with a name: "[We] fished, and caught more than we could eat. We named our quarters Sunfish, and rested well and undisturbed." Sometimes unusual natural features or particularly beautiful landscapes received names: "We called our quarters Horned Tree, because near to our tent there stood a tree, in which the antlers of a deer had been laid, and now appeared to have grown into it," and "we pitched our tents in a very beautiful cleared spot. *Gallichwio* [Cammerhof] was so delighted with it that he called it *Mon Plaisir in the Wilderness* [My Pleasant Place in the Wilderness]."[113]

Some Moravian place names and maps had rather prosaic or utilitarian purposes. Johannes Ettwein's map of the north branch of the Susquehanna from the Wyoming Valley to the mission of Friedenshütten, completed after the founding of the mission in 1765, ascribes names to the locations of important natural resources or dangerous areas along the river: A Great Tract of Oak Forest, Cherry Islands, Sage Bush, Long Bottom, Bad Water Swamp, or Dangerous Corner. Ettwein's names are interspersed with important written details such as "*Hier gibt es Ginsengwurzel* [here there are ginseng roots]," or "*Oppening wo es viel Pateto's gibt* [Opening where there are many potatoes]."[114] These useful notes and names would have likely been appreciated by Moravian travelers (fig. 1.5).

On his journey in the fall of 1742, Zinzendorf sometimes concerned himself simply with recording the beauty of the natural environment. As he rode through the forests that bordered the Susquehanna River north of Shamokin, he wrote about the vibrant and varied colors of the autumn trees: "The foliage of the forest at this season of the year, blending all conceivable shades of green, red, and yellow, was truly gorgeous, and lent a richness to the landscape that would have charmed an artist. At times we wound through a continuous growth of diminutive oaks, reaching higher than our horses' girth, in a perfect sea of scarlet, purple, and gold, bounded along the horizon by the gigantic evergreens of the forest."[115] During his

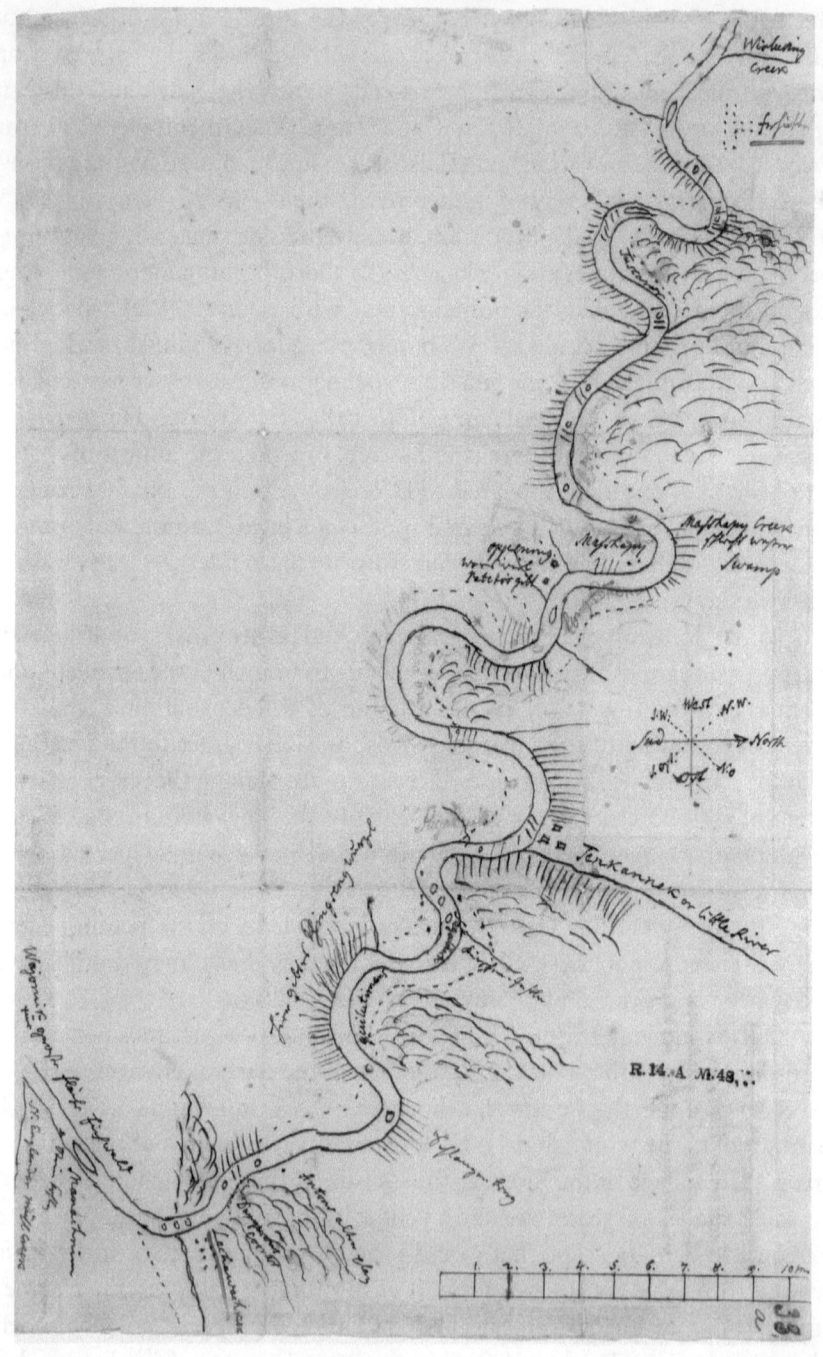

Fig. 1.5 Johannes Ettwein's Map of the Upper Susquehanna. TS Mp.212.15, UA Herrnhut.

journey eastward along the Susquehanna, after leaving Otstonwakin on the West Branch of the river, Zinzendorf and his companions followed the Great Warrior's Path that led from the Great Island (Lock Haven) along the river toward the Wyoming Valley, called by the Moravians, "Wajomick." It was during his stay in Wajomick that he would compose several new hymns that preserved in song the experiences, people, and places of his Pennsylvania journeys.

Wajomick (Wyoming): October, 1742

It was a brilliantly sunny day in mid-October 1742. Zinzendorf sat on the floor of his tent, pitched on a sandstone bluff above the Susquehanna River in the Wyoming Valley. As he hummed the words to a new hymn that he had improvised that morning, he surveyed the broad waters of the river and the three-mile expanse of valley floor that stretched to the south. Below his camp lay the town of Wajomick, home to Shawnee and Chickasaw refugees from the Carolinas, Georgia, and Florida, and Mohicans displaced from the New York Colony. After visiting with Shikellamy and his community at Shamokin, Zinzendorf had traveled to Wajomick via Otstonwakin. He was accompanied by Sattelihu (Andrew Montour), the son of Madame Montour from Otstonwakin, and some of his closest Moravian friends: Anna Nitschmann, Martin Mack, and Jannetje Mack. Together, they had intended to bring the Christian message to the residents of Wajomick and their Delaware and Nanticoke neighbors in the "great Deserts of Skehantowanno."[116] So far, things had not gone according to plan.

Sitting in the sun, fifty-two miles north of Bethlehem, Zinzendorf mused on their venture. The little group of five had been continually surrounded after their arrival at the town by Shawnee warriors. Zinzendorf had attempted to speak with the chief and sachem, Kackawatcheky, with Sattelihu serving as interpreter. But perhaps Zinzendorf and Sattelihu were not persuasive enough, for the chief seemed uninterested in the prospect of allowing the Moravians to live and practice their religion in the valley. Such matters of spirituality belonged to the Europeans, not Native people, he had said. Zinzendorf offered his shirt buttons and shoe buckles as gifts to the chief and other residents. Martin Mack would later state in his diary that Zinzendorf had given away so many buttons from his clothing, he had to fasten his underclothes and shoes with string. For the past ten days, the Moravians had lived on the boiled beans they had brought with them from

Shamokin, which they apportioned sparingly three times per day. Still the residents of Wajomick appeared uninterested in speaking with them about Christianity or feeding them, and had little regard for the Moravians' buttons. They already had numerous, more expensive European trade goods—iron kettles and pots, long rifles, and beautiful translucent glass beads in hues of white and blue. These beads were especially cherished and often became grave goods that accompanied the burial of loved ones on the sandstone bluff above the town.

After a few days in Wajomick, Zinzendorf decided to move his tent east of the town onto that high sandstone outcropping overlooking the river. He was likely unaware of the burial ground beneath. Martin Mack's diary records the events of their stay on the sandstone bluff. Zinzendorf apparently occupied himself primarily with preparing the hymn manuscripts for the eleventh and twelfth supplements to the main Moravian hymnal, the *Herrnhuter Gesangbuch*. He also composed several new hymns that recorded the events of their stay in the area of Wajomick. One of these hymns celebrated a Mohican woman that Jannetje Mack had met as she was walking through Wajomick. Jannetje could speak a little of the Mohican language, and the woman told her that she had heard of the Christian God and was curious to learn more.[117] After this encounter, Zinzendorf had also gone with Jannetje to meet the woman, and had spoken to her of "the love of Jesus . . . in terms of most persuasive tenderness."[118] The woman offered fresh beans and cornbread to Zinzendorf and Jannetje. Returning to his tent, Zinzendorf was apparently overjoyed that this woman would be "*der erste Schawanohs sich gläubt* [the first of the Shawnee to believe]," perhaps mistaking her tribal affiliation in his haste to record the encounter in a hymn text. He was now hopeful that the Christian message would resonate with some residents in Wajomick: "*Darüber wurden eins zwey, drey, und denken itzt noch einerley* [Now where there is one, there will be two, three, and I think not just one only]."[119] Although it would be another twelve years before the Mohican woman would request baptism by the Moravians, receiving the Christian name of Marie on July 28, 1754, Zinzendorf carefully recorded this first encounter with her in song.

Later that day, Zinzendorf would have another fateful encounter worthy of committing to song. As he was immersed in his work on the hymnal supplements, he failed to notice two large blowing adders, called by the

Moravians *Blaseschlangen* (blowers or blowing snakes), enjoying the warm afternoon sun at the entrance of his tent. His companions did not fail to notice the snakes. In a panic, Martin Mack sprang up from the campfire to scare them away from the tent entrance. Unfortunately, rather than directing the snakes away from Zinzendorf, Mack startled the snakes and they glided quickly into the tent, where they slipped over Zinzendorf's legs and disappeared into the hymn manuscripts. In all the confusion, Zinzendorf somehow had not been bitten, and had miraculously survived his encounter with the "very tempters of the Garden of Eden." It was a miracle that would become legendary among Moravian missionaries. The awe and wonder of the story was fueled by the subsequent discovery of an entire nest of snakes just outside the entrance of the tent.[120] Zinzendorf had clearly been the beneficiary of divine Providence, and a chance encounter with two snakes became the story of a "miracle" in which the Moravian leader had been snatched from Satan's jaws by an all-powerful God. The sandstone bluff, site of the temporary and precarious encampment of Zinzendorf and his companions, now became Moravian sacred ground, and was committed into song and iconography (fig. 1.6).[121] Zinzendorf, Nitschmann, and the Macks would also learn from Wajomik's residents that their tent was pitched on already existing sacred ground—the site of hundreds of Native burials along the top of the sandstone bluff. Long before the Shawnee and other current residents had sought refuge in the Wyoming Valley in the early eighteenth century under the protection of the Six Nations, people had been laid to rest there with their bodies facing the east and the rising sun, beautiful glass beads scattered around their graves.

As Zinzendorf finished his next two hymns and recorded them into manuscript form, he began the process of communicating the travails and blessings of his stay in Wajomick, including his encounter with the snakes, to Moravian congregations across the Atlantic World. These hymns would be entered into the eleventh and twelfth supplements of the *Herrnhuter Gesangbuch* as hymns 1853 and 1902 (fig. 1.7a–b).

HG 1853

Wir dachten an die hirten-treu	We thought upon the Shepherd-faithful,
des Jesuah Jehovah	Jesus Jehovah
In der betrübten wüsteney,	In these afflicted wastes
mit namen Skehantowa.	known as Skehantowa.

Fig. 1.6 Detail of Nikolaus Ludwig von Zinzendorf, Anna Nitschmann, and several Shawnee warriors in the Wyoming Valley as they appear in an oil-on-canvas study by Johann Valentin Haidt, 1752. PC 37, MAB.

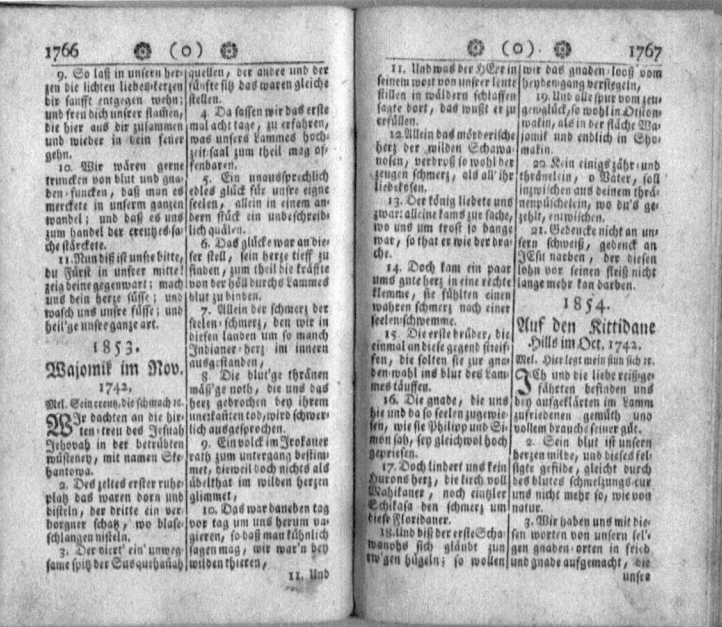

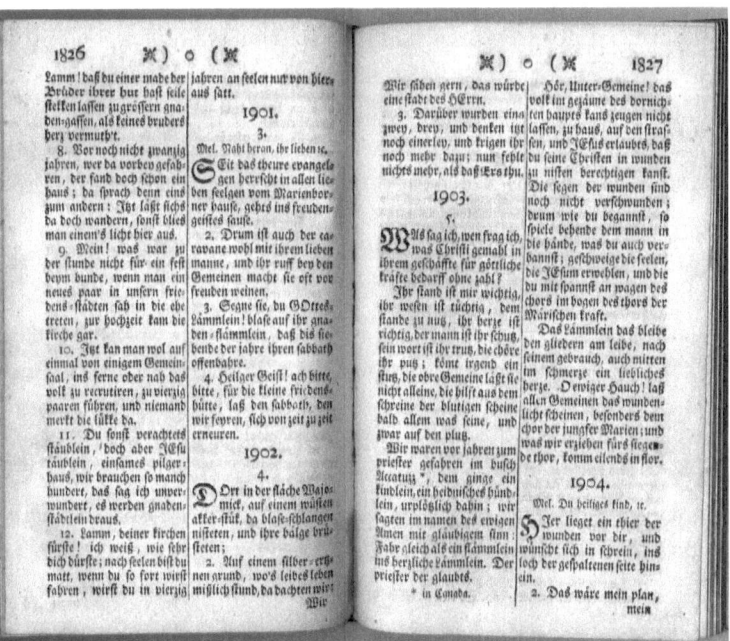

Fig. 1.7a–b Hymns written in the Wyoming Valley from the 11th and 12th supplements of the *Herrnhuter Gesangbuch* (HG 1853 and HG 1902). Cea 54, Anh. 11–12, MAB. Photographs by author.

Des zeltes erster ruhe-platz	The tent's first resting-place
das waren dorn und disteln,	was thorns and thistles,
Der dritte ein verborgner schatz,	The third place was a dangerous treasure,
wo blase-schlangen nisteln.	where blowing adders nested.
Der viert' ein' unwegsame spitz	The fourth was an impassable place
der Susquehannah quellen,	at the source of the Susquehanna,
Der andre und der fünfte sitz	The other and the fifth place
das waren gleiche stellen.[122]	it really was just the same.

HG 1902

Dort in der fläche Wajomick,	There in the plains of Wajomick,
auf einem wüsten akker stük,	on a little acre of waste-land,
Da blase-schlangen nisteten,	There blowing adders nest,
und ihre bälge brüsteten.[123]	and puff out their bellows.

When Zinzendorf returned to Bethlehem from Wajomick, he crafted his diary notes into an official record of *Johanans Reisen unter die Heyden* (*Johanan's Trips to the Heathen*) and produced a small map of his travels that could be sent to Germany and England (fig. 1.7). He also wrote the preface to the hymnal supplements he had prepared in Wajomick, and signed the document with his Native name, Johanan, inserting the Native place-names that attended its preparation: "*Aus dem Zelte vor Wayomik, in der grossen Ebene Skehandowana*, in Canada, am 15. Oct. 1742. *Euer unwürdiger Johanan* [From the wilderness of Wajomick, in the great flats of Skehandowana, in Canada, on the 15th of October, 1742. Your unworthy Johanan]."[124]

A Moravian Pennsylvania

The hymns that Zinzendorf composed in Wajomick became powerful symbols of a distinctly Moravian version of Pennsylvania, its landscape, and its people. By the time future Moravians visited the Wyoming Valley, these hymns were well known and served as sonic markers of Moravian history, place, and identity. In October 1748, Baron Johannes de Watteville, Zinzendorf's son-in-law, noticed traces of Zinzendorf still marked upon the landscape of Wyoming: a tree that had been carved with the intials "J" for Johanan and "C" for Conrad Weiser. Watteville added his own markings to the tree, carving an "A" for Anna Nitschmann, Zinzendorf's second wife, and the years "1742" and "1748." He also visited the unmarked but still well remembered spot where Zinzendorf's tent had first been pitched, and the

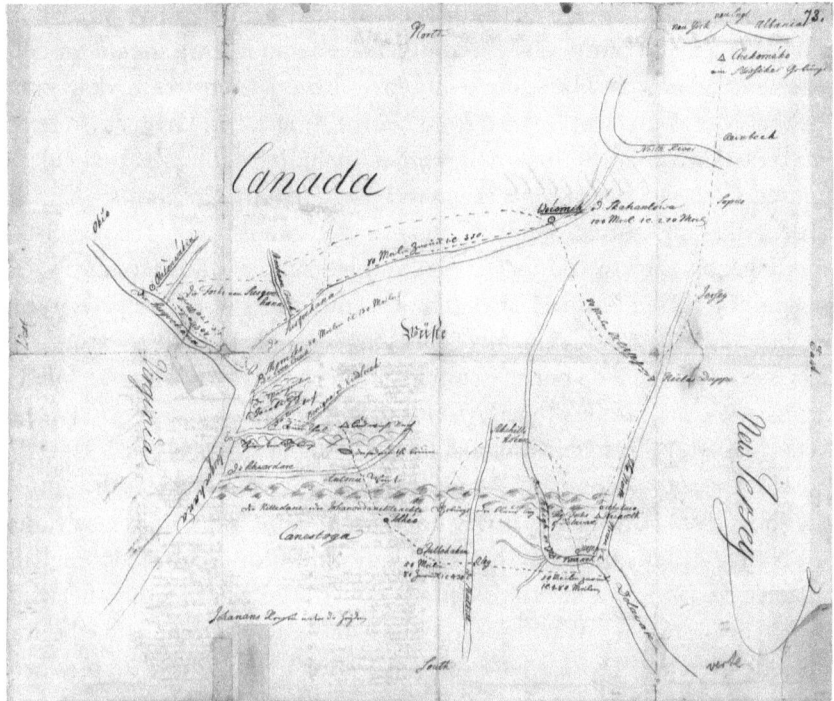

Fig. 1.8 Sketch map entitled "Johanans Reisen unter die Heiden [Johanan's Travels to the Heathen]." TS Mp.212.10, UA Herrnhut.

sandstone outcropping along the river where God had delivered Zinzendorf from the fangs of the adders.[125] These places now firmly resided in the topography of Moravian history and imagination.

However, it was not just Moravian songs, diaries, and maps that captured the American landscape and carried it across the Atlantic. Moravian missionaries collected plant, bird, and animal specimens throughout the forests of Pennsylvania, New York, and Ohio to send back to Germany. By 1758, the Moravian seminary at Barby had a robust and famous natural history collection, a "cabinet of curiosities" or *Naturaliencabinet* that was used by students and remarked on by travelers.[126] In 1776, Johann Wolfgang von Goethe visited Barby and viewed the cabinet. His interest in the American wilderness was further piqued by a song shared with him by missionary and musician Christian Gregor, originally written by Gregor for his daughter as an eleventh birthday present in 1771. The song commemorated Gregor's journey from Bethlehem to Friedenshütten, along the northern branch of the

Susquehanna, and celebrated the distinctive qualities of Pennsylvania's flora and fauna. Gregor humorously recommended camping by a stream for both water and a bath, and building temporary shelters from tree bark, a practice that Moravians had learned from Native Americans. Gregor also referenced Pennsylvania's animals and reptiles, including a line that particularly amused Goethe: "*Item Klapperschlangen und der Art Geschwänz* [namely, rattle snakes and similar kinds of tails]," as well as a listing of the deer, bear, woodpigeons, eagles, wild cats, fox, and wolves, and all those creatures that "*sonst noch Tatzen hat zum Tischgebrauch* [that otherwise have paws to use at table]." Goethe would remember this song for more than thirty years, and referenced it in his 1808 poem, "Zum 21. Juni. An Silvie von Ziegesar," which begins with the famous line: "*Nicht am Susquehanna, der durch Wüsten fließt* [Not on the Susquehanna that flows through empty wastes]."[127]

However, the Moravian Pennsylvania of the mid-eighteenth century was short lived. Ultimately, it was not a Moravian vision of Pennsylvania, or a Native American vision that would prevail. It was a distinctly English-inflected version of American identity that would arise in the aftermath of the Revolutionary War. By the time Goethe wrote his poem, the war was long over, and Thomas Jefferson was ending his term as third president of the United States of America. The physical and cultural landscapes of Pennsylvania's river valleys had also shifted dramatically. Native American towns and villages had long since been removed westward after the Six Nations signed away lands belonging to the Delaware and other tribal communities, beginning with the Treaty of Fort Stanwix in 1768. And there were no longer Moravian missions in Pennsylvania. As the western border of European colonial settlements shifted dramatically westward, other European immigrants, primarily Scots-Irish, poured into Pennsylvania's forests by the hundreds and thousands. These new immigrants transformed the eastern parts of Penn's Woods into miles of cleared agricultural land, and deforested thousands of acres in the Delaware, Ohio, and Susquehanna Valleys. When Charles Dickens sailed down the Allegheny in the early 1840s, in what had been the heart of Native country in the eighteenth century, he observed hundreds of new Scots-Irish farmsteads and settlements crowding along the riverbank, and the "wounded bodies" of trees like "murdered creatures":

> Then there were new settlements and detached log-cabins and frame-houses, full of interest for strangers from an old country: cabins with simple ovens, outside, made of clay; and lodgings for the pigs nearly as good as many of the

human quarters; broken windows, patched with worn-out hats, old clothes, old boards, fragments of blankets and paper; and home-made dressers standing in the open air without the door, whereon was ranged the household store, not hard to count, of earthen jars and pots. The eye was pained to see the stumps of great trees, thickly strewn in every field of wheat.... It was quite sad and oppressive, to come upon great tracts where settlers had been burning down the trees, and where their wounded bodies lay about, like those of murdered creatures, while here and there some charred and blackened giant reared aloft two withered arms, and seemed to call down curses on his foes.[128]

Even the nineteenth-century novelist James Fenimore Cooper, who studied the Moravian mission histories and diaries of Johann Heckewelder and David Zeisberger to create his best-selling novel *The Last of the Mohicans: A Narrative of 1757*, did not advocate for the preservation of the American forest. Writing of his youth, Cooper, like Dickens, noted the same "dark and charred stumps that had, in the preceding season, supported some of the proudest trees of the forest." But, unlike Dickens, he decried the "deep forests" and "dreary and dark wood[s], where the rays of the sun could but rarely penetrate," and praised the pastoral transformation wrought upon the landscape by European colonists: "What had been a timeless, wild land of pure nature, [became] schools, academies, churches, meeting-houses, turnpike roads, and market towns . . . neat and comfortable farms . . . and beautiful and thriving villages."[129] These new "beautiful and thriving" communities boasted soundscapes that could be heard for several miles through the forest. The sounds of industries and towns overtook the soundscapes of the forest, its plants, trees, animals, birds, and reptiles.

For Native people, this environmental transformation was also a slow process of sonic and spatial loss. The dreams and visions of Native Christians, as recorded by missionaries, serve as a potential way to understand the mental toll that colonial expansion must have extracted from Native communities. One night, Moses Tatemy, Zinzendorf's host on his first journey into Native country, had a terrible vision: "There seem'd to be an impassable Mountain before him. He was pressing towards Heaven as he thought, but his Way was hedg'd up with Thorns that he could not stir an Inch further. He thought if he could but make his Way thro' these Thorns and Briers, and clime up the first steep Pitch of the Mountain, than then there might be Hope for him, but no Way or Means could he find to accomplish this." Tatemy's vision is very similar to another dream experienced by the very first Native Moravian, the Shekomeko resident, Shabash (baptized as Abraham). One evening in 1738, Shabash drank himself into unconsciousness and had a vision of a "roar of gushing water [that] filled his ears and he saw before him a group of Indians drunk and naked and unable to escape the onrushing water. A voice told him he must give up all wickedness. As the vision continued, a strong light shone all about him, and he heard 'a noise like the blowing of a pair of bellows' and a 'violent blast of wind which dispersed the Indians into the air.'"[130] According to Moravian historian Rachel Wheeler, Tatemy and Shabash's visions may have articulated the transformation of once-familiar natural environments into disabling forces that rendered the seers helpless.[131]

As Tatemy and Shabash may have sensed, the natural environment surrounding their communities was shifting, and with this shift came the dissolution of a vision of a shared Pennsylvania that had briefly existed in the mid-eighteenth century. Conflicts over the use of the land and its natural resources, and the increasingly racially inflected rhetoric that dominated the discourse of all sides during the Seven Years' War, would result in destructive acts of violence and physical removal for the Native population of Pennsylvania. In the 2000 census taken by the United States government, Pennsylvania ranked last in the reported percentage of its population classified as "Native American." There are no reservation lands or federally recognized tribes within the modern borders of Penn's Woods. The last land grant made to a Native community, the Cornplanter Tract, was subsumed by the Kinzua Dam in the 1960s, forcing the relocation of the two hundred people who lived there, despite promises that it would belong to them and their descendants "until the end of time." Still, in recent decades, there have been new efforts to recognize Pennsylvania's Native American past. Much of this activity has been centered around the area that is now Bethlehem, with the founding of the Lenni Lenape Historical Society and the Museum of the American Indian in Allentown, and the Lenape Nation of Pennsylvania Cultural Center and Trading Post in Easton.[132] These cultural organizations and museums are important reminders that the Lehigh River Valley was a traditional center for Delaware culture long before the arrival of the Moravians and other settlers. And the Pennsylvania landscape is still embedded with traces of its Native and colonial pasts—names, burials, songs, maps, and stories that record the many histories of its valleys, mountains, and forests.

Plymouth (Wyoming, Wajomick): June, 1905

Where today are the Pequot? Where are the Narragansett, the Mohican, the Pocanet and other powerful tribes of our people? They have vanished before the avarice and oppression of the white man, as snow before the summer sun . . . Sleep no longer, O Choctaws and Chickasaws . . . Will not the bones of our dead be plowed up, and their graves turned into plowed fields?

—Tecumseh, 1811

In early June 1905, several Polish mine workers were excavating a basement for a new building on Bead Street in Plymouth, a mining community located near the site of a former Shawnee village on a sandstone bluff overlooking the Susquehanna River in the Wyoming Valley. It was early evening, as the men painstakingly hacked away at the soft bedrock underneath the site. They had already worked full shifts in the nearby anthracite mines and were anxious to return home. Suddenly, a splinter of bone emerged from the rough earth underneath the workmen's pickaxes. As they dug further, they discovered more bones—an entire body, carefully laid to rest in a shallow pit surrounded by a complete row of stones set on their edges. Underneath the body was a floor of smooth stones. She was a Native woman, laid to rest with her feet to the east. Nearby, and about three feet to the north, was the body of a Native man. His grave was also carefully bordered with an edge of upright stones and he lay facing east. Around him, the workers found more than two hundred faceted dark-blue glass beads. His chest was layered with beadwork comprised of hundreds of small blue-and-white glass beads. From his left shoulder to his wrist were more than one hundred brass rings, many inlaid with seals of transparent or semi-transparent glass. Two glass decanters and a baked clay pipe lay buried nearby. As the workers continued to dig the cellar, a third body was unearthed, a man encircled in upright stones. This man had no grave goods, but his body faced to the east.

Perhaps the Polish mine workers, exhausted and hot, were surprised to find three bodies in their newly dug basement. Or perhaps they were aware that the area around Bead Street, the top of the sandstone bluff now known locally as "Plymouth Rock," was a Native American burial ground containing hundreds of ancient graves. Some of these burials were rumored to be so old that even the eighteenth-century Shawnee residents who had built a town four hundred yards to the west of the bluff along the banks of the Susquehanna regarded them as ancient, since they "were there when they came into the country."[133] The Shawnee town had itself long since disappeared and was named now only as a curiosity on nineteenth-century American maps as the "Old Shawanese Town," or remembered in the archival records of the Moravian Church in Bethlehem, Pennsylvania, by its German name, "Wajomick."[134]

Gone were the homes of the town's chief, Kackawatcheky, and Marie, the Mohican woman who had requested baptism by the Moravians. Gone was the tent of Zinzendorf and his companions, pitched on top of the sandstone bluff. Gone were the snakes. The Moravian Church had long ago ceased to sing the hymns written by Zinzendorf to record the events that transpired during his visit to this place. But the burials had remained. And the beads. Hundreds of glass beads that were plowed up in farmers' fields and in construction sites all along the northern bank of the Susquehanna. So many beads that this place had become known as "Bead Street."

By the time darkness fell that evening in June 1905, a large crowd had gathered to view the bodies. By morning, the beads, the glass decanters, the baked clay pipe, the bits of bone and stone were gone, vanished into nearby homes and businesses as objects of curiosity and historical interest. Now there was only a name, "Bead Street," marking the many histories of this place.

Notes

1. HG 2251, verse 4.
2. The Meadowcroft Rockshelter in Washington County is currently the earliest known archaeological site in Pennsylvania, and was occupied perhaps as early as 16,000 years ago. Randall M. Miller, William Pencak, and Pennsylvania Historical and Museum Commission, eds., *Pennsylvania: A History of the Commonwealth*, A Keystone Book (University Park: PA: Pennsylvania State University Press; Pennsylvania Historical and Museum Commission, 2002), ix–x. For further discussion of Pennsylvania's Paleo-Indian history, see David Jay Minderhout, *Native Americans in the Susquehanna River Valley, Past and Present* (Lanham, MD: Bucknell University Press, 2013); Kurt W. Carr and J. M. Adovasio, *Ice Age Peoples of Pennsylvania* (Harrisburg: PA: Pennsylvania Historical and Museum Commission, in cooperation with the Pennsylvania Archaeological Council, 2002); Kurt W. Carr and Roger W. Moeller, *First Pennsylvanians: The Archaeology of Native Americans in Pennsylvania* (Harrisburg: PA: Pennsylvania Historical and Museum Commission, 2015); Daniel K. Richter, "A Framework for Pennsylvania Indian History," *Pennsylvania History: A Journal of Mid-Atlantic Studies* 57, no.3 (1990): 236–261; and Daniel K. Richter, *Native Americans' Pennsylvania* (University Park, PA: Pennsylvania State University Press, 2005).

3. Jean O'Brien has argued that the process of European settlement often involved the renaming and claiming of Native places in an attempt to erase their earlier histories. Settler histories of New England also perpetuated myths of a landscape that was unpeopled and vast, to legitimize land claims. O'Brien, *Firsting and Lasting: Writing Indians Out of Existence in New England*, Indigenous Americas series (Minneapolis, MN: University of Minnesota Press, 2010), 1.

4. Simon Schama, *Landscape and Memory* (New York: A. A. Knopf, 1995), 61; quoted in Belden Lane, *Landscapes of the Sacred*, Isaac Hecker Studies in Religion and American Culture (New York: Paulist Press, 1988), 239.

5. Tim Cresswell, *Place: A Short Introduction,* Short Introductions to Geography (Malden, MA: Blackwell Publishing, 2004), 11.

6. Cresswell, *Place: A Short Introduction*, 10.

7. Barry Truax, *Acoustic Communication*, 2nd ed. (Westport, CT: Ablex Publishing, 2013), 392; Bernard L. Krause, *The Great Animal Orchestra: Finding the Origins of Music in the World's Wild Places* (Boston: Back Bay Books Little Brown, 2013), 27–28.

8. Daniel K. Richter, "First Pennsylvanians," in *Pennsylvania: History of the Commonwealth*, 10.

9. Pennsylvania Department of Conservation of Natural Resources, "Penn's Woods: A History of Pennsylvania's Forests," accessed July 28, 2016, www.dcnr.state.pa.us/cs/groups /public/documents/document/dcnr_009325.pdf.

10. John L. Cotter, Daniel G. Roberts, and Michael Parrington, *The Buried Past: An Archaeological History of Philadelphia* (Philadelphia: University of Pennsylvania Press, 1992), 6. Other common trees included larch, walnut, ash, tulip poplar, sugar maple, wild cherry, cedar, and sycamore.

11. Loskiel, *History of the Mission of the United Brethren*, I, VI, 68.

12. Loskiel, *History of the Mission of the United Brethren*, I, VII, 78–80.

13. The coastal flats of the Susquehanna and Delaware Estuary were home to numerous species of fish: pike, eel, mackerel, salmon, carp, catfish, blackfish, whitefish, rockfish, trout, sucker, and shad, as well as oysters, crabs, and even harbor seals. Several early Moravian documents, including Loskiel's mission history, David Zeisberger's mission diary, and Johannes Ettwein's travel journal of 1765, record that harbor seals were seen in the north branch of the Susquehanna near Wyoming. David Zeisberger, *Schoenbrunn Story; Excerpts from the Diary of the Reverend David Zeisberger, 1772–1777, at Schoenbrunn in the Ohio Country* (Columbus: Ohio Historical Society, 1972); John Ettwein and John Jordan, ed., "Rev. John Ettwein's Notes of Travel from the North Branch of the Susquehanna to the Beaver River, Pennsylvania, 1772," *The Pennsylvania Magazine of History and Biography* 25, no. 2 (1901): 208–219.

14. For a more extensive list of birds, see Loskiel, *History of the Mission of the United Brethren*, "Birds," I, VII, 89–94.

15. For a detailed history of trails in Pennsylvania, see Wallace, *Indian Paths of Pennsylvania*.

16. Joy Harjo and Stephen Strom, *Secrets from the Center of the World*, Sun Tracks 17 (Tucson, AZ: University of Arizona Press, 1989), 1; quoted in J. Scott Bryson, *The West Side of Any Mountain* (Iowa City: University of Iowa Press, 2005), 62.

17. Yi-Fu Tuan, *Space and Place: The Perspective of Experience* (Minneapolis: University of Minnesota Press, 1977), 119.

18. James Hart Merrell, *Into the American Woods: Negotiators on the Pennsylvania Frontier* (New York: Norton, 1999), 20. Jon Parmenter, *The Edge of the Woods: Iroquoia, 1534–1701* (East Lansing: Michigan State University Press, 2010), xxvii.

19. Records of the Moravian Missions to the American Indians [MissInd], 164.6, 4/19/1804, Moravian Archives, Bethlehem [MAB].

20. Delaware Tribe of Indians, Lenape Talking Dictionary, accessed April 7, 2018, http://talk-lenape.org/stories?id=72#1694.

21. MissInd 171.6, 4/21/1800, MAB.

22. MissInd 177.5, 1/28/1803, MAB.

23. MissInd 177.6, 2/23/1804 and 3/18/1804, MAB.

24. Peter Charles Hoffer, *Sensory Worlds of Early America* (Baltimore: Johns Hopkins University Press, 2003), 28.

25. MissInd 135.1, 6/6/1768, and 175.1, 1/3/1817, MAB.

26. Loskiel, *History of the Mission of the United Brethren*, II, XIV, 191.

27. Dreams were an important way of receiving spiritual communications, for both Europeans and Native Americans. See Ann Marie Plane, Leslie Tutle, and Anthony F. C. Wallace, eds., *Dreams, Dreamers, Visions: The Early Modern Atlantic World* (Philadelphia: University of Pennsylvania Press, 2013); and Pointer, *Encounters of the Spirit*, 139–140.

28. Loskiel, *History of the Mission of the United Brethren*, I, VII, 76–77.

29. Loskiel, *History of the Mission of the United Brethren*, 42.

30. Paul A. W. Wallace and William A. Hunter, *Indians in Pennsylvania*, 2nd ed. Anthropological Series 5 (Harrisburg: Pennsylvania Historical and Museum Commission, 2005), 123. Loskiel, *History of the Mission of the United Brethren*, I, VII, 75–76.

31. MissInd 171.2, 11/20/1798, MAB.

32. Hoffer, *Sensory Worlds of Early America*, 34.

33. For an extensive list of Munsee and Unami names for plants and animals, see August C. Mahr "Delaware Terms for Plants and Animals," *Anthropological Linguistics* 4, no. 5 (1962): 1–48.

34. Elma E. Gray, *Wilderness Christians: The Moravian Mission to the Delaware Indians* (Ithaca, NY: Cornell University Press, 1956), 8.

35. Jane T. Merritt, *At the Crossroads: Indians and Empires on a Mid-Atlantic Frontier, 1700–1763* (Chapel Hill: University of North Carolina Press, 2003), 42.

36. August C. Mahr and John Gottlieb Ernestus Heckewelder, "A Canoe Journey from Big Beaver to the Tuscarawas in 1773: A Travel Diary of John Heckewaelder," *Ohio State Archaeological and Historical Quarterly* (1952): 294.

37. Hoffer, *Sensory Worlds of Early America*, 34.

38. Robert Warrior, "Conclusion: Intellectual Trade Routes," in *The People and The Word: Reading Native Nonfiction*, Indigenous Americas (Minneapolis: University of Minnesota Press, 2005): 182–183. Many of Pennsylvania's modern highways follow the routes of Native pathways. Daniel Richter has asserted that the southern portion of US Highway 522 follows a transportation corridor between the Cumberland Valley and the Juniata River that has been in use at least since 6,500 B.C. as a route for transmitting rhyolite—a volcanic substance used for stone tools. In the eighteenth century, this path was known as the Frankstown Path (Richter, "First Pennsylvanians," in *Pennsylvania: A History of the Commonwealth*, 12). Also see Wallace, *Indian Paths of Pennsylvania*, 2.

39. Wallace, *Indians in Pennsylvania*, 44.

40. According to Loskiel's history, directional information could be gleaned from the position of the sun and moon and the growth of moss on the northern sides of trees: Loskiel, *History of the Mission of the United Brethren*, I, II, 30.

41. Loskiel, *History of the Mission of the United Brethren*, I, II, 23.
42. Frederick C. Johnson, *Count Zinzendorf and the Moravian Occupation of the Wyoming Valley* (Wilkes-Barré, PA: 1904), 26; and Katherine Faull, "From Friedenshütten to Wyalusing: Johannes Ettwein's Map of the Upper Susquehanna (1768) and an Account of His Journey," *Journal of Moravian History* 11 (2011): 90.
43. Loskiel, *History of the Mission of the United Brethren*, I, II, 25.
44. Loskiel, *History of the Mission of the United Brethren*, I, II, 25.
45. Wallace, *Indian Paths of Pennsylvania*, 15, 38n; and "Spangenberg's Journey to Onondaga, 1745," in *Moravian Journals Relating to Central New York, 1745-1766* ed. William Martin Beauchamp (Bowie, MD: Heritage Books, 1999), 11. For further information on warrior's trees, see Wallace, *Indian Paths of Pennsylvania*, 10.
46. "Better than a mile from the town [Pettquotting] was a peeled tree, on which, was written with charcoal in Delaware the whole verse, 'The Saviour's blood and righteousness.' This made us a good road sign." Paul A. W. Wallace, *Thirty Thousand Miles with John Heckewelder, Or, Travels among the Indians of Pennsylvania, New York & Ohio in the 18th Century*, The Great Pennsylvania Frontier Series (Lewisburg, PA: Wennawoods Publishing, 1998), 248; quoted from "Abraham Steiner's Account of His Journey with Johann Heckewelder from Bethlehem to Pettquotting on the Huron River near Lake Erie, and Return, 1789," MAB.
47. Hoffer, *Sensory Worlds of Early America*, 43; also see Gray, *Wilderness Christians*, 20.
48. Wallace, *Indians in Pennsylvania*, 42.
49. Although the modern usage of the word "swamp" indicates a marshy area, in the eighteenth century a "swamp" was a lowland area overgrown with thickets of bushes and trees. Loskiel, *History of the Mission of the United Brethren*, I, I, 9.
50. Wallace, *Indian Paths of Pennsylvania*, 6; quoting from A. G. Spangenberg, "Spangenberg's Notes of Travel to Onondaga in 1745," translated in *The Pennsylvania Magazine of History and Biography* 2, no. 4 (1878): 424-432.
51. John Bartram, *Travels in Pensilvania and Canada*, March of America Facsimile Series, No. 41 (Ann Arbor: University Microfilms, 1966), 13, 37. Bartram was the first European American botanist. He collected specimens for the new Royal Society in London and Carl Linnaeus's Swedish Academy of Sciences, and worked with London merchant Peter Collinson to ship American "curiosities"—Native American pottery and arrowheads, fossils, seeds, and plants, birds and insects—to English collectors.
52. Wallace, *Indian Paths of Pennsylvania*, 7.
53. Ettwein, "Rev. John Ettwein's Notes of Travel, 1772," 208-209.
54. Lewis Evans's map of 1755 was the first good map of Pennsylvania and adjoining areas, and was used extensively during the Seven Years' War by General Edward Braddock.
55. Thomas Pownall, *A Topographical Description of the Dominions of the United States of America* (Pittsburgh: University of Pittsburgh Press, reprint 1949), 23.
56. Pownall, *A Topographical Description of the Dominions of the United States*, 30.
57. Pownall, *A Topographical Description of the Dominions of the United States*, 31.
58. Mary Rowlandson, *Narrative of the Captivity and Restauration (1682)*. Quoted in Roderick Nash, *Wilderness and the American Mind* (New Haven: Yale University Press, 2001), 28.
59. Ettwein, "Rev. John Ettwein's Notes of Travel, 1772," 213-214.
60. Bartram, *Travels in Pensilvania*, 70-71.
61. Bartram, *Travels in Pensilvania*, 70 and 72.
62. August Gottlieb Spangenberg, David Zeisberger, John Martin Mack, "Diary of Brother David Zeisberger's and Henry Frey's Journey and Stay in Onondaga from April 23rd

to November 12th, 1753," in *Moravian Journals Relating to Central New York, 1745–1766*, ed. William Martin Beauchamp (Bowie, MD: Heritage Books, 1999), 158.

63. Loskiel, *History of the Mission of the United Brethren*, III, II, 29.
64. Bartram, *Travels in Pensilvania*, 77.
65. Bartram, *Travels in Pensilvania*, 29.
66. Bartram, *Travels in Pensilvania*, 44.
67. See "Narrative of a Journey to Shecomeco, in August of 1742," in *Memorials of the Moravian Church*, 62, for a short biography of Conrad Weiser; and "Zinzendorf's Narrative of a Journey from Bethlehem to Shamokin in September of 1742," in *Memorials of the Moravian Church*, ed. William Cornelius Reichel (Philadelphia: 1870), 83, for a short biography of Shikellamy.
68. Bartram, *Travels in Pensilvania*, 31.
69. Wallace, *Indian Paths of Pennsylvania*, 12.
70. "Diary of a Journey Made by the Brethren Grube and Rundt to Wajomik, 1754," in Johnson, *Count Zinzendorf and the Moravian Occupation of the Wyoming Valley*, 175.
71. Michael N. McConnell, *A Country Between: The Upper Ohio Valley and Its Peoples, 1724–1774* (Lincoln: University of Nebraska Press, 1992), 39–40. Although violins and flutes may have been added to existing Native musical practices and may have represented changing performing traditions, they probably did not alter the underlying social and religious significance of community rituals and events.
72. James Hart Merrell, "Shamokin, 'The Very Seat of the Prince of Darkness,'" in *Contact Points: American Frontiers from the Mohawk Valley to the Mississippi, 1750–1830*, ed. Andrew R. L. Cayton, Fredrika J. Teute, and the Omohundro Institute of Early American History & Culture (Chapel Hill: University of North Carolina Press, 1998), 27; quoted in Katherine Faull, "The Experience of the World as the Experience of the Self: Smooth Rocks in a River Archipelago," in *Re-Imagining Nature: Environmental Humanities and Ecosemiotics*, ed. Alfred K. Siewers (Lewisburg, PA: Bucknell University Press, 2014), 197–214, 199.
73. Katherine Faull, "The Experience of the World as the Experience of the Self," 197–199; also see Jane T. Merritt, *At the Crossroads: Indians and Empires on a Mid-Atlantic Frontier*, 2–3. Cynthia Radding has similarly argued that frontier spaces provided opportunities for dialogue between settlers and Native communities, and provided the possibility for new cultural ways to emerge in as much as was possible in a colonial context. Cynthia Radding, "Human Geographies and Landscapes of the Divine in the Northern Mesoamerican Borderlands," in *Re-Imagining Nature: Environmental Humanities and Ecosemiotics*, 215; and Radding, "Borderlands of Knowledge about Nature: Crossing and Creating Boundaries in Early America," *Early American Studies: An Interdisciplinary Journal* 13, no. 2 (April, 2015): 503–510.
74. Faull, "The Experience of the World as the Experience of the Self," 207.
75. "Mack to Onondaga, 1752," in *Moravian Journals Relating to Central New York*, 120.
76. "Mack to Onondaga, 1752," in *Moravian Journals Relating to Central New York*, 145.
77. "Zeisberger and Frey to Onondaga, 1753," in *Moravian Journals Relating to Central New York*, 180.
78. "Diary of Br. John Martin Mack's and Christian Fröhlich's Journey to Wayomick and Hallobanck," in Johnson, *Count Zinzendorf and the Moravian Occupation of the Wyoming Valley*, 37.
79. See Chief Luther Standing Bear, "Indian Wisdom," in Michael P. Nelson and J. Baird Callicott, *The Great New Wilderness Debate* (Athens, GA: The University of Georgia Press,

2008); Melanie Perreault, "American Wilderness and First Contact," in Michael L. Lewis, *American Wilderness: A New History* (New York: Oxford University Press, 2007); and Bill Devall and George Sessions, *Deep Ecology: Living as If Nature Mattered* (Salt Lake City: G. M. Smith, 2007).

80. Faull, "From Friedenshütten to Wyoming," 87.

81. "Count Zinzendorf's Review of His Experience among the North American Indians," in *Memorials of the Moravian Church*, 134.

82. "Mack to Onondaga, 1752," in *Moravian Journals Relating to Central New York*, 350.

83. Loskiel, *History of the Mission of the United Brethren*, III, II, 29.

84. Loskiel, *History of the Mission of the United Brethren*, VI, III, 106.

85. "U. es ist ihm auch recht wol, so wol in Josuas als Augustus Hütte unter ihnen gewisen. Ihre Abdzeit haben sie mit nüzlichen Discursen zu gebracht u. offt an die erstaunl. lieb des Hlds u. ihrer Gnadenwohl gedacht. Morgens u. Abends singen sie versgen. Im jagen sind sie zieml. glückl." *Gnadenhütten Diary*, 11/24/1754, MissInd 117.3, MAB. On another occasion, Schmick noted in a letter to Spangenberg that Joshua, Anton, and Jacob had preached and sung verses of the Savior in the woods. "Jacob, Josua u. Anton werden die Aufsicht über die Brr. heben, u. ihnen Worte u. Verse von Hland sagen u. singen." Schmick to Spangenberg, 7/20/1755, MissInd 118.6, MAB. Also see Wheeler, *To Live upon Hope*, 109–11.

86. *Report of the Brown Brethren in Gnadenhütten, 1753*, 4/9/1753, no. 31, MissInd 319.4.17, MAB.

87. Wheeler, *To Live Upon Hope*, 103.

88. Wheeler, *To Live Upon Hope*, 103.

89. Loskiel, *History of the Mission of the United Brethren*, II, IV, 66. Also see Seidel, MissInd 117.3, 9/4/1752, MAB.

90. *The Bethlehem Diary* (Bethlehem, PA: Archives of the Moravian Church, 1971), vol. I, 47. Johanan was Zinzendorf's Native name.

91. Reichel, ed., "Zinzendorf Among the Delawares, July 24–August 2, 1742," in *Memorials of the Moravian Church*, 28.

92. "Zinzendorf Among the Delawares, July 24–August 2, 1742," in *Memorials of the Moravian Church*, 30.

93. Sometime during the time period between July 28 and July 30, Zinzendorf also sent another hymn to the congregation in Bethlehem: HG 1796, "At Christ's Manger," a hymn named in reference to Christ's birth in the town of Bethlehem. The hymn reached Bethlehem on July 31, and was sung in a Singstunde that evening. *Bethlehem Diary* I, 50.

94. The meaning of the name "Sikihillehocken" is unknown, but it likely referred to the region of land west of the Schuylkill.

95. Seuse reportedly heard angels singing the text of "*In dulci jubilo*," and was inspired to join them in a dance of worship and to record the text. The tune first appears in Codex 1305, a manuscript dating to the turn of the fifteenth century, currently owned by the Leipzig University Library.

96. For the complete text of the hymn, see *Memorials of Moravian Church*, 39–44 (German); and *Bethlehem Diary* I, 53–57 (English).

97. Shikellamy and other Haudenosaunee officials had been in Philadelphia to negotiate with the Pennsylvania governor for the final removal of Delaware settlements within the areas of the Walking Purchase. The Tulpehocken Path from Shamokin to Tulpehocken Creek was used by the Haudenosaunee coming from Onondaga to the Tulpehocken Valley and the Philadelphia region. Those going north from Tulpehocken referred to the pathway as the Shamokin Path.

99. This belt is believed to be the same one that is currently owned by the Völkerkundemuseum Herrnhut ("Irokesen Wampum-Schnüre," H.05 ES 404 b, UA Herrnhut). For more information on the wampum belt in Herrnhut, see Alexander Glitsch to John W. Jordan, 6/2/1885, MissInd 211.20, MAB. Also see Johnson, *Count Zinzendorf and the Moravian and Indian Occupancy of the Wyoming Valley*, 45; Loskiel, *History of the Mission of the United Brethren*, II, 30–32; and Christian Feest, "Wampum from Early European Collections, Strings, Belts, and Bracelets," *American Indian Art Magazine* (Spring 2014): 32–41, 71.

99. *Bethlehem Diary*, I, 64–65 (HG Appendix XI, No. 1836). This hymn also referenced two pre-existing Moravian hymns. See HG Appendix VIII No. 1267, and Appendix VII No. 1254.

100. *Bethlehem Diary* I, 90.

101. *Bethlehem Diary* I, 90.

102. *The Moravian Hymnbook* I, hymn 2086.

103. Jacob Bürstler was a Palatine immigrant who had settled in Oley.

104. "Count Zinzendorf's Narrative of a Journey from Bethlehem to Shamokin," in *Memorials of the Moravian Church*, 76. This hymn had been composed by Zinzendorf on June 13, 1742.

105. These places were named for people close to Zinzendorf: Ludwig (Zinzendorf), Erdmuthe (Zinzendorf's first wife), Anna (Zinzendorf's second wife), Benigna (Zinzendorf's daughter), and Jacob (Jacob Bürstler). Moravian place names were occasionally chosen by lot, including the names of the Ohio mission communities. See *John Ettwein and the Moravian Church during the Revolutionary Period*, 259–260, 279.

106. "Count Zinzendorf's Narrative of a Journey from Bethlehem to Shamokin," in *Memorials of the Moravian Church*, 81. According to Katherine Faull, these Moravian place names are no longer used, except for Benigna's Creek Winery on the Susquehanna Heartland Winery Trail. See Faull, "The Experience of the World as the Experience of the Self," 206.

107. For further discussion on the relationship between Christian missions and early efforts to map the Western Reserve (Ohio), see Amy DeRogatis, *Moral Geography: Maps, Missionaries, and the American Frontier*, Religion and American Culture (New York: Columbia University Press, 2003), 1–4, 30–31.

108. John W. Jordan, "Spangenberg's Notes of Travel to Onondago in 1745," *Pennsylvania Magazine of History and Biography* 2, no. 4 (1898): 426; quoted in Faull, "The Experience of the World as the Experience of the Self," 205.

109. DeRogatis, *Moral Geography*, 4.

110. Judith Ridner, "Building Urban Spaces for the Interior: Thomas Penn and the Colonization of Eighteenth-Century Pennsylvania," in *Early American Cartographies*, ed. Martin Brükner (Chapel Hill: University of North Carolina Press, 2011), 332.

111. "Spangenberg to Onondaga, 1745," in *Moravian Journals Relating to Central New York*, 12.

112. "Spangenberg to Onondaga, 1745," in *Moravian Journals Relating to Central New York*, 15.

113. "Cammerhof and Zeisberger to the Five Nations, 1750," in *Moravian Journals Relating to Central New York*, 27–31. This place is also referenced in "Zeisberger and Frey to Onondaga, 1753," in *Moravian Journals Relating to Central New York*, 158, except it is called *Mon Plaisir in the Desert*. Cammerhof and Zeisberger also named additional places on their journey such as Hill of Peace, Rose Meadow, Joseph's Heights, Haven of Peace, David's Castle, and Deer Pasture.

114. Faull, "From Friedenshütten to Wyoming," 88–89.

115. "Count Zinzendorf's Narrative of a Journey from Bethlehem to Shamokin," in *Memorials of the Moravian Church*, 94.

116. Oscar J. Harvey and Ernest G. Smith, *A History of Wilkes-Barré, Luzerne County, Pennsylvania from Its First Beginnings to the Present Time* (Wilkes-Barré, PA: Raeder Press, 1909), 208; quoting "J. Martin Mack's Recollection of a Journey from Otstonwakin to Wyoming, in the Wilds of Skehandowana," in *Memorials of the Moravian Church*, 100.

117. HG 1853, verse 18: "Und bis der erste Schawanohs sich gläubt zun ew'gen hügeln / so wollen wir das gnaden-looß von diesem gang versiegeln." Despite Zinzendorf's assertion that the woman was Shawnee, she was actually Mohican. Although her Native name was not recorded, she was baptized as Marie.

118. "J. Martin Mack's Recollection of a Journey from Otstonwakin to Wyoming, in the Wilds of Skehandowana," in *Memorials of the Moravian Church*, 104.

119. HG 1902, verse 3.

120. "J. Martin Mack's Recollection of a Journey from Otstonwakin to Wyoming, in the Wilds of Skehandowana," in *Memorials of the Moravian Church*, 106.

121. This encounter with the snakes was subsequently embellished and retold with added details. Some versions stated that the incident occurred at night, or happened while several Shawnee warriors were hiding in the dark, intending to kill Zinzendorf. In those versions, the warriors supposedly abandoned their plan after observing Zinzendorf emerging from the tent unharmed by the snakes. For an alternative version of this story, see John Hill Martin, *Historical Sketch of Bethlehem*, 2nd ed. (Philadelphia, 1873), 138–139.

122. HG 1853, verses 1–3. Marked with the heading, "Wajomik im Nov. 1742."

123. HG 1902, verse 1. Marked with the heading, "Wyomik im Nov. 1742."

124. Zinzendorf referred to the lands north of the Susquehanna as "Canada."

125. It was Zinzendorf's son-in-law who would hold the first communion service in the Wyoming Valley when he returned in 1748: "Watteville faithfully proclaimed the Gospel, and on the 7th of October was celebrated the Lord's Supper, the first time the holy sacrament was administered in the Wyoming Valley. The hymns of the little company swelled solemnly through the night, while the Indians stood listening in silent awe at the doors of their wigwams. And when they heard the voice of the stranger lifted up in earliest intercession, as had been Zinzendorf's voice in that same region six years before, they felt that the white man was praying that they might learn to know his God." Johnson, *Count Zinzendorf and the Moravian Occupation of the Wyoming Valley*, 43.

126. Johann Jakob Bossart, the collection's second curator, wrote a well-known natural history treatise: *Kurze Anweisung Naturalien Zu Sammeln*. Barby's *Naturaliencabinet* eventually evolved into Völkerkundemuseum Herrnhut. Bethlehem and Nazareth had collections too. Bethlehem resident, Lewis David de Schweinitz, was the first American Moravian botanist, and produced beautiful drawings and descriptions of plants and trees in eastern Pennsylvania. See Moravian Archives, Bethlehem, *This Month in Moravian History*, October 2008: accessed November 11, 2016, www.moravianchurcharchives.org/publications/month-moravian-history/.

127. I would like to thank Katherine Faull for sharing this information with me.

128. Charles Dickens, *American Notes and Pictures from Italy (1842)*, Oxford Illustrated Dickens (Oxford: Oxford University Press, 1987), 152–153. By the time of Dickens' American journey, there were an estimated 128,000 self-sufficient settler farms in Pennsylvania.

129. Hoffer, *Sensory Worlds of Early America*, 76. Despite the dramatic environmental shifts of the eighteenth and nineteenth centuries, Pennsylvania today retains much of its precolonial environment. Thanks to the work of environmentalists such as Joseph Trimble Rothrock, who advocated for forest preservation and created conservation agencies such as the Pennsylvania Forestry Association and the Association of Forest Landowners, more than 60 percent of Pennsylvania is currently covered by forest land: accessed May 7, 2017, http://news.psu.edu/story/293182/2013/10/29/sustainability/professor-pennsylvanias-forest-cover-remains-stable-59. There are thirteen different tracts of virgin forest still extant in Pennsylvania, such as the Alan Seeger Natural Area in Huntingdon County and Cook Forest State Park in Clarion, Forest, and Jefferson Counties. These places still offer the chance to experience Pennsylvania as it may have been before European settlement. In this way, modern-day Pennsylvania differs from many other mid-Atlantic regions where large-scale ecological transformations had already impacted the natural environment dramatically by the nineteenth century.

130. Wheeler, *To Live Upon Hope*, 68.

131. Wheeler, *To Live Upon Hope*, 95.

132. The Council of Three Rivers American Indian Center in Pittsburgh is another important organization devoted to recognition of Pennsylvania's Indigenous people. For more information on modern-day Native American communities and tribes in Pennsylvania, see Minderhout, *Native Americans in the Susquehanna River Valley, Past and Present*; and David J. Minderhout and Andrea T. Frantz, *Invisible Indians: Native Americans in Pennsylvania* (Amherst, NY: Cambria Press, 2008).

133. Christopher Wren, "Description of Indian Graves on Bead Hill, Plymouth, Pennsylvania," in *Proceedings and Collections of the Wyoming Historical and Geological Society, 1911–12*, vol. 12 (Wilkes-Barre, PA: E. B. Yordy), 204.

134. Wren, "Description of Indian Graves on Bead Hill," vol. 12, 201. Travelers later in the eighteenth century, such as Spangenberg in 1767 and Johannes Ettwein in 1768, noted that the Shawnee town was already deserted.

BETHLEHEM

The first time I saw Bethlehem, Pennsylvania, was a summer evening in June 2002. My father and I had recently discovered a published translation of a late eighteenth-century travel diary kept by our ancestor Johann Jacob Eyerly Jr. in the periodical stacks in the basement of Penn State's Pattee Library. We knew nothing about Johann Jacob other than that we shared the same last name. According to Pennsylvania historian Paul Wallace, translator and publisher of Eyerly's diary, the original document resided in the Moravian Archives in Bethlehem. And so we made the four-hour car trip from our farm in central Pennsylvania in search of the manuscript.

I clearly remember driving into Bethlehem from the west along US Route 22 with our car windows rolled down. As we descended into the little valley at the confluence of Monocacy Creek and the Lehigh River, a brightly lit Moravian star shone from the hillside on the other side of the river. At the far end of Main Street, I could just glimpse the four-story stone façade of the Single Brothers' House and the spire of the Central Moravian Church. Candles shone in every window—a sight I recognized from the Pennsylvania "Deutsch" towns of Lancaster County. In the distance, the warm and humid air carried the sounds of car horns, a train whistle, and wind. Beyond that, an almost imperceptible hum of air conditioners, streetlights, and gas pumps mixed with the engine-whine of a plane and amplified rhythms projected from car radios. As we checked into the Hotel Bethlehem, a historic art deco hotel built by the Bethlehem Steel Corporation in the 1920s, a further glimpse of Bethlehem's past emerged. On the wall hung a picture of a log cabin—Bethlehem's first house—the site of which was now covered by the hotel parking lot.

Late that night, my father and I walked down the hill from the hotel to a springhouse along the banks of Monocacy Creek. Signs pointed to its historic significance as a stopping place for Native Americans traveling along a trail that passed to the side of the hotel. It was a "never-freezing" spring and had been counted on for hundreds of years to provide water in any season. Along the creek, crickets and peepers called incessantly from the reeds at the water's edge. In the background, I could just hear the thump of double bass and piano from the swanky 1741 on the Terrace supper club, and the four-part harmony of the hymn "All Glory Be to God on High" from the choir rehearsal at the Central Moravian Church across the street.

In the morning, we met Vernon Nelson, archivist for the Moravian Church, Northern Province. He quickly retrieved the tattered manuscript of Eyerly's journal from the vault. Clad with white gloves, we turned its pages gently. I had never touched anything so old. We learned that Johann Jacob had served the nearby Moravian settlement of Nazareth as blacksmith and church organist and was one of the very first members of the Pennsylvania legislature. He also represented the interests of the Moravian Church to the American government in Philadelphia where he worked with the US Congress and President George Washington to establish a Bureau of Housing and Land Tax. The journey that spurred him to keep a diary as he walked across Pennsylvania in the summer of 1794 was undertaken to survey lands he had long worked to secure for the church in the Erie Triangle.

But, only five years later, Eyerly died at the age of forty-three, after being sanctioned by the church for becoming "too worldly." All that remained were his words on the page, and his detailed descriptions of the acoustic and natural ecology of the Native lands north of Fort Pitt. Still, it was enough to envision a past landscape of trees, plants, animals, wind, and water, and intricate pathways carved across the Allegheny Front. What I did not understand at the time was how persistently those pathways still connected Pennsylvanians both historic and contemporary across geography, time, and place.

2

FRIENDS & STRANGERS

SElige Gemeine!	Blessed community!
fahr ins Lammes blut,	Go into the Lamb's blood,
bade dich fein reine,	Bathe yourself beautifully pure,
bleib ein Lammes-gut,	Remain a Lamb's-possession,
schließ dich in das fächlein,	Close yourself in the little compartment,
das der speer gemacht,	That the spear made,
labe dich am bächlein,	Refresh yourself at the brook,
das dir heil gebracht.[1]	That brought you salvation.

IN THE SUMMER OF 1754, TWELVE YEARS AFTER Zinzendorf's journey to the communities of Shamokin and Wyoming, a young Englishman by the name of Thomas Pownall set out to explore Pennsylvania. His destination was Bethlehem, a "curious and remarkable" town on the western branch of the Delaware River. In Bethlehem, Pownall was told, he could expect to find a settlement of more than nine hundred people living in a multiethnic religious commune founded by the Moravian Church. Bethlehem, a Christian Utopia hacked and quarried from the forests and hills that surrounded it, had been quickly peopled by successive waves of *See-Gemeinen* (sea congregations) seeking religious freedom and missionary work in the frontier spaces of Penn's Woods, and Native Christians seeking new forms of spiritual and communal power and sustenance. As Pownall reached the Lehigh River Valley one summer evening and looked down on Bethlehem from the top of the southern mountains, he was amazed to see a small European-style town:

> Coming out from amidst a wilderness of woods through which I had been travelling some daies all at once at the top of a hill & viewing hence this cultivated populous settlement & its cluster of College like buildings large & spacious all of stone; with the grounds all around planted with orchards; & varied with tillage in all its forms of culture; & border'd on the banks of the river . . . My Eye was struck with unexpected pleasure. The Place itself makes a delightfull

Fig. 2.1 Thomas Pownall, "A View of Bethlehem, the great Moravian settlement in the province of Pennsylvania, sketch'd on the spot by his excellency, Governor Pownal; painted & engraved by Paul Sandby," London: Published according to Act of Parliament by Thos. Jefferys, the corner of St. Martin's Lane, 1761. Library of Congress. Public domain.

landskip but found, & thus seen in the center of a wilderness.... I made here a Sketch of it after a drawing from which an engraving has been made & published.[2]

Although idealized, Pownall's sketch of Bethlehem represented the viewpoint of many mid-eighteenth-century travelers who passed through the town (fig. 2.1). These strangers understood little of the vibrant religious culture of the Moravian Church and its large network of mission communities. Therefore, it was almost always surprising to find such an orderly and pious community strikingly built in the midst of the "wilderness" at the borders of European colonial settlements and neighboring Native American territories. But Bethlehem in 1754 existed on borders that were not just spatial, but also sonic.[3] Moravians shared a belief in the power of sound to shape religious community. All community members were required to sing hymns throughout the day and to improvise new hymns as a demonstration of the Holy Spirit's action in their lives. The acoustic environments of Bethlehem framed the lives and relationships of Moravian Christians.

What you could hear and how you could sing were important aural markers of communal belonging. Sound was an expression of religious, social, and spatial identity.

Just as early Moravian missionaries had sought to imprint distinctly Moravian-inflected places onto the Pennsylvania landscape, newer Moravians also sought to create acoustic order within their own communities. For the first twenty years after its founding in 1741, Bethlehem was a strictly governed communal church economy. Moravian Christians did not own property and lived in segregated group housing determined by age and gender. Only those who had undergone an intense two-year process of initiation could fully participate in the religious life of the community. Moravians understood their place within the *Gemeine* (community) or their separation from it through the articulation of social rules that governed relationships, gender roles, and hierarchies of race, class, and religion. Moravians assigned meaning to acoustic space. Carefully constructed "strangers' areas" around Bethlehem existed outside of the sound boundaries of the social and religious practices that defined Moravian identity. These spaces created a barrier that separated Moravians of all ethnic backgrounds from other neighboring communities. Whether those communities were Native or European, they nevertheless existed outside of the Gemeine. But, as the conflicts of the Seven Years' War and the American Revolution overtook Bethlehem, understanding who was a friend and who was a stranger became complex and often unresolvable quandaries.

Sounding Bethlehem

Du weißt wir geben alles hin, nur eins nicht, die Gemeine.[4]
[You know we will give up everything, except one, the community.]

HG 1204, verse 2

Bethlehem was the first settlement to be planned and built entirely according to Moravian architectural principles on this side of the Atlantic.[5] Larger Moravian towns all followed a predictable schema that was drafted on-site and then sent for approval from the church leadership in Europe. Heerendijk (1736), Pilgerruh (1737), Herrnhaag (1738), and Bethlehem (1741) were *Anstalt* (institution, establishment) communities that consisted initially of a main building, called the *Gemeinhaus* or *Anstalthaus*, around which additional buildings were grouped.[6] Bethlehem was to be a central hub for missionary activities, as well as the financial and the industrial center for

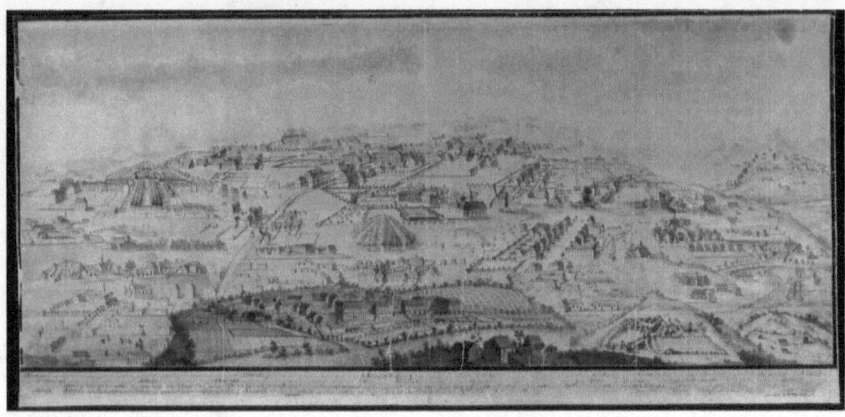

Fig. 2.2 International Settlement Scene, May 12, 1758. Unknown artist, pen and ink. Moravian Historical Society, Nazareth, Pennsylvania.

North America. The Bethlehem community was responsible for supporting the spiritual and economic activities of the outlying missions in Pennsylvania, New York, and Connecticut. As an industrial center, Bethlehem's residents worked at crafts and industries that turned raw goods into finished products for themselves and the nearby agricultural Moravian settlements of Nazareth, Nain, Lititz, Emmaus, Lebanon, and Hope. Each community was an important part of this larger network. In addition, Moravians in Bethlehem recognized that they belonged not only to a network of communities in eastern North America, but also to a spiritually linked worldwide Gemeine, as portrayed in this painting from the Moravian Historical Society in Nazareth that depicts Moravian missions across the Atlantic World contained within a single geographic space (fig. 2.2).

From 1741 to 1762, Bethlehem also operated as a cooperative, communal society referred to as the Economy. Under this system, individuals did not own their own land or businesses. The profit of trades and individual work was committed to a common church account.[7] This social and economic system dictated the way people lived, studied, worked, worshipped, and sang, but it also served other purposes. The strict scheduling of worship and work sustained early economic growth in the new and sometimes-difficult environment of the Pennsylvania colony. The Economy also helped to define the social relationships of those who lived in the Gemeine. The community was divided into two groups: the missionaries' or pilgrims' community, called the *Pilgergemeine*, and the home community, called the *Hausgemeine*.[8]

This division of labor allowed the members of the pilgrim's community to fulfill their missionary duties without the worry of childcare or financial support. Those tasks were fulfilled by the Hausgemeine.[9] All members of the Economy were supplied with food, housing, an education, community support, and a place to worship. Individuals, therefore, possessed little personal property. The majority of individual needs were provided by the church.

The Gemeine was also subdivided into communal housing groups called "choirs" that were defined by age, gender, and marital status.[10] Moravians believed that differences in gender and life-stage changed the way that each individual perceived and approached their religious faith.[11] Children were divided into choirs for nursing babies, infants, and then into gendered choirs for five to twelve year olds. At puberty, they were initiated into the sacrament of Holy Communion, and transferred to the Older Boys' and Older Girls' choirs. At age twenty, young adults moved into the choir houses of the Single Brothers and Single Sisters until they married.[12] Married adults lived within the Married Mens' and Married Womens' choirs. Widows and widowers shared communal quarters again with those of their gender. Members of the same choir ate, worked, worshiped, and slept in communal houses together. This living arrangement strengthened the unity of the choirs as a whole because members relied on their choirs for support, rather than on their biological families.

The hierarchy of people within a choir was also strictly defined, from the *Hausaufseher* (house overseer) or *Pfleger* (pastoral leader), to the *Stubenaufseher* (room overseer) or *Diener* (financial leader), who was responsible for the spiritual and physical needs of each choir. Sleeping places were assigned to choir members by the lot. Every member of the Gemeine had a distinctive place within the community, but this place was not supposed to be defined by social standing. Even from the founding of the church at Herrnhut, Moravians had desired to create a Christian society free of class restrictions. Clothing was to be utilitarian, designed for warmth and health, not exhibition of status. Eighteenth-century visitors to Moravian towns such as Bethlehem and Herrnhut noted "ideal Christian communities" and "working moral Utopias" marked by uniformity of dress and familiarity among members of all social classes.[13] Although Zinzendorf was a nobleman, he would have worn the same garments as any other male Moravian, in neutral colors of black, brown, gray, and blue.[14]

Women were allowed a bit more freedom in their attire: they wore collarless bodices laced with ribbons and a white neckcloth. The color of the ribbons that laced their bodices visually represented their status in the community: married or single.[15] The color of their skirts was not predetermined, but brown was usually worn for everyday use and for services of Holy Communion. On Sundays, women were permitted to wear brightly colored or patterned skirts. A picture owned by the Moravian Archives in Bethlehem portrays a young Moravian woman dressed in a patterned bodice, so individual taste may have determined wardrobe, to a certain extent (fig. 2.3).[16] The most distinguishing feature of a woman's clothing was a close-fitted white hood or bonnet that was worn over the hair and ears. The hood was tied with a ribbon that matched the color of the ribbon used to lace the bodice. Children dressed in the same manner as adults, but their clothing could be brightly colored.[17]

Moravian clothing visually reflected the spatial distribution of men, women, and children into separate communal houses. This included the separation of the sexes for the majority of worship services. There was originally no church building in Bethlehem, because each of the choir houses included its own *Saal* (worship hall), which meant that men and women worshipped separately except for community gatherings in the Gemeinhaus. The community was also divided spatially into separate walking paths for men and women, including the graveyard (*Gottesacker*, God's Acre), which was also divided by gender.[18]

Life in Bethlehem was therefore characterized by a carefully maintained social and spatial order that ascribed religious meaning to the organizing of people and the construction and layout of streets, residences, community buildings, and industrial sites. Like other colonial settlements throughout North, Central, and South America, Bethlehem reflected the town planning practices that settlers brought with them from Europe. Often, these detailed plans for new communities were designed to provide structure and impose order on what settlers viewed as contested frontier spaces.[19] These new communities imposed spatial planning with little regard for topography or culture, controlling both lands and people by ordering them into defined spaces.[20] As Geoffrey Baker has written of Spanish urban planning in Cuzco, Peru, "the physical concord of the European town plan was a spatial ideal created in the abstract, a bold attempt to construct a harmonious world from scratch."[21]

Sound was also a part of European settlement and colonization efforts throughout the Americas. Sound had marked conceptions of civic time and space in European towns and cities, and so it naturally became a basic

Fig. 2.3 Johann Valentin Haidt, "Portrait of a Young Girl," eighteenth century. Oil on canvas. PC 12, MAB.

principle of the design and the marking of social order in colonial areas.[22] The soundscapes of settler places (bells, musical instruments, town criers, industrial machinery) could, in theory, impose instant sonic order.[23] Bethlehem was no different from many settler communities in this respect. The gendered divisions that characterized Moravian life had important implications for the way that Bethlehem was constructed as a spatial community, as well as its constitution as a sounded Gemeine. A complex soundscape of human and nonhuman sounds underlay communal and social life in Bethlehem. Hearing and interpreting sound was not just important to understanding internal relationships within the Gemeine but also defined the external boundaries of the Gemeine. These internal and external boundaries followed the regulated practices of Moravian worship and the spatial plan of Bethlehem's built environment.

At the heart of the physical community lay the central common area.[24] This common area was bordered on the south by the Single Brothers' House and on the east by the Doctor's House, medical *Laboratorium*, and Apothecary, as well as the *Kinderanstalt* (Older Boys' School). The northern side of the common area was bordered by the farm and stable complex, called the *Hof* (courtyard). These agricultural buildings contained horse stables, a barnyard, cow and pig stables, the stablehands' house, and other buildings dedicated to animal husbandry. Near the Hof was the First House (the first building constructed in Bethlehem), and various industrial buildings, including the wheelwright, hatter, joiner, currier, nailsmith, weaver, and cooper. Adjacent to the Hof, running along the bluff above Monocacy Creek and forming the western side of the common area were other trade buildings of the industrial quarter: the blacksmith, locksmith, potter, cabinetmaker, and turner. Although the smithy was a stone building, the metal tools, bellows, anvil, and fire of the smithy were foregrounded and acoustically powerful sounds, which could be heard almost perpetually over a long distance. The whirring of the potters' wheel, and the lathes, iron molds, and finishing tools of the cabinetmaker and turner, although much quieter, nevertheless resonated beyond the walls of the wooden buildings that housed them. The blended soundscape of the common area incorporated various human and nonhuman sounds: swallows and house sparrows, sheep, chickens, cows, roosters; wind in the branches of the trees; the potters' wheel, the blacksmith's hammer, and the apothecary's mortar and pestle; the murmur of people talking, playing, working, singing. This was the core soundscape of Bethlehem. 🌐 **See website chap2.1, Interactive sound map: "Bethlehem in 1758."**

Along the Minisink Path, which wound down to the Monocacy Creek between the blacksmith and the pottery, was the Spring House, built over the original spring near the now-abandoned Delaware village of Menagachsuenk. The creek and spring provided an important water source for Bethlehem's other industries and trades: the water works, oil mill, fulling mill, grist mill, tannery sheds, slaughter house, dye house, and brewery. Along the Lehigh River was a separate industrial area near *Wunden Eiland* (Wound Island) with a flax house, a *Brecher* (building for breaking and drying flax), soap-boiling house, saw mill, and the Single Sisters' laundry. The Single Brothers' laundry was to the west along the Lehigh. The trades and industries practiced by both men and women were important contributions to the communal economy, and the building of these industrial spaces was completed at the same time as the first residential buildings. By 1747, there were thirty-two different trades and industries operating in Bethlehem. This would grow to sixty-two separate occupations by 1762. The layered sounds of mills, smithies, carpentries, and tanneries formed a perpetual wall of sound along the creek, echoing up into the town during daylight hours. The grinding machinery of the grist mill and the rotating wheel of the water mill created a continuous rhythmic ostinato that underlay almost all activities in Bethlehem.[25]

To the east of the common area was a street named Sisters' Lane. This was the location of the remaining communal choir houses and related structures: the Single Sisters' House and Kitchen, *Mädgenanstalt* (Older Girls' School, also called the Bell House), Gemeinhaus, Gemeinsaal (also called the Old Chapel), the Nursing Womens' or Married Womens' House, and the Married Mens' House. There was a small common area in front of the Bell House, which was bounded on the eastern side by the Single Sisters' House, and on the western side by the Gemeinsaal. To the south, this common area faced Sisters' Lane, the Single Sisters' Kitchen, and the terraced communal gardens stretching down to the Lehigh River, forming a bounded soundscape of communal life. Within this space, house and farm birds (house sparrows, finches, and swallows) and garden insects such as bees were blended with the low murmur of people at work or play. Even in the private rooms of the Bell House and Single Sisters' House, the singing of hymns carried through walls and floors, creating a continuous sense of sounded prayer and meditation. All choir houses also had indoor bells that arbitrated a sense of communal time. Choir house bells were rung throughout the day to signal changes in work routines and worship times and to summon residents to their meals.[26] Three large outdoor bells on the top of the Bell House and the Gemeinhaus were used to mark the quarter-hours and to announce community-wide events.[27]

The timber frame structures and hardwood floorboards of the communal houses and the individual worship halls of each house were especially suited to amplify the sound of the singing voice. Although the Moravians had simply built Bethlehem from available natural materials, they were aware of the particular acoustic properties of their communal houses. Throughout the day, there were times reserved for worship and singing in each choir Saal, as well as daily worship services in the Gemeinsaal. During worship, hymns could be sung into the floorboards, while lying face down around the edges of the room. Done in a large group, this produced a particular effect of transmitting the vibrations of many voices through the boards of the floor. The wooden boards also amplified the sound of the singing, and could transmit the vibrations throughout the entire structure and outside of its walls. This practice of singing was known as "prostration," and was an important part of community worship. There are numerous references to prostration in the *Bethlehem Diary*.[28] Johann Friedrich von Heinitz also observed prostration in Herrnhut and wrote a detailed account of the practice:

> Now one began this holy ceremony with several liturgical hymns, which were sung very slowly, quietly, and melodically. . . . After this address all brethren and sisters lay down with their faces on the ground in order to worship the immaculate Lamb. They remained in this position as long as the "Te deum laudamus" continued, which they sang together in two choirs, that is the brothers began one verse and the sisters finished it. They sang this "Te Deum" in such a melodic and quiet manner that one thought one would hear the voices of the angels, arch-angels, and cherubim laying prostrate before the Lamb in heaven. The posture of the brethren and the sisters, who all lay prostrate with their face on the ground, and this melodic singing enchanted me in such a way and sank into the depth of my heart, so that I could not hold back my tears of joy for seeing, after all, on this earth a church of Christians faithful to Jesus, who were giving Him the honor that is due to Him.[29]

This type of spiritually directed singing was also practiced in the Gottesacker. Special services were held throughout the year in the graveyard, accompanied by the frequent singing of hymns. Moravians believed that the vibrations of music had the power to transcend the boundaries of the spiritual world and call to deceased members of the community.[30] Therefore, hymn singing was not confined to the worship halls, but was integrated into daily life, enlivening even the most mundane activities. According to Moravian historian Helmut Erbe: "The sisters met together in the common room in the afternoon.... These sessions were always quite lively and animated; someone cobbled, another tailored, a third one prepared powder for the apothecary, a fourth copied, several peeled carrots, a few knitted, others stretched, sewed, etc., and there was a cordial and free discussion about love during all of which the most beautiful verses on the blood [of Christ] were sung."[31] When any important event happened in the community, a song was composed. Even unborn children heard the voices of their mothers singing special hymns for infants.[32]

The boundaries of the Gemeine were therefore manifested in song, and the sounds of hymns and community bells formed an ever-present soundscape that characterized life in the Bethlehem community.[33] Moravian choir life included a daily and weekly schedule of worship and devotional services involving the entirety of an individual's waking and sleeping hours.[34] There was a set schedule for eating, working, sleeping, waking, praying, and singing. The day began at four o'clock in the morning with the ringing of a small bell in each of the choir houses. At five o'clock, a resident sang several hymn verses in the hallway of the sleeping quarters as a wake-up call.[35] Following the morning bells and hymns, residents heated milk and water in the choir house kitchens. The day officially began at six o'clock with the service of Morning Benediction. At this service, the Losung for the day was read, accompanied by the singing of hymn verses. The service typically lasted fewer than fifteen minutes. This first presentation of the Losung at the service of Morning Benediction was the foundation of the day's schedule, since the text was an overarching theme for the entire day. The Losung was also paired with a hymn stanza appropriate to its theme, as well as a devotional collect. Congregants were instructed to meditate on all three and to sing the hymn verse throughout the day.

After the Morning Benediction, residents breakfasted in the *Speisesaal* (dining hall). After breakfast, each resident went to work, and remained at their trade until noon. Children attended a *Kinder-Stunde* (children's meeting) each weekday from eight thirty until nine o'clock. Older boys and girls attended this service on occasion, as well as adults whose occupations allowed them time to attend. At noon, residents gathered in the dining halls of the choir houses for the main meal of the day. Before sitting down to eat, they joined in the singing of a hymn verse. When lunch was ended half an hour later, they also sang a hymn verse, and then returned to work. The choir house bells rang again at six o'clock to signal the evening meal—typically tea, buttered bread, and soup.

Around seven o'clock in the evening, a short community-wide devotional meeting was sometimes held in the Gemeinsaal, called the *Gemein-Stunde* (community meeting). This service opened and closed with singing of hymn stanzas. On other days of the week, between seven thirty and nine o'clock, a *Chorviertelstunde* (choir quarter hour) was observed in the worship halls of the individual choir houses. At the choir services, several hymn stanzas were sung, followed by a litany specific to the choir, or a short sermon on a devotional theme relating to the mission of the choir. At nine o'clock, choirs also held an *Abend-Stunde* (evening meeting) or Singstunde. Following this, residents went to bed, but communal and ritual time was maintained during the night through night watches (*Nachtwache*).[36] Typically, two members of each choir were assigned to pray at every hour of the day and night for "whatever anyone heard of joy or sorrow, from near or far, every concern of this or that people, this or that community, this or that individual person."[37] Those assigned to the night watches were known as *Beter* (intercessors) or *Nachtwächter* (night-watchers).[38] The hours assigned to each intercessor were chosen by lot. The intercessors also sang hymn verses in addition to praying silently, and there were special hymns devoted to intercessors and night-watchers (fig. 2.4).

The intercessors were not the only "watchers" who sang hymn verses throughout the night. Four or more watchmen were typically employed to guard the community during the night, and all men older than sixteen and younger than sixty were required to take their turn at this occupation.[39] Since no member of the community was permitted to be outside after ten in the evening, if emergencies arose, such as a need for the doctor, community members could alert the watchmen to summon aid by placing a lamp in their window. In 1756, after the beginning of the Seven Years' War, a series of watchtowers and palisades was constructed along the borders of the town, including a watchtower near the Strangers' Store. Moravian brethren took turns at posts along the palisade and towers throughout the night.[40] Whether it was a time of peace or hostility, it was also the duty of a watchman to sing a hymn stanza at the turning of each hour:[41]

> The clock is eight! to [Bethl'em] all is told,
> How Noah and his seven were saved of old.
>
> Hear, Brethren, hear! the hour of nine is come!
> Keep pure each heart, and chasten every home!

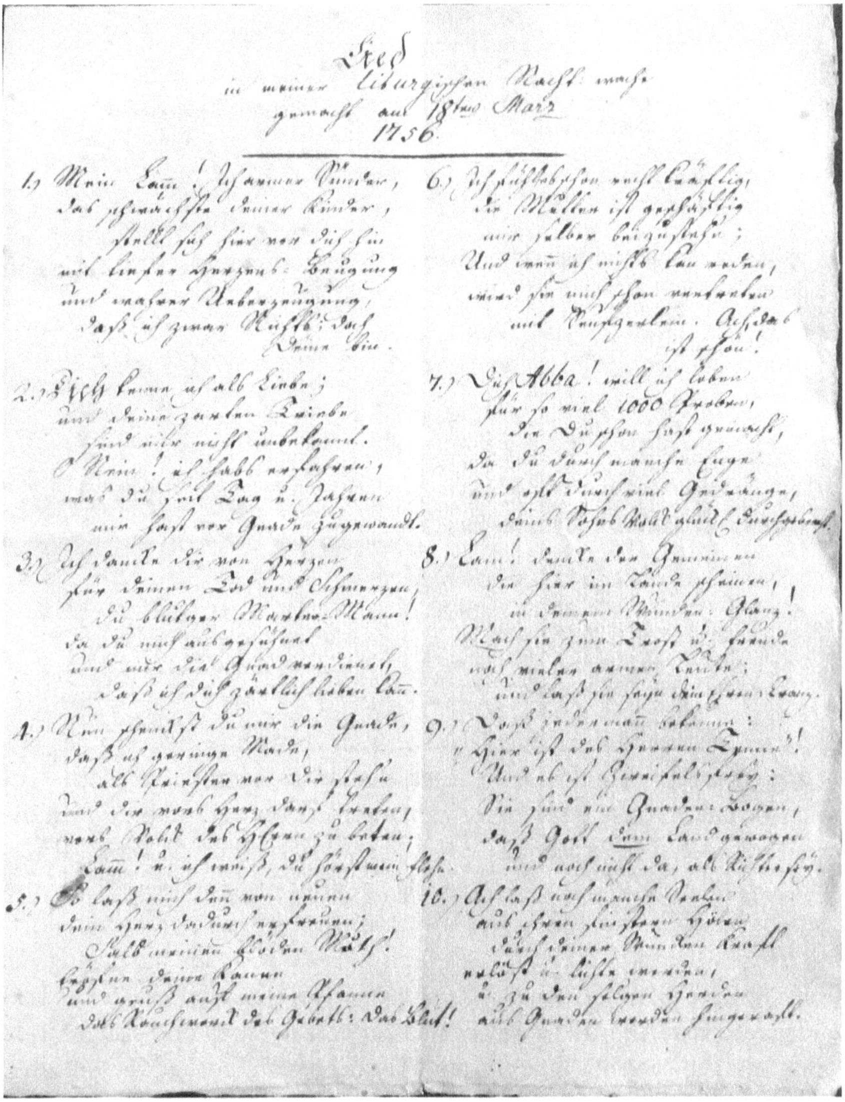
Lied
in meiner liturgischen Nacht-wache
gemacht am 18ten März
1756.

Fig. 2.4 Johann Daniel Sydrich, hymn for the night-watch titled "Lied in meiner liturgischen Nachtwache," 1756. Spangenberg Poetry Collection, folder 2, no. 8, MAB.

Hear, Brethren, hear! now ten the hour-hand shows;
They only rest who long for night's repose.

The clock's eleven, and ye have heard it all,
How in that hour the mighty God did call.

It's midnight now, and at that hour you know,
With lamp to meet the Bridegroom we must go.

The hour is one; through darkness steals the day;
Shines in your hearts the morning star's first ray?

The clock is two! who comes to meet the day,
And to the Lord of days his homage pay?

The clock is three! the Three in One above
Let body, soul and spirit truly love.

The clock is four! where'er on earth are three
The Lord has promised He the fourth will be.

The clock is five! while five away were sent,
Five other virgins to the marriage went!

The clock is six, and from the watch I'm free,
And every one may his own watchman be![42]

Even in the intervals between the hours, watchmen sang other bits of sacred song.[43] Moravians believed that these hymns were both consciously and unconsciously perceived by the sleeping community: "In the evenings, we . . . fell asleep beneath the song of a group on watch. It was not unusual to awake at night hearing again from somewhere the hymn verse of the night watchman and falling asleep again with this thought."[44] Sometimes hymns were also played on musical instruments during the evening or nighttime hours. On Sunday, July 12, 1747, the *Bethlehem Diary* records: "After the other events, several brethren and sisters made some sweet music with flutes around Bethlehem and rocked the Brethren and sisters to sleep with juicy stanzas about the little birds in the cross's atmosphere."[45]

The image of the Moravian Gemeine as a community manifested in sound and song was further reinforced through the textual content of hymns. The verses of "Was macht ein Creuz-Luft-Vögelein, / wenns sich schwingt zu dem Lämmelein" ("Was Does a Cross-Air-Bird, / When It Swings about the Little Lamb"), a popular hymn in the 1740s, detailed the cycle of various life activities in a Moravian community, such as getting up in the morning, taking a walk, working at a trade, or going to sleep:

Morgen-Versel Was macht ein Creuz-Luft-Vögelein, wenns raus fliegt aus dem bettelein?	Morning verses What does a cross-air-bird, When it flies around out of its little bed?
Tage-Versel Wie aber wenns Creuz-Luft-Vögelein gesperrt ist in ein bäuerlein?	Daytime verses How does the cross-air-bird, When it is working in the little farmyard?
Wenn spazieren geht Was macht ein Creuz-Luft-Vögelein, . . . dem schönsten parc rum gehn?	When going for a walk What does a cross-air-bird, . . . Going around the beautiful park?
Wenns melancholisch ist Was macht ein Creuz-Luft-Vögelein, wenn man es klopft aufs schnäbelein?	When feeling a little melancholy What does a cross-air-bird, When it has been knocked on its little head?
Abend-Versel Was macht ein Creuz-Luft-Vögelein, wenns sich legt in sein bettelein?	Evening verses What does a cross-air-bird, When it lays itself down in its little bed?
Wenns ans heimgehn komt Wie machts ein Creuz-Luft-Vögelein, wenns raus will aus dem hüttelein?[46]	When it is time to die How does a cross-air-bird, When is it time to go out of its little body?

Hymn texts such as "Was macht ein Creuz-Luft-Vögelein" reflected the larger ritual cycles of Moravian daily life, and also helped to regulate mental health and to assist Moravians through particularly difficult times of illness or death. Even children were to be reminded of their place within the communal order through singing, and learned to celebrate their membership in a larger community of past, present, and future Christians. Children were taught to accept difficulties and hardships, and to view death as an intended and even welcome end to a Christian life, as exemplified in Friedrich Cammerhof's report on the instruction of children in Bethlehem: "We also told them something about the cross-air-birds, and we sang songs about them. That pleased them very much, and they were overjoyed. It is their greatest joy to think and talk with each other about going to the Lamb and kissing his wounds. And indeed they have no other concept of death than this one. What is more blissful on earth than being such a child?"[47]

In addition to participating in a daily cycle of rituals and hymns, Moravians of all ages also celebrated special occasions and feasts with hymns and instrumental music. Singstunden and Liebesmahlen were held every day, but on special feast days these services were even more elaborate.[48] Each choir had a yearly festival day that typically began at seven thirty in the morning with the playing of hymns specific to the particular choir by the trombone choir.[49] The choir's hymns would be sung again at a Lovefeast at two o'clock in the afternoon, and again at night in an open-air service.

Choir hymns also served as sonic markers of the passage from life into death. Deaths within the Gemeine were marked with the playing of three chorales by the trombone choir from the Bell House tower.[50] The first chorale tune announced the death. The second tune represented the choir of the deceased, and the third tune repeated the death announcement.[51] The *Bethlehem Diary* and other community records contain hundreds of descriptions of burial services that depict the basic importance of singing hymns or playing them instrumentally during the ritual of committal. The burial of Peter Bartholet in 1744 is generally representative of the role of music in Moravian deathways: "Towards evening the wasted tabernacle left behind by Br. Peter Bartholet was taken with singing and instrumental music to our Hill of Rest. There—following a short and pleasant address by Br. Boehler based on the watchword for the day, Healthy in body and soul, and a heartfelt prayer—his body was committed to the Savior in preparation for the day of resurrection and to the earth for rest. Singing with instrumental accompaniment, the brethren and sisters returned home in a procession following the same order as that in which they had come."[52]

The entire week before Easter was also celebrated with special services to commemorate deceased members of the community, culminating in an Easter procession to the Gottesacker. The trombone choir typically initiated the procession at three o'clock in the morning by playing an Easter hymn:

> Christ is risen from the dead,
> Thou shalt rise too, saith my Savior,
> Of what should I be afraid?
> I with him shall live forever,
> Can the head forsake his limb,
> And not draw me unto him?[53]

The Moravian Easter service was one of the most important events of the calendar year. At this quiet early morning service (no trades or industries operated on Sundays), worshippers could likely hear the rushing waters of the Lehigh River, and the dawn chorus of spring birds, forming a natural counterpoint to the Easter hymns and the chorale tunes played by the trombone choir. Throughout the year in Herrnhut and Bethlehem, music was frequently performed out-of-doors and in consequence was a distinct aural feature of the soundscapes of Moravian towns. In Bethlehem, there were numerous natural spaces around the community that served as performance spaces. In Herrnhut, Singstunden were sung at a special *Lindenwald* (linden wood) outside of the town.[54] And on Saturday evenings in Bethlehem, instrumental serenades were played throughout the town. The finale of the serenade procession ended in the graveyard with the singing of improvised hymns.

Since little distinction was made between secular and religious activities in Moravian communities, the singing of hymns was also a part of daily work life in Bethlehem. There were special hymns for each occupation: stablehands, watchmen, reapers, spinners, and tailors, among others, each had their own hymns (see fig. 2.5 for a hymn dedicated to spinners). Women often cooked or sewed to tune of a hymn.[55] Men sang while harvesting a field, erecting a barn or mill, excavating a cellar, or felling timber.[56] As Friedrich Cammerhof reported to Zinzendorf in Europe: "Bethlehem blooms and grows greener every day. I cannot express how lovely and inspiring a sight it is to see a corps of upwards of thirty reapers move off to the fields under the sound of sweet music."[57] It was common for both male and female harvesters to hold a worship service in the fields, or to process to work with flutes, French horns, drums, and cymbals, along with their sickles. Many of these musical occasions are mentioned in the *Bethlehem Diary*: "The Single Brethren returned at 9 pm from the harvest in Gnadenthal, during extremely heavy rain and thunder, and entered Bethlehem with some lovely music with the Waldhorns."[58] As nineteenth-century Moravian music historian Rufus Grider poetically stated in his treatise on musical life in Bethlehem, *Historical Notes on Music in Bethlehem, Pennsylvania*: "They [the Moravians] contrived to form a frame of the Beautiful around their toil and labor; an idea which does not suggest itself to this age of progress."[59]

Like many other Christian communities, including the Shakers, Moravian *Arbeitslieder* (work songs) celebrated and eased the daily tasks of life—milking, washing clothes, and threshing grain—placing manual labor into the realm of the spiritual:

Du süßter Herzbezwinger,	You sweetest heart-conqueror,
Die Melker, Wäscher, Schwinger,	The milkers, launderers, and threshers,
Die sehen jetzt auf Dich	They look now to you
Und warten mit Verlangen,	And wait with longing
Um Segen zu empfangen	To receive blessing
Aus Deinem blutgen Seitenstich.	From your bloody side-wound.

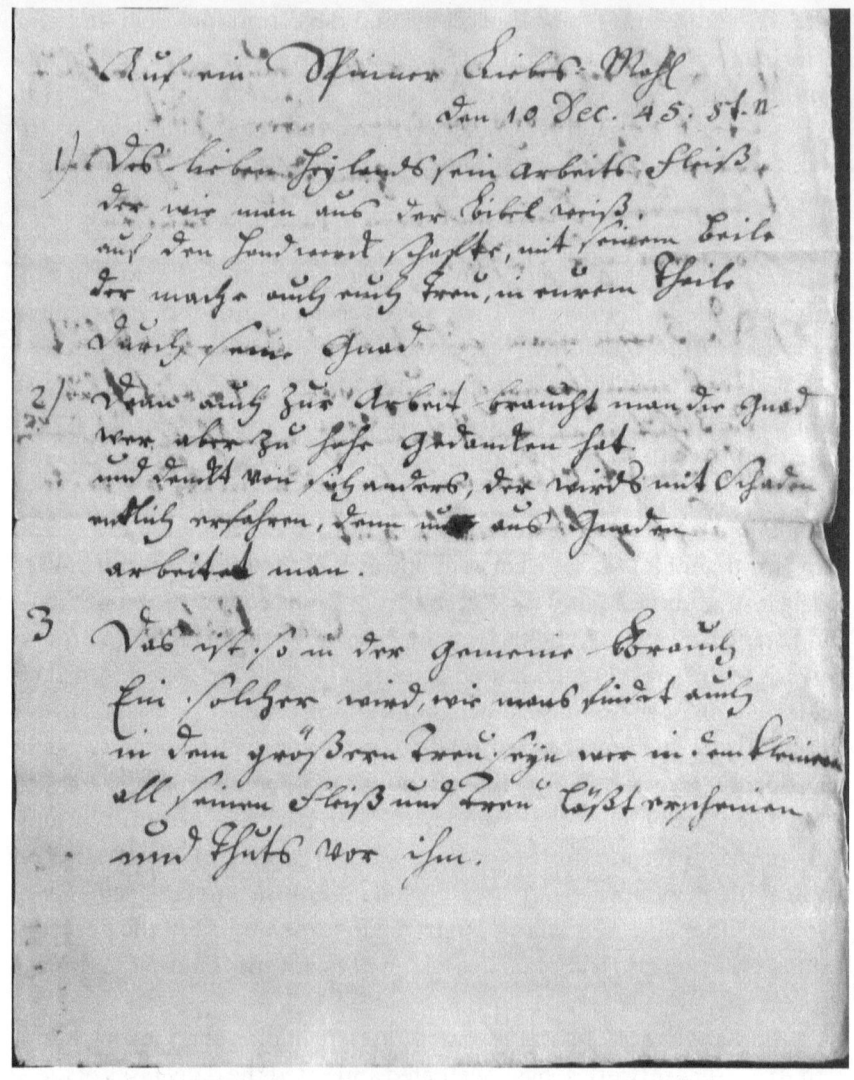

Fig. 2.5 Hymn for the spinners' Lovefeast, December 10, 1745. Spangenberg Poetry Collection, folder 12, MAB.

Du bist bei allen Dingen, In all things,
Beim Milken, Wäschen, Schwingen, In the milking, laundering, and threshing,
Das einzge Augenmerk. You are the only focus.
Dir leben wir auf Erden, We live for you upon the earth,
Bis wir Dich sehen werden, Until we will see you,
Dir tut man jedes Tagewerk.[60] Everyone will do their daily work for you.

Arbeitslieder also celebrated Christ's physical manifestation on Earth, linking Moravians with the "holy stories" of Christ's human family: Mary, Joseph, and his brothers and sisters. In this hymn by Zinzendorf, the Single Brothers were reminded that both Jesus and his father Joseph earned their living at the trade of carpentry. Hard work, and practicing a trade, were important components of Moravian life well-lived:

Mich deucht ich sehe Es	I think I see It
in vater Josephs hüttel	In father Joseph's little hut,
so handwerks-volk gemäß	Like all craftsmen,
bald in dem akker-kittel;	Now in the farming garb
bald gräbts nach einer wurz;	Now digging for a root,
bald schafft so was fürs haus;	Now doing something for the house,
bald nimmts den zimmer-schurz;	Now taking the apron,
bald mit der geissel naus.[61]	Now out with the whip.

In addition to work life, play in Moravian communities could also be framed by hymns. Young Moravian men and women, especially those who were unmarried, were encouraged to use time not reserved for worshipping or working at trades for playing musical instruments and singing hymns, or playing card games with hymn verses written on the cards, in addition to reading and playing chess, checkers, and other games like *Mühle* (nine-men's morris) and billiards.[62]

Individual Moravians also used hymns as an alternative to casting lots, especially in times of personal distress. When Anna Böhler's husband died, leaving her with two small children, her "indescribable pain" led her to open the hymnbook at random. There, she found the verse: "God has given you your lot with His own hand because your capacity is known to Him; and he will also not forget always to embrace and support you with faithfulness, love, and mercy." She was comforted in despite of "sleepless nights and dark days."[63] Anna's marriage would have also originally been consecrated and consummated in song. Husbands were to sing hymn verses while engaging in sex with their wives as a way to remind them of their spiritual duties to the community, to each other, and to God.[64] And so it was perhaps only logical that Anna should find comfort in the end of her marriage through a hymn. The writings of individual Moravians, such as Lebensläufe and diaries, generally express great satisfaction with participating in this particularly sonic version of communal life: "Beginning with the *Singstunden*, we sang the whole day through. We greeted each other with a hymn verse, we sang while working, we prayed and sang when the time for the prayer hour had come.... As we strove to live according to the word of the scripture, so, too, did we live in song."[65] Through song, the ordinary spaces and activities of Bethlehem became sacred.

Freunde und Fremde [Friends and Strangers]

Hymns served as important aural markers of belonging for those who lived within the Economy and choir system. But hymns were also important to defining relationships with those who lived outside of the Gemeine. The soundscapes of Bethlehem's communal and industrial buildings, gardens, orchards, and fields assumed great importance in defining social relationships. Sound defined the external boundaries of the community, and these boundaries followed the spatial plan of Bethlehem's built environment. There were separate spaces for travelers and visitors (*Fremde*, strangers) and friends and neighbors (*Freunde*, friends). Some Moravians lived outside of the communal choir houses but were still considered a part of the Gemeine. This included people living in settlements near Bethlehem built specifically for Native Christians, such as Nain, and Moravian farms such as Burnside Plantation, both of which were sited along Monocacy Creek. Settlers living in the vicinity of Bethlehem sometimes attended Moravian worship services, or visited its taverns or store, and were considered "neighbors." 🌐 **See website chap2.2, Interactive and static maps: "Sound Boundaries of Bethlehem."**

Studying the sound boundaries of the Bethlehem community helps to elucidate the relationships between Native and European Moravians. In the Pennsylvania State Archives, a census of the Bethlehem Economy dated November 29, 1756, records eighty-two Native American members of the Gemeine. Some of these Native members lived within the choir houses in Bethlehem, such as Martha, Theodora, Maria, and Christina, who were members of the Single Sisters' Choir.[66] Other Native Moravians lived in Nain, a community along the Monocacy Creek north of the industrial quarter and Burnside Plantation. Daily life in the community of Nain, which was also a part of the *Kmeende* (the Mohican translation of the word *Gemeine*), was very similar to daily life in Bethlehem. Although sonically distant from the core soundscapes of Bethlehem, Moravian historian Katherine Faull asserts that, "The diaries of the mission village [Nain] reveal a life that centered around the services and liturgical life of many other *Gemeinen*. Devotional paintings depicting

Christ's life were displayed in the chapel, festival days were marked by trombone choirs, scriptures were read, hymns were sung, lovefeasts celebrated, in Mohican, some in Delaware and some in German. Services were conducted bilingually, sometimes tri-lingually in Mohican, Delaware and German."[67] Although Nain was spatially removed from the core of communal buildings in Bethlehem, it was liturgically and spiritually a part of the Bethlehem Gemeine.

But some nearby spaces reserved for Native people were not considered to be within the boundaries of the Gemeine. This was the case with the lodgings built for traveling or visiting Native Americans who were not members of the Moravian Church: the Indian Saal, Indian Hotel (*Indianer Haus*) and Indian Kitchen, located near a fording place in the Monocacy Creek along the Minisink Path. Although very close to Bethlehem's industrial area on the other side of the creek, and certainly within the outer range of the communal soundscapes of Bethlehem's core, the Indian Hotel and its subsidiary buildings were "strangers' areas" and not considered a part of the community. Native travelers could hear the ringing of the Bell House bells and other sounds from the industrial quarter and the communal choir houses. But the sounds of food preparation from the kitchens and the chopping of firewood for the kitchen fires in the Indian Kitchen would have also blended with the edge of the natural boundaries of the fields and forests across the Monocacy. The murmur of the creek, forest birds, animals, and insects would have dominated the soundscape around the Indian Hotel.

European travelers were similarly accommodated at separate lodgings for "strangers," including the Crown Inn, or the Tavern Across the Water, located at a fording place along the Lehigh River.[68] The new King's Roads that connected Bethlehem to Philadelphia and Easton at the Forks of the Delaware were especially important in bringing these strangers into contact with Bethlehem. The Crown Inn and a second lodging house, the Sun Inn, constructed in 1758, all hosted travelers on the King's Roads.[69] Like the Indian Hotel, the fording places of the Lehigh lay on the edges of Bethlehem's communal soundscape. Although the location of the ford near the Crown Inn was close to the central part of Bethlehem near the Single Brothers' House, the natural sounds of wind, water, marsh and river birds, dominated the soundscape of the ford, connecting travelers or passing residents with the natural boundaries of the landscape that surrounded Bethlehem. These spaces were clearly crafted as "strangers' areas," and did not constitute a part of the Gemeine either spatially or sonically. 🌐 **See website chap2.1, Interactive sound map: "Bethlehem in 1758."**

A King's Road that ran to the east of Bethlehem also allowed travelers and nearby settlers to access a *Fremden Laden* (Strangers' Store) on the *Ladengasse* [Store Road]. Before the opening of the store, Moravians had already been trading in cash with people outside of the community for the products of Bethlehem's industries and trades as well as imported goods such as linen, powder, shot, locks, snuff, tobacco, blankets, and barrels. The Moravians were also purveyors of spices and whalebone, silk, coffee, tea, sugar, and chocolate, and various household items such as combs and dishes. But these business activities also created a problem for Bethlehem's residents. They necessitated the presence of strangers in the community. An official Strangers' Store was opened for business in 1753 in an extension of Timothy Horsfield's house on the edge of the town.[70] Timothy Horsfield and his family were Moravian, but they did not belong to the choir system and were thus ideally suited to run the store.[71] Town builders constructed a house for the family on other side of the Gottesacker, and this structure also housed the Strangers' Store. The Horsfield house was therefore half communal and half private, and the Horsfields were the first people to inhabit a private space on lands owned by the Moravian Church in Pennsylvania. Near to the store, there was also a row in the Gottesacker reserved for the burial of strangers. This was intended for use in emergencies—guests who died while staying at the inns, for instance—or for interments of people who lived on nearby farms.[72]

Bethlehem's elders carefully monitored relationships between Moravians and outsiders. A special "Commission of the Brethren" was constituted to oversee the presence of strangers in the town. The most basic task of the commission was to establish a spatial system that designated appropriate places for visitors, such as the Strangers' Store and the inns. Church elders also established ways to welcome visitors to Bethlehem without disturbing religious ceremonies or the activities of the women's choirs in particular. Curious visitors were greeted by a *Fremdendiener* (Stranger-servant), an older man appointed to give tours, handle questions, and supervise unknown individuals. Strangers were allowed into communal areas or the communal choir houses for special "outreach" worship services, but the Commission of the Brethren also oversaw behavior within the community as a way to monitor the relationships of Moravians with outsiders.[73] Only adult men and married women could hold positions in businesses that welcomed customers from outside of the Gemeine, such as the inns. Children and unmarried women were not permitted to work in these locations.

The Moravians' plan for handling strangers also included limited access to certain religious teachings that were only appropriate for members of the Gemeine. For instance, when a Swedish visitor to Bethlehem inquired about an unusual text on the front side of the Single Brothers' House—"Vater und Mutter und lieber Mann, habt Ehr vom Jünglings Plan [Father and Mother and dear husband, be honored by the young men's plan]"—he was told that the meaning of the inscription would be kept among "the Brethrens' secrets."[74]

Becoming a part of the *Gemeine* and gaining access to its "secrets" was an intensive process, and there was a well-planned application process for prospective Moravians. Candidates were first interviewed to verify that they had no debts, bad marriages, or imprisonments.[75] They also had to display significant religious commitment, and Moravians were well aware that their life-style was not appropriate for everyone. A particularly revealing passage from the *Bethlehem Diary* details the hesitance of church elders to admit new members, especially young children or those who were deemed not suitable to living within the strictures imposed by choir life and the Economy:

> Br. Boehler explained the rationale of the congregation with regard to people coming to live in the congregation and particularly children who are still under their parents' control: That we are not to encourage anyone by a single word to live among us; we do not do it that way. They also should not be irritated if in spite of repeated requests they receive an unfavorable reply. It is true that we speak a thousand encouraging words to people, urging them to come to the Savior, to let themselves be saved, etc. But when people come to the *Gemeine* and even perhaps get themselves admitted, then they must no longer do as they wish. And when we receive children and then the parents later get the idea, "I could now use my child to greater advantage at home," and change their minds, then indeed we remain concerned for the children and do not like to let them go at the wrong time and to the hurt of their souls; therefore it causes us extraordinary pain to see such children taken away.[76]

Many people applied for membership in the Bethlehem Gemeine during the mid-eighteenth century, either for themselves or their children, and were turned away. Some of those denied were allowed to live nearby and to worship in Bethlehem, such as the Horsfields, but were not given a place in the communal system.[77]

While it was very difficult to be admitted to the Gemeine, and Moravians actively supervised the activities and presence of strangers in their communities, Moravians were certainly allowed to interact with the outside world. According to historian Katherine Carté Engel, over time this resulted in the blending of Bethlehem into the larger context of the Pennsylvania colony as "travelers, acquaintances, and friends all passed through [Bethlehem]."[78] Maintaining friendships with those outside of the Gemeine was not forbidden. In fact, cultivating ties outside of the community was an important strategy for helping outsiders to understand the Moravian version of religious life. According to Engel, "members of the Oeconomy lived under a very different set of rules from those of their non-Moravian neighbors, but they did not turn their backs on them. Instead, Bethlehem's Moravians moved comfortably between their communal homes and their wider neighborhood, just as many of them slipped easily between German and English."[79]

Every interaction with a stranger was a chance to spread the message of Christianity. Strangers frequently visited Bethlehem and attended worship services, and there were regular English and Native-language services for this purpose. Strangers were also welcome to attend worship in missions outside of Bethlehem. While there were borders to the Gemeine, Bethlehem also existed within the larger circles of Pennsylvania society. Moravians' lives revolved around their daily spiritual routines and missionary work in a way that was distinct from other settlers and Native American communities, but they still existed as a part of, rather than in opposition to, their "neighbors." The boundaries of the Gemeine permitted outside friendships, business relationships, worship opportunities, and kinship ties.[80]

This meant that "strangers" and "neighbors" also experienced Bethlehem's communal, industrial, and religious soundscapes. During the eighteenth century, many travelers visited Moravian towns and settlements on both sides of the Atlantic. Not surprisingly, what these strangers most often observed while visiting communities like Bethlehem were the sounds—especially those sounds that pertained to worship and the singing of hymns. Travel journals and other accounts by outside observers often contain descriptions of the sound of Moravian singing and are remarkably consistent in mentioning its softness, sanctity, and communal nature.[81] There was a particular way to sing and to worship for those in the Gemeine. Learning how to sing was especially important for those who wished to become Moravian, and a necessary part of sonically demonstrating true conversion. Those who sang as Moravians were Moravians. The particular style of Moravian singing was audible to insiders and outsiders, sounding a musical boundary between those were "friends" and those who were "strangers."[82]

In writing about Moravian singing, outside observers and Moravians themselves tended to focus on how Moravian singing was different from the singing of "others," namely Lutherans, Pietists, and other Protestant denominations. Moravian singing was to be a foretaste of the beautiful and harmonious singing of the heavenly church, and Moravian sources tended to describe the singing of other churches as "shouting."[83] A blending of voices and timbres, and a quiet, meditative attitude were all necessary ingredients for appropriate singing. According to the Single Brothers' diary for 1759: "The beauty of communal singing consists of natural beauty, that is, the divine simplicity and gravity of the text that makes the ears feel. The art of worldly music, as well as the so-called church music, is contrary to the purpose of communal music."[84] Throughout the eighteenth century, church records continued to reiterate the same aesthetic considerations as necessary for beautiful singing: that hymns be communally, rather than individually rendered, with a performance practice favoring soft dynamics, simple rhythms and structures, and heartfelt emotion. The minutes of the General Synod of 1775 provide the following description: "It is also a part of the external beauty in the meetings which at the same time has an influence on the heart, that therein *many* sing and perform music as if there were only *one*, and thus besides binding the hearts to a single purpose also directs the necessary attention of each single individual upon the whole."[85]

Moravians envisioned hymns as effective ways to attract people to the church and to communicate ideas that were important to Moravians more effectively both inside and outside of the community. Strangers who attended worship in the Gemeinsaal were confronted with a prominent painted border around the ceiling that displayed the words of a hymn: "The Savior's blood and righteousness my beauty is, my glorious dress." This visual, painted border was thought to be especially important for communicating with visitors, and related the basic message of Moravian Christianity succinctly in a single hymn verse.[86] Public services were often performed in various Native languages, although most frequently in Mohican and Delaware, and hymns figured prominently in those services, as well.[87] A public service attended by Native Moravians and visiting Native Americans is mentioned in the *Bethlehem Diary* in February 1744: "After five o'clock a signal for rising was provided by music played in the Saal, and after the ringing of the bell the congregation assembled for the holy Meal of the Soldiers of the Lamb. It was conducted by Br. Boehler with an ineffable sense of the presence of our wounded and tormented Head and of His slain body and shed blood. The dear Indians were also present, to the heartfelt delight of us all, namely, Gideon, Josua, Lucas, Thomas, Esther, and Mary, the daughter of Gideon."[88] Singing hymns in Native languages was an important part of these public worship services, as well as worship services for Native Moravians within the Gemeine. The composition of hymns in Mohican was commemorated each year in August. According to the *Bethlehem Diary* for August 1744, "The watchword for this day, *The Mahikand tribe sings Lamb*, on which the dear Br. Anton conducted a significant Singstunde, was most appropriate, since just two years ago yesterday the hymn, *He is it indeed, etc.*, was composed and sung in Schecomeko [Shekomeko] in the presence of seventy Mahikand Indians as well as a number of white people."[89]

Strangers' services were also offered in English for the benefit of settlers living around Bethlehem or for visiting Native Americans who could speak English: "Today many strangers were here, visiting us. Pyrlaeus preached in German and Utley in English. Hagen spoke on the text in the afternoon. Buttner conducted the large Gemeinstunde. We sang the *Te Christum laudamus* with special blessing and abasement."[90] These services happened according to a set schedule, so people living in the vicinity of Bethlehem knew when they could enter the community to attend worship, and the services were generally well attended: "Br. Antes conducted the service for strangers at ten o'clock; Pyrlaeus conducted the English service and Br. Boehler the service for strangers in the afternoon. . . . There were many strangers present."[91] Even away from Bethlehem, those in missionary service made an effort to learn English to communicate as widely as possible with English-speaking settlers. Along the Bald Eagle creek in 1772, Johannes Ettwein gave a service in English for Scots-Irish residents in the area: "By request, I [Ettwein] preached in English to a goodly audience of assembled settlers from the Bald Eagle creek and the south shore of the West Branch [of the Susquehanna River]."[92]

Strangers were also invited to participate in many different musical activities sponsored by the Moravians, including the annual Easter service. In 1743, the *Bethlehem Diary* recorded the presence of "many strangers here [for the Easter service], German and English. Br. Boehler preached to them in both languages. In the afternoon Br. Buttner spoke on the text with unction and blessing. Br. Boehler, who conducted the Gemeinstunde, sang a hymn extemporaneously. This day closed by our prostrating ourselves and singing the *Te Jesum laudamus*."[93] Strangers also observed or even participated in musical processions: "Following the hourly intercession, in which the *Te Christum laudamus, etc.*, was sung, the Single Brethren's Choir, together with many strangers who were present, went about Bethlehem singing, with full musical accompaniment."[94] As these passages attest, strangers were invited to attend various worship services and musical events within the Bethlehem community. Apparently, these experiences were often moving and affective: "At six o'clock in the morning the brethren and sisters were awakened by music. Then they gathered in the *Saal* for Speaking. At ten o'clock the regular preaching service took place, which Br. Böhler conducted with great power and grace. There were many strangers present, who felt the words as power from God."[95]

These diary passages also highlight the issue of language and multilingualism in Moravian communities. Moravians sometimes employed language differences to exclude outsiders and maintain internal control, but they also made an effort to deemphasize these barriers and encourage communication.[96] Languages themselves formed important markers of sonic identity in a frontier environment where people of differing backgrounds inhabited a single geographic space. Most Moravian community gatherings entailed an accommodation involving language. At the wedding of Samuel and Mary, two Native Moravians, English served as a lingua franca for Native American visitors of many different tribal backgrounds:

> Election by grace was strongly impressed upon the hearts of the congregation, and toward the [elder] applied this to the Indian couple, Samuel and Mary, who were present and who were to be married. It brought us all grace. At the close of this address we sang in English: *Most worthy Spirit, etc.* Next came a short address in English concerning the importance of the married state. Then Samuel and Mary were blessed for their future married life in the name of our Lamb, His Father, and of the Holy Spirit. Following the stanza *Grant in the Bottom of their, etc.*, the benediction was pronounced in German, and the congregation dispersed. A special stirring of grace was evident in connection with this marriage. An hour later a wedding lovefeast was held.[97]

The wedding of Samuel and Mary was typical of many frontier gatherings. The sheer number of dialects and languages spoken in the Pennsylvania colony made most public events difficult.[98] In the eighteenth century, Native North Americans spoke 221 mutually unintelligible languages, fractured into myriad dialects. Europeans, too, spoke many different dialects and languages. A settler from France might speak Breton, Norman, or Basque, rather than Parisian French. A Spanish merchant might speak Castilian, Galician, or Catalonian, and a British immigrant, East Anglian or Manx.[99] The soundscape of

German dialects in Pennsylvania was similarly diverse: Silesian, Low Saxon, Swabian, and Frisian, among many others. Even members of the Gemeine may have sometimes chosen to express themselves in a combination of languages as evidenced in a remarkable bilingual letter from Rachel Post (a Wampano Moravian) to her spiritual mentor, Maria Spangenberg, originally written in German and Mohican, and then recopied into English. In the letter, Rachel writes, "I can't express how it was with me when I received that Blood, *muchree wekanisese pekachkanon* [very sweet little bloods] . . . and when Br. Joseph gave it to me, my heart glowing and filled with the Sap of Life and I thought, *Muchree onewe onewe onewe pachtomawas* [very much, I thank thee, I thank thee, I thank thee, O my dear Savior very heartily]."[100] Rachel's text is beautifully rendered in dual languages, lending a depth of meaning derived from the orthography, grammar, vocabulary, and sounds of each language (figs. 2.6a–c).

Given the prominence of language in establishing relationships or even subverting them, Native-language hymns were critical to the success or failure of the Moravian missions. Mohican and Delaware hymns affirmed the multicultural nature of the Gemeine through what Joanne van der Woude has termed "lived intercultural performance," or the performance of hymns by singers from different cultural or ethnic backgrounds.[101] As early as 1747, the Moravian Johann Christoph Pyrlaeus, began a Mohican and Mohawk language school in Bethlehem. Apparently, the language curriculum included hymn writing and also learning new words by listening and repeating, or "getting an Indian ear," so that the pronunciation would be correct.[102] The Native-language hymns written by Pyrlaeus and other missionaries, as well as Native spiritual leaders such as Joshua Sr. and his wife, Bathsheba, were used in a variety of contexts in Bethlehem and other missions, and they offered spiritual and social access to the Gemeine for those who were not fluent in German.[103] Native-language hymn texts also sometimes made connections with Native spiritual traditions, as represented in a Mohican hymn verse attributed to Joshua and Pyrlaeus and sung to the German chorale tune "Seelen Bräutigam." In this verse, the spiritual sustenance of Christ's blood is described in the form of a *beson*, a Native spiritual medicine that had the power to heal both physical and spiritual illnesses:

Paquaik assanaik	Strong wounds
awanetsch mnata	Whoever drinks
watsche sogachnak	The sap is the one
watachgäik ghackak	Who gets healthy.
natsch onammanisso.[104]	

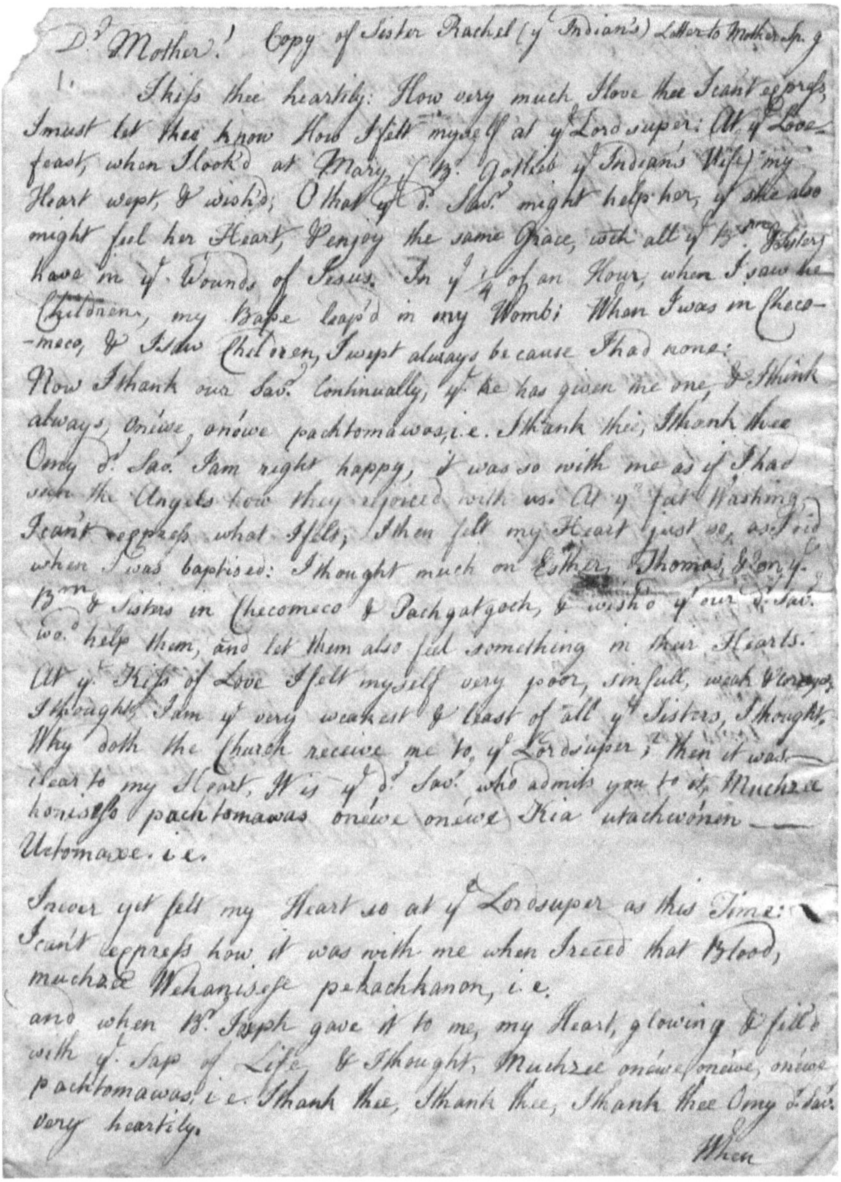

Dr. Mother! Copy of Sister Rachel (ye Indian's) Letter to Mother Sp. 9

I kiss thee heartily: How very much I love thee I can't express, I must let thee know How I felt myself at ye Lord super: At ye Love-feast, when I look'd at Mary (Br. Gottlieb ye Indian's Wife) my Heart wept, & wish'd; O that ye d. Sav. might help her, yt she also might feel her Heart, & enjoy the same Grace, with all ye Br. & Sisters have in ye Wounds of Jesus. In ye ¼ of an Hour, when I saw ye Children, my Babe leap'd in my Womb: When I was in Checo-meco, & I saw Children, I wept always because I had none: Now I thank our Sav. continually, yt he has given me one, & think always, Onive, onewe pachtomawas; i.e. I thank thee, I thank thee Only d. Sav. I am right happy, it was so with me as if I had seen the Angels how they rejoiced with us. At ye feet Washing I can't express what I felt; I then felt my Heart just so, as I did when I was baptised: I thought much on Esther, Thomas, & on ye Br. & Sisters in Checomeco & Pachgatgoch, & wish'd ye our d. Sav. wo. help them, and let them also feel something in their Hearts. At ye Kiss of Love I felt myself very poor, sinfull, weak & loving, I thought, I am ye very weakest & least of all ye Sisters, I thought, Why doth the Church receive me to, ye Lords super; then it was clear to my Heart, It is ye d. Sav. who admits you to it, Muchee honesse pachtomawas onewe, onewe Kia utachwonen Utomaxe. i.e.

I never yet felt my Heart so at ye Lords super as this Time: I can't express how it was with me when I received that Blood, muchee Wehanisse pakachkanon, i.e. and when Br. Joseph gave it to me, my Heart, glowing & fill'd with ye Sap of Life, & I thought, Muchee onewe, onewe, onewe pachtomawas; i.e. I thank thee, I thank thee, I thank thee O my d. Sav. very heartily.

When

Fig. 2.6a–c English and German copies of a letter from Rachel (Rahel) Post to Eva Maria Spangenberg, ca. 1746. MissInd 319.2, MAB.

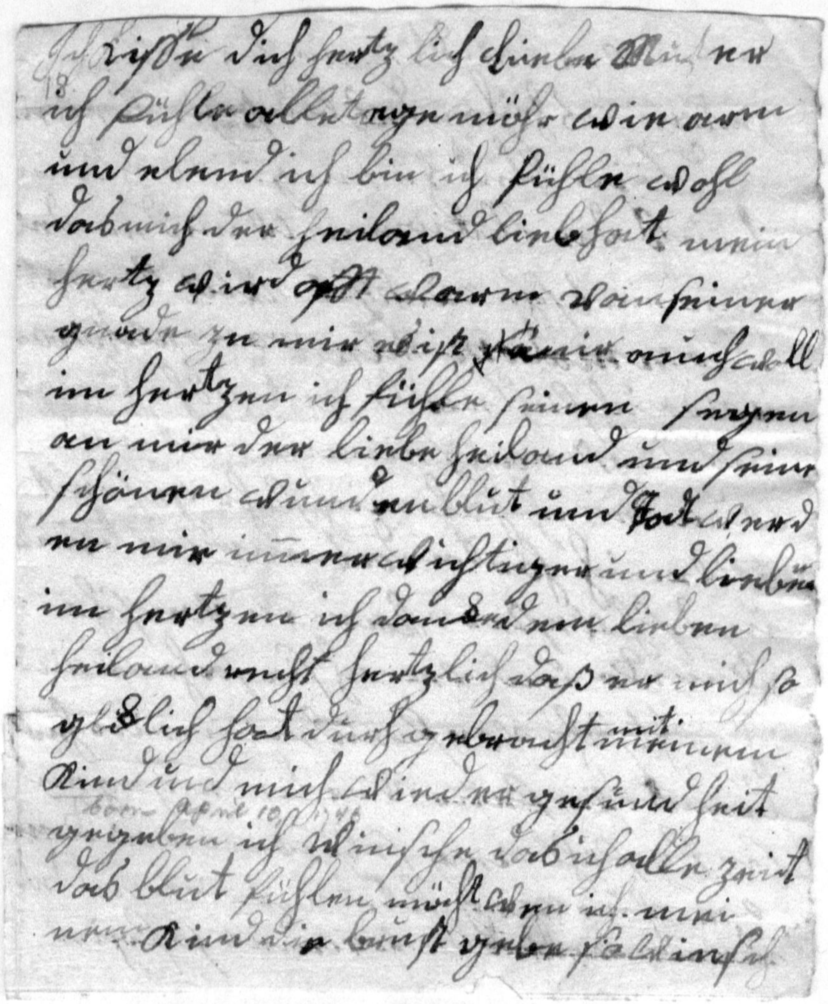

Fig. 2.6a–c *(continued)*

Polyglot singing was an especially important part of worship in Bethlehem and other communities when people of various nationalities and language backgrounds were present.[105] On August 21, 1745, the same hymn verse was sung simultaneously in English, German, Swedish, Danish, and Jewish-German (Yiddish) at a Moravian service. And on September 4, 1745, thirteen languages were sung together in polyglot harmony: Bohemian,

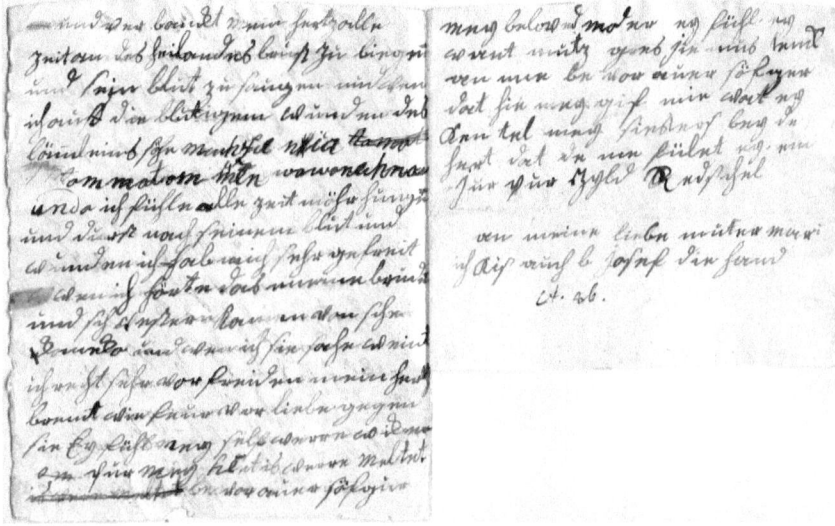

Fig. 2.6a–c (continued)

Dutch, English, French, German, Greek, Irish, Latin, Mohawk, Mohican, Swedish, Welsh, and Wendish. Apparently, this could have been a sixteen-language performance, but the singers, Matthew Reuz (Denmark), Matthew Hancke (Poland), and Christopher Baus (Hungary), refused to sing.[106] The *Bethlehem Diary* also records a particularly "heavenly" worship service in the summer of 1749 when a group of Moravian Inuit from Greenland visited Bethlehem. On June 9, 1749, the Native Moravian Gemeine from Gnadenhütten traveled to Bethlehem to attend a farewell service for the Greenlanders in the Saal of the Single Brothers' House. Also present at the service was an Arawak boy from Berbice, named John Renatus. Inuit, Delaware, Mohican, and Wampano representatives, as well as John Renatus, each sang several hymns in their own languages. Then, everyone sang hymn verses simultaneously in different languages, including English and German, accompanied by wind and stringed instruments. The diarist recorded that this service was "something totally heavenly, and a concert without compare."[107] Moravian archival collections also preserve some of the hymns that may have been sung on these occasions. The hymn "Lamm, lamm, o lamm!" is recorded in more than twenty-five different languages in various manuscripts from Bethlehem, including Hebrew, Arabic, Gaelic, Arawakan, and Mohican (figs. 2.7a–c).

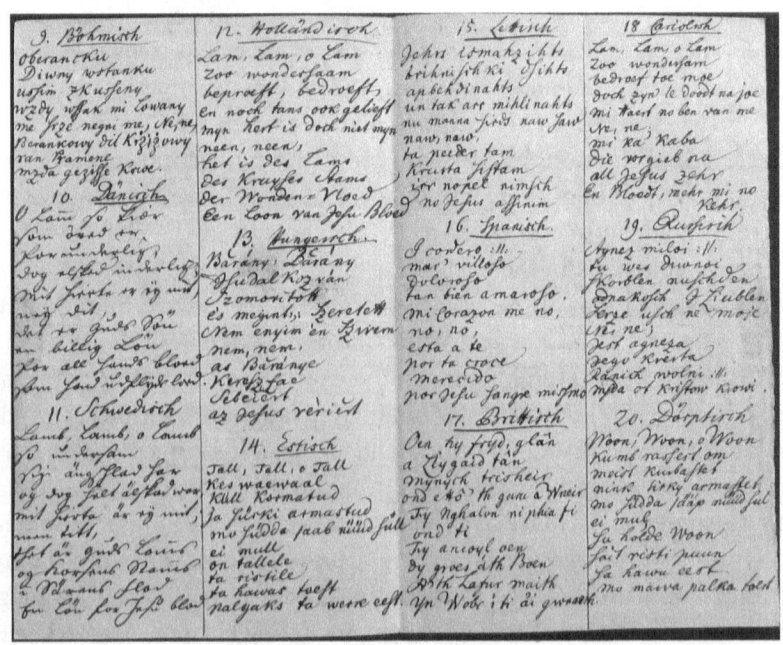

Fig. 2.7a–c Manuscript hymn verse "Lamm, Lamm, O Lamm," represented in twenty-five different languages, n.d. Poetry Collection, box 2, MAB.

Singing Moravian hymns regardless of language was clearly important. But it was also important to have the proper performance practice. Native members of the Gemeine were expected to participate in the Moravian ideal of soft, harmonious, and prayerful singing. When European Moravians wrote about the singing of Native Moravians, they emphasized the ways in which Native singing succeeded in sounding "Moravian": Johannes Hagen remarked on the "angelic voices" of the gathered Mohican singers in Shekomeko, and an anonymous diarist at the Wechquetank mission noted that Nathanael, a mission resident, sang "his German and Indian verses so beautifully and clearly that one could feel that the Savior with his glistening wounds was near to him."[108] On another occasion, a diarist in Gnadenhütten commented on the "beautiful sounds coming through the thick woods and over the high mountains" as he heard the Native men returning from hunting and singing hymn verses.[109] At a Lovefeast held to celebrate Joshua Sr.'s appointment as overseer of Gnadenhütten, the *Bethlehem Diary* records that the hymns and prayers were performed so beautifully by Johannes and Nathanael that the "birds sang and fish sprang. . . . [and] all the brethren and sisters received a genuine feeling."[110] Outside observers also praised the beauty of the singing of Native Moravians. When the English missionary David McClure visited the Moravian community at Lagundo-Utenünk (Friedensstadt) on the Beaver River in western Pennsylvania, he heard morning and evening services conducted in Delaware. He was especially interested in the singing, and wrote in his diary that he "could not praise highly enough the reverence shown by old and young, and the beautiful singing."[111]

Instruction in both vocal and instrumental music was part of the training received by all Moravians in the Single Sisters' and Single Brothers' Houses, the Girls' and Boys' Houses, and outlying missions such as Gnadenhütten. Musical lessons and instruction were free of charge, and everyone was expected to participate in playing and singing music. Musical "Practicings" and *Dichterschulung* (poet schools) were a part of daily life, and accessible to all Moravians, including Native members of the Gemeine. According to the minutes of the Fifth Moravian Synod: "In December 1750 the school matters in Gnadenhütten were organized according to the plan formed long ago, and a beginning was made so that even the smallest children of 3, 4 years old could hardly wait to come into the classes. They learned to read and to sing Indian and German verses. On Sabbath days various older brothers and sisters come together in order to have a singing exercise in their Indian verses. During these exercises they also always learn some German."[112] The Bethlehem diarist recorded in 1756 that the Native children at Nain were enthusiastic about singing hymns and learning to read: "The children came into our huts, sat down, and wanted their ABC book, and we held a nice little Singstunde with German and Indian stanzas."[113]

Some Native Moravian parents sent their children to attend Moravian boarding schools. Joshua Sr. enrolled his son, Joshua Jr., in the German boarding school on the farm of Henry Antes in Fredericktown, Pennsylvania. At the German school, the curriculum for Joshua and other boys would have included three areas of education: religion, intellectual inquiry and philosophy, and vocational training. Music was a significant part of the training in religion, and the hymnal was used for study and meditation along with the Bible. The children were also taught to memorize hymn verses in German and English. There were frequent opportunities to make music throughout the day, and impromptu sessions of chamber music in the evenings. Joshua also took lessons on the organ and spinet. These keyboard lessons, especially on the organ, were intense apprenticeships lasting for six months to one year in length. During this time, prospective keyboardists memorized around four hundred chorale tunes and their harmonizations. These had to be played in all twelve keys, for the purpose of accompanying improvised singing. They also received rigorous training in sight-reading so that they could accompany the *Collegium Musicum* (instrumental ensemble) in Bethlehem, Lititz, or Nazareth, or other instrumental ensembles in the outlying missions.[114]

Joshua's time at the German school allowed him to become fluent in European musical practices. He was already fluent in Native musical practices, and could speak German, English, Delaware, and Mohican. Heckewelder would describe Joshua as a man who had been "brought up in the fear of the Lord, and had from his childhood, been within the pale of the Society. He had a genious [sic] for learning, both languages, and the mechanical arts; was a good cooper and carpenter; could stock a gun nicely and no one excelled him in building a handsome canoe ... He was fond of reading, in the Bible, hymn book, and other religious books."[115] Joshua would become the most noted Native musician of the eighteenth century.

As a child, Joshua's instruction at the Antes farm and the Moravian school at Gnadenhütten enabled him to fully participate in the Moravian Gemeine. What this meant for Joshua and other Native Moravian children is difficult to assess from the mission records. Certain passages hint that

Native children came to enjoy singing Moravian hymns: "In February 1758, Brother Mack visited the Indians across the Lehigh River, as he often did. Little four-year old Martha, an Indian girl, stopped him and asked if he was going to have a meeting today. When he said no, she must have been disappointed because she liked to sing. She sat down on a bench and started to sing one of the hymns she had learned from the Moravians: 'Ach, mein herrlich Jesulein, mach mir ein sanft Bettlein [Ah, my dear little Jesus, make for me a soft bed].' After she finished her song in German, she sang it in the Indian language, and then she walked away."[116]

Hymns, processions, instrumental music, and worship services created opportunities for intercultural exchange of musical and religious ideas, and it is tempting to wonder what role Native Moravians assumed in these exchanges. Native Moravians certainly took on leadership roles and exercised some musical and religious agency in Moravian communities such as Gnadenhütten, Nain, and Bethlehem. Native musicians and church elders built musical instruments, composed and copied music, and added their own touches to musical manuscripts, instruments, and compositions. Although the Moravian Church intended to create a transatlantic network of mission communities connected through an idealized and uniform singing style, the result was probably very different from this ideal. Moravian musical traditions can certainly be regarded as colonial structures that attempted to standardize, indeed to colonize, Native soundscapes and traditions, but the kinds of adaptations and borrowings of cultural traditions that were common results of colonization might be better understood as a process of transitioning, rather than an erasure of previous traditions. Native musicians likely actively shaped Moravian hymnody to suit their own traditions and musical styles.

Although performances of Native-language and German-language hymns used German chorale tunes, there was some flexibility in the realization of the harmony. Even in German Moravian communities, this was an inherently social form of music-making, depending on multiple voices to realize the harmonies in ever-changing ways that responded to spiritual inspiration. So it is possible to speculate that Native Moravians may have sung hymns in ways that reflected performance practices drawn from their own musical traditions. This is hinted at in a passage from the *Bethlehem Diary*: "Sixteen Indians from Checomeko had a lovefeast with the Bethlehem ones and some from Nazareth.... The Indian brothers and sisters sang many verses in Mahican in quite lovely and for them hardly to be expected harmony."[117] What surprised the diarist was not that the singing sounded beautiful, but that it was done in German-style harmony. The Moravian singers likely changed their approach to the chorales in response to singing for the larger gathering of people in Bethlehem.

In outlying mission communities, Moravian hymns might have been performed monophonically or heterophonically, as opposed to the homophonic harmony of the German-Moravian style. But the difficulty of reconstructing specific performance practices, since most diary passages or records of Native singing focused on the aesthetics of singing rather than the musical performances, makes it especially important to consider how and to what extent the Moravian records can be used uncover the voices of Native Moravians and their contributions to the larger musical traditions of the Moravian Church.[118] What we can speculate is that the process of becoming a "friend," rather than a "stranger," was permeable, and may have allowed space for the creative agency and contributions of Native musicians and singers. This reveals something important about the Gemeine itself as a sonic construction. In reality, the boundaries of Moravian communities encompassed differing forms of musical expression, despite the larger ideals of the Moravian Church. It is imperative, then, that we listen carefully for the many ways in which Moravians understood and limited their communities based on musical participation, aesthetics, and religious commitment, because these boundaries may demonstrate not only the dominant narratives of the Moravian Church, but also the persistence and survivance of the cultural aesthetics, practices, and values of Native Moravians.

Notes

1. HG 2018, verse 1.
2. Pownall, *A Topographical Description of the Dominions of the United States of America*, 102.
3. See Rath, *How Early America Sounded*; and Hoffer, *Sensory Worlds of Early America*, for further detailed discussions of sounded spaces in early America.
4. Quoted in the Lebenslauf of Johann Jungmann. See Katherine Faull, "The Life of Johann Georg Jungmann (1720–1808)," in *The Distinctiveness of Moravian Culture: Essays and Documents in Moravian History in Honor of Vernon H. Nelson on his Seventieth Birthday*, ed. Vernon H. Nelson, Craig D. Atwood, and Peter Vogt (Nazareth, PA: Moravian Historical Society, 2003), 189.
5. See Craig D. Atwood, *Community of the Cross: Moravian Piety in Communal Bethlehem*, Max Kade German-American Research Institute Series (University Park, PA:

Pennsylvania State University Press, 2004) for a detailed history of Bethlehem and religious life in the settlement. Outlying missions were also planned using pre-approved structures. See Merritt, *At the Crossroads*, for information on the physical structures of outlying mission settlements. Also see Carola Wessel, "'We Do Not Want to Introduce Anything New': Transplanting the Communal Life from Herrnhut to the Upper Ohio Valley," in *In Search of Peace and Prosperity: New German Settlements in Eighteenth-Century Europe and America*, ed. Renate Wilson, Hermann Wellenreuther, and Hartmut Lehmann (University Park, PA: Pennsylvania State University Press, 2000).

6. Peucker, *A Time of Sifting*, 21.

7. Joseph Mortimer Levering, *A History of Bethlehem, Pennsylvania, 1741–1892: with Some Account of Its Founders and Their Early Activity in America* (Bethlehem, PA: 1903), 182–183. For detailed descriptions of the Economy system, see Katherine Carté Engel, *Religion and Profit: Moravians in Early America* (Philadelphia: University of Pennsylvania Press, 2009), 35–38, and chapter 2; and Katherine Carté Engel, "The Strangers Store: Moral Capitalism in Moravian Bethlehem 1753–1775," *Early American Studies: An Interdisciplinary Journal* 1, no. 1 (October 2007): 99–100.

8. Gray, *Wilderness Christians*, 26.

9. Engel, *Religion and Profit: Moravians in Early America*, 48.

10. The term "choir" denotes the separation of individuals into groups, and does not refer to a musical choir.

11. Zinzendof, *Berliner Reden* (1738). Quoted in Katherine Faull, "Faith and Imagination: Nikolaus Ludwig von Zinzendorf's Anti-Enlightenment Philosophy of Self," in *Anthropology and the German Enlightenment: Perspectives on Humanity*, ed. Katherine Faull (Lewisburg, PA: Bucknell University Press, 1995), 31.

12. Some sources indicate that seventeen was the age necessary for entry into the Single Sisters' and Single Brothers' Choirs.

13. James D. Nelson, *Herrnhut: Friedrich Schleiermacher's Spiritual Homeland* (PhD dissertation, University of Chicago, 1963), 90.

14. Nelson, *Herrnhut: Friedrich Schleiermacher's Spiritual Homeland*, 26, 28.

15. Women wore pink ribbons before marriage and blue after marriage. If widowed, they changed their ribbon color to white. Married women also wore purple ribbons after marriage and before the consummation of the marriage (Paul Peucker, "In the Blue Cabinet: Moravians, Marriage, and Sex," *Journal of Moravian History* 10 (2011): 27). For a detailed description of Moravian dress, see Elisabeth Sommer, "Fashion Passion: The Rhetoric of Dress within the Eighteenth-Century Moravian Brethren," in *Pious Pursuits: German Moravians in the Atlantic World*, ed. Michele Gillespie and Robert Beachey (New York: Berghahn Books, 2007).

16. "Portrait of a Young Girl," Johann Valentine Haidt, PC 12, MAB.

17. Nelson, *Herrnhut: Friedrich Schleiermacher's Spiritual Homeland*, 26–28. Also see August Spangenberg, "Something of Bodily Care for Children," accessed February 8, 2017, http://bdhp.moravian.edu/education/spbodilycare/germanbod/translation/gbodtrans01.htm. This treatise is also available in a Delaware translation.

18. Atwood, "Sleeping in Arms of Christ: Sanctifying Sexuality in the Eighteenth-Century Moravian Church," *Journal of the History of Sexuality* 8, no. 1 (1997): 25; and Peucker, "In the Blue Cabinet," 15. Married people may have sometimes gone into woods secretly to spend time together (30).

19. Baker, *Imposing Harmony: Music and Society in Colonial Cuzco*, 22–23. Baker describes this process as it applied to the building of new Spanish towns in the Americas.

20. Judith Ridner, "Building Urban Spaces for the Interior: Thomas Penn and the Colonization of Eighteenth-Century Pennsylvania," in *Early American Cartographies*, ed. Martin Brükner (Chapel Hill: University of North Carolina Press, 2011), 326.

21. Baker, *Imposing Harmony: Music and Society in Colonial Cuzco*, 23.

22. See Baker, *Imposing Harmony: Music and Society in Colonial Cuzco*, 26–27, 49–50, and 31–35.

23. Baker, *Imposing Harmony: Music and Society in Colonial Cuzco*, 49. For more information on the musicology of urban spaces, see Reinhard Strohm, *Music in Late Medieval Bruges* (Oxford: Clarendon Press, 1985); Fiona Kisby, ed., *Music and Musicians in Renaissance Cities and Towns* (Cambridge: Cambridge University Press, 2006); and Egberto Bermudez, "Urban Musical Life in the European Colonies: Examples from Spanish America, 1530–1650," in *Music and Musicians in Renaissance Cities and Towns*, ed. Fiona Kisby 167–180 (Cambridge: Cambridge University Press, 2006), 167–180.

24. See William J. Murtagh, *Moravian Architecture and Town Planning: Bethlehem, Pennsylvania, and Other Eighteenth-Century American Settlements* (Chapel Hill: University of North Carolina Press, 1967); Beverly Prior Smaby, *The Transformation of Moravian Bethlehem: From Communal Mission to Family Economy* (Philadelphia: University of Pennsylvania Press, 1988); and Gollin, *Moravians in Two Worlds*, for an overview of Moravian town structures and development. For a discussion of worship practices in early Bethlehem, see Atwood, *Community of the Cross*. The physical descriptions of Bethlehem in this section of the chapter, and in the online maps, represent the year 1758.

25. Engel, "The Strangers' Store," 91.

26. Martin, *Historical Sketch of Bethlehem*, 154.

27. Gray, *Wilderness Christians*, 28–29. Before these outdoor bells were placed in position, the hours were marked by a bell suspended from a large tree. Bells provided a type of temporal architecture and communal time-keeping at a time when public and private clocks were rare (Corbin, "Identity, Bells, and the Nineteenth-Century French Village," 191). Also see Corbin, *Village Bells: Sound and Meaning in the Nineteenth-Century French Countryside*.

28. "Then, during the singing of the stanza, *Lord Jesus Christ, Your death*, the congregation fell to the ground and commended itself and its new members to the wounds of the Lamb and closed with the stanza, *Spread out both wings, etc.*" See *Bethlehem Diary* I, 18–19; II, 54; and I, 49 for additional descriptions of Prostration.

29. "Nun fing man diese heilige Handlung mit mehreren Gesängen an, die sehr langsam, leise und melodisch gesungen wurden. . . . Nach dieser Rede legten sich alle Brüder und Schwestern mit dem Gesicht auf die Erde, um das unbefleckte Lamm anzubeten und blieben in dieser Stellung so lange, wie das 'Te deum laudamus' dauerte, das sie zusammen in 2 Chören sangen, d.h. die Brüder fingen eine Strophe an und die Schwestern beendigten sie. Sie sangen dieses 'Te deum' in einer so melodischen und leisen Weise, daß man die Stimmen der Engel, der Erzengel und Cherubinen zu hören glaubte, die sich vor dem Angesichte des Lammes im Himmel niedergelegt haben. Die Stellung der Brüder und Schwestern, die alle auf dem Boden lagen, das Angesicht auf der Erde, und dieser melodische Gesang haben mich derart entzückt und sich mir bis in die Tiefen meines Herzens eingeprägt, so daß ich meine Freudentränen nicht zurückhalten konnte, weil ich noch eine Kirche auf der Erde sah von jesustreuen Christen, die ihm die Ehre geben, die ihm gegeben werden muß." Hickel, "Das Abendmahl zu Zinzendorfs Zeiten," 29–30. Quoted in Peter Vogt, "Listening to 'Festive

Stillness': The Sound of Moravian Music according to Descriptions of Non-Moravian Visitors," *Moravian Music Journal* 44, no. 1 (Spring 1999): 16.

30. In the early 1740s, the Gottesacker was laid out in the space behind the Laboratorium and Gemeinsaal, and its complete plan was finalized on 14 August, 1748, and sent to Europe for approval. This graveyard was a striking representation of the early population of the Gemeine, and demonstrates the cultural and ethnic diversity of Bethlehem. According to Jon Sensbach, who has studied early race relations in Moravian communities: "Side by side lie the graves of Mahicans, Mandinkas, and Marburgers; of Delawares and Dresdeners; of Brothers and Sisters from Guinea, Gnadenhütten, and Gnadenthal." Jon Sensbach, "Race and the Early Moravian Church: A Comparative Perspective," *Transactions of the Moravian Historical Society* 31 (2000): 1. See *Memorials of the Moravian Church* for a list of Indian residents buried in Bethlehem's *Gottesacker*. Also see Augustus Schultze, *Guide to the Old Moravian Cemetery of Bethlehem, PA, 1742-1897* (Bethlehem, PA: The Comenius Press, 1898).

31. "Die Geschwister kamen nachmittags auf dem Gemeinsaal zusammen, um herzlich und ungebunden miteinander von dem zu diskursieren, so der Heiland von Zeit zu Zeit geben würde, z. B. von der Kinderzucht. Dabei ging's recht munter und lebendig zu; der eine schusterte, der andere schneiderte, der dritte machte Pulver für die Apotheke, der vierte kopierte, einige schälten Rüben, einige strickten, andere spannen, nähten usw., und dabei wurde von der Liebe recht herzlich und frei diskuriert und mitten drunter die schönsten Blutverse gesungen." Hellmuth Erbe, *Bethlehem, Pa: Eine kommunistische Herrnhuter Kolonie des 18. Jahrhunderts* (Stuttgart: Ausland und heimat Verlagsaktiengesellschaft, 1929), 92.

32. Wilhelm Bettermann, *Theologie und Sprache bei Zinzendorf* (Gotha: L. Klotz, 1935), 19.

33. In the German community of Herrnhaag, there was even a carillon in the spring house at the center of the community's square that played Moravian chorales throughout the day (Peucker, *A Time of Sifting*, 21).

34. For a detailed discussion of the daily liturgical cycle in Moravian communities, see Nicole Schatull, *Die Liturgie in der Herrnhuter Brüdergemeine Zinzendorfs* (Tübingen: Francke, 2005).

35. "At five o'clock in the morning the brethren and sisters were awakened by music and summoned by the bell to arise." *Bethlehem Diary* II, 47.

36. Moravian Church historians have traditionally contended that the intercessions were practiced without ceasing for the first one hundred years after the founding of the church in 1727. For some modern Moravians, this achievement is considered one of the most important accomplishments of the early Moravian Church, and is viewed as a reflection of the general desire of both historic and contemporary Moravians to maintain a continuous connection with God.

37. "Was man zu Freude oder Leid aus der Nähe und Ferne vernahm, all besondern Anliegen dieses oder jenes Volkes, diese oder jener Gemeine, dieser oder jener einzelnen Person." Kölbing, *Die Gedenktage der erneuerten Bruderkirche*, 152–54.

38. Nelson, *Herrnhut: Friedrich Schleiermacher's Spiritual Homeland*, 196.

39. J. E. Hutton, *A History of the Moravian Church*, 2nd ed. (London: Moravian Publication Office, 1909), 218.

40. Martin, *Historical Sketch of Bethlehem*, 17. Also see "Commission of the Brethren, 7th February, 1752," for the appointment of new nightwatchmen: accessed April 7, 2017, http://bdhp.moravian.edu/community_records/meeting_minutes/journal/1752commjournal.html.

41. "The Watchers are to sing a verse from a suitable hymn, at the change of the successive hours in the night, with a view to encourage and edify the Congregation." *Brotherly Union*

and Agreement at Herrnhut (1727). Published in *Der Deutsche Socrates*, 289–296. Translated in *Pietists: Selected Writings*, 330.

42. Hutton, *History of the Moravian Church*, 218.

43. Hutton, *History of the Moravian Church*, 218. "Thus passes the watchful night, to be broken as morning nears by the ringing of the bells, the singing of the brethren of a chorale, the entry of the brothers from the kitchen with steaming pots of water for tea, and finally by the Morning Blessing at which is sounded the Watchword for the new day in a Brethren congregation" (Nelson, *Herrnhut: Friedrich Schleiermacher's Spiritual Homeland*, 218).

44. "Wenn die Zeit des Stundengebets an einen gekommen war, . . . nicht ohne beim Aufwachen nachts wiederum von irgendwoher den Liedvers des herumgehenden Nachtwächters zu hören und mit diesem Gedanken wieder einzuschlafen." Hanns Joachim Wollstadt, *Geordnetes Dienen in der christlichen Gemeinde dargestellt an den Lebensformen der Herrnhuter Brüdergemeine in ihren Anfängen. Mit 4 Kunstdrucktafeln* (Göttingen: Vandenhoeck und Ruprecht, 1966), 84–85.

45. *Bethlehem Diary*, December 7, 1747, MAB.

46. HG 2251, verses 2–6, 8. See Peucker, *A Time of Sifting*, 64, for a discussion of the hymn.

47. Peucker, *A Time of Sifting*, 68–69; quote of letter from Cammerhoff to Zinzendorf, May 22–24, 1747, PP CJF 2, MAB.

48. See Rufus A. Grider, *Historical Notes on Music in Bethlehem, Pennsylvania from 1741 to 1871* (Bethlehem, PA: J. Hill Martin, 1873), 34, for a complete list of Moravian festivals.

49. For a representative sample of hymn verses corresponding to Moravian choirs, see the choir chapters in the manuscript, *Auszug für Singstunde*, UA NB.IV.R1 120b, Unity Archives, Herrnhut: "Das XXXII Capitel: Von den Männern," "Das XXXIII Capitel: Von den Weibern," "Das XXXIV Capitel: Von den Wittwen," "Das XXXV Capitel: Von den ledigen Brüdern," "Das XXXVII Capitel: Von den großen Knaben," "Das XXXVIII Capitel: Von den ledigen Schwestern," "Das XXXIX Capitel: Von den großen Mädgen," and "Das XL Capitel: Von den Kindern."

50. Grider, *Historical Notes on Music in Bethlehem, Pennsylvania*, 3.

51. Grider, *Historical Notes on Music in Bethlehem, Pennsylvania*, 31, lists the chorale tunes that are played for various choirs to announce a death.

52. *Bethlehem Diary* II, 129. Peter Bartholet was buried on September 3, 1744.

53. English translation from *A Collection of Hymns for Use of the Protestant Church of the United Brethren* (1826).

54. Near the *Lindenwald* is also a well which is known by local residents as *Zinzendorfs Ruh* (Zinzendorf's Rest). Zinzendorf apparently composed hymns and poems at the site of the well in order to draw inspiration from the natural setting.

55. Maurer Maurer, "Music in Wachovia, 1753–1800," *The William and Mary Quarterly* 8, no. 2 (1951): 217.

56. Levering, *A History of Bethlehem, Pennsylvania*, 184.

57. Gray, *Wilderness Christians*, 28.

58. *Bethlehem Diary*, July 17, 1747, MAB.

59. Grider, *Historical Notes on Music in Bethlehem, Pennsylvania*, 4.

60. Hymn composed by August Spangenberg. Quoted in Erbe, *Bethlehem, Pa. Eine kommunistische Herrnhuter Kolonie des 18 Jahrhunderts*, 93. In 1746, Spangenberg offered the following comment on the laborers at the agricultural farms in Nazareth: "Never since the creation of the world were there made and sung such lovely and holy shepherds, ploughing, reapers, thrashing, spinners, knitters, sewers, washers and other laboring hymns, as by these

people. An entire farmers' hymn book might be made by them." Martin, *Historical Sketch of Bethlehem*, 47.

61. *Synode Protokoll* (1745), 142. Later published in the *Herrnhuter Gesangbuch* as hymn 2085, verse 7.

62. Peucker, *A Time of Sifting*, 65. Playfulness was seen as an important way to approach spiritual life in a childlike manner, and games and singing helped to cultivate this.

63. Katherine Faull, *Moravian Women's Memoirs: Their Related Lives 1750–1820*, Women and Gender in North American Religions (Syracuse, NY: Syracuse University Press, 1997), 73.

64. Peucker, "In the Blue Cabinet," 26.

65. "Wir werden uns heutzutage kaum eine Vorstellung davon machen können, in welcher Weise diese auswendig in die Herzen gesungenen Lieder mit ihren leicht eingehenden Melodien im Gemeinleben gewirkt haben. Von den Singstunden aus sang man durch den Tag. Man begrüßte sich mit einem Liedvers, man sang zur Arbeit, man betete und sang.... Wie man nach dem Wort der Schrift zu leben sich mühte, so lebte man im Lied." Wollstadt, *Geordnetes Dienen in der christlichen Gemeinde*, 84–85.

66. Martin, *Historical Sketch of Bethlehem*, 14–16. The census did not record Native Americans who lived in the vicinity of Bethlehem, or who stayed within the community but were not part of the Gemeine. From 1742 to 1746, the Native congregation lived in Friedenshütten, near Wunden Eiland. In 1746, the majority of the congregation left to build the new community of Gnadenhütten. After the massacre at Gnadenhütten and the destruction of that mission, the congregation relocated to the vicinity of Bethlehem, and Nain was constructed and occupied from 1758 to 1763.

67. Faull, "Stories of the Susquehanna: The Nain Indian House," accessed October 13, 2016, http://storiesofthesusquehanna.blogs.bucknell.edu/2014/01/24/the-nain-indian-house/.

68. The Crown Inn was the location of the first bookstore in the American colonies, and managed by innkeeper Samuel Powell (Levering, *A History of Bethlehem, Pennsylvania*, 191). This bookstore is still open and is currently called "The Moravian Bookshop." It is located on Main Street in Bethlehem at the site of the Doctor's House (Laboratorium) and Kinderanstalt.

69. According to a survey conducted by Moravian cartographers Nicholas Garrison and George Golkowsky on October 2, 1760, a King's Road crossed the Lehigh River at a fording place near the Bethlehem community. Please see the online map point, Ford of the Lehigh, to view this survey. ⊕ See website chap2.1, Interactive sound map: "Bethlehem in 1758." The Moravians also operated a ferry service at the ford for many years. See Levering, *A History of Bethlehem, Pennsylvania*, 161, for a history of the ferry.

70. See Murtagh, *Moravian Architecture and Town Planning*, 69–73, for a discussion of the Strangers' Store; also see Engel, "The Strangers' Store," and Engel, *Religion and Profit*.

71. Eventually, Timothy Horsfield's children and his slave Josua (Ybo) did become members of the Gemeine.

72. Levering, *A History of Bethlehem, Pennsylvania*, 191.

73. Engel, *Religion and Profit*, 62.

74. Peucker, *A Time of Sifting*, xi–xii. The inscription comes from HG 2157, verse 10, which was written by Zinzendorf on 19 September, 1745, for the birthday of his son, Christian Renatus.

75. Engel, *Religion and Profit*, 66.

76. *Bethlehem Diary* II, 91 (June 21, 1744).

77. Engel, "The Strangers' Store," 100. Also see Engel, *Religion and Profit*, chapter 2, "Bethlehem's Neighborhood."

78. Engel, *Religion and Profit*, 67.

79. Engel, *Religion and Profit*, 67–68.

80. Engel, *Religion and Profit*, 39–40.

81. Vogt, "Listening to 'Festive Stillness': The Sound of Moravian Music according to Descriptions of Non-Moravian Visitors," 15, 21–22. Further descriptions of the sound of Moravian singing are provided in chapter 3.

82. Moravians were well aware of the particular qualities of Moravian singing, and often wrote about it: "What is the *spirit*, what is the *form*, and what is the *practice* of Moravian song? The absorbing feeling is the love of Jesus, and this spirit of Moravian song has gone with the United Brethren wherever they have pitched their tents. They have cultivated it at all times and seasons; over the couch of infancy, and around the bed of the sick and dying; in the morning and at eventide, at the table of the Holy Communion, as well as on their journeys and travels by land and sea. This spirit of Moravian teaching and song, is the grand spirit which holds us together; it is this which marks us as somewhat peculiar and distinct from other churches." Grider, *Historical Notes on Music in Bethlehem, Pennsylvania*, 20; quoting Rev. Lewis West, "The Music of the Sanctuary in the Moravian Church," published in February 1858, *Fraternal Messenger*.

83. "Religions, except for ours, are not liturgical, for example their singing is so chaotic that everyone shouts out what he wants.... With the pious people of the world, e.g. the Methodists, singing usually means just loud shouting. In contrast, we sing beautifully due to our intimacy, of course, not so clearly because when you are into it with all your heart, you don't think about the individual words. When the individual words are pronounced clearly, then it is really a sermon. [Die Religionen außer uns find nicht liturgisch, z.B. geht es bei ihrem Singen so unordentlich zu, daß ein jeder drein schreit, wie er will.... Bei frommen Leuten in der Welt, z.B. bei den Methodisten, bedeutet das Singen meist nur laut schreien. Wir singen dagegen schön wegen unserer Innigkeit, allerdings dabei nicht deutlich, denn wenn man mit ganzem Herzen dabei ist, bedenkt man die einzelnen Worte nicht. Wenn man die einzelnen Worte deutlich ausspricht, ist es schon mehr eine Predigt]." *Synode*, September 3, 1753, September 15, 1746, and June 13, 1740. Quoted in Uttendörfer, *Zinzendorfs Gedanken über den Gottesdienst* (Herrnhut, 1931), 54.

84. "Die Schönheit der Gemeinmusik besteht darin, daß sie die naturellen Schönheiten, das ist die göttliche Simplizität und Gravität des Textes, den Ohren gefühlig macht. Die Kunst sowohl der Weltmusik als der sogenannten Kirchenmusik ist dem Zweck der Gemeinmusik ganz entgegen." *Jünger-Haus Diarium*, April 21, 1759, UA Herrnhut.

85. "Verlass der vier Synoden der evangelischen Brüder-Unität, von den Jahren 1764, 1769, 1775, und 1782," Paragraph 917, MAB.

86. Levering, *A History of Bethlehem, Pennsylvania*, 145.

87. Martin, *Historical Sketch of Bethlehem*, 21.

88. *Bethlehem Diary* II, 35. This service took place on February 15, 1744.

89. *Bethlehem Diary* II, 125. This service took place on August 26, 1744.

90. *Bethlehem Diary* II, 49 (March 22, 1744).

91. *Bethlehem Diary* II, 94 (June 28, 1744).

92. Jordan, "The Diary of John Ettwein, 1772," 211.

93. *Bethlehem Diary* II, 54 (April 5, 1744).

94. *Bethlehem Diary* II, 122 (August 16, 1744).

95. *Bethlehem Diary* II, 149 (October 25, 1744).

96. Brian W. Thomas, "Inclusion and Exclusion in the Moravian Settlement in North Carolina, 1770–1790," *Historical Archaeology* 28, no. 3 (1994): 23.

97. *Bethlehem Diary* II, 36 (February 16, 1744). In Native communities such as Onondaga, John Bartram heard a multiplicity of different languages being spoken: English, French, German, Iroquoian languages, Shawnee, Delaware, and Mohican (Bartram, *Travels in Pensilvania*, 48). He also attended a council meeting where representatives of different tribal affiliations could not understand each other, so they requested that Conrad Weiser interpret the proceedings in English (61).

98. Due to these communication difficulties, at the wedding of Samuel and Mary some communications between "friends" and "strangers" may have actually been inaudible: signs, body signals, touching or sitting together, sharing a meal, singing, or simply extending hospitality to guests or finding points of congruence in a cultural event such as a wedding.

99. James Axtell, "Babel of Tongues: Communicating with the Indians in Eastern North America," in *The Language Encounter in the Americas, 1492–1800: A Collection of Essays*, ed. Edward G. Gray and Norman Fiering, European Expansion and Global Interaction (New York: Berghahn Books, 2000), 16–18.

100. Manuscript letter, Rachel Post to Maria Spangenberg, undated. MissInd 319.2.1. Letter dictated in English. I would like to express my thanks to Christopher Harvey for his help in translating the Mohican text. See also Rachel Wheeler, "Women and Christian Practice in a Mahican Village," *Religion and American Culture: A Journal of Interpretation* 13, no. 1 (2003): 44.

101. Woude, "Polyglot Harmony: Moravians Among the Indians"; also see Erben, *Harmony of the Spirits*.

102. John Gottlieb Ernestus Heckewelder, *History, Manners, and Customs of the Indian Nations: Who Once Inhabited Pennsylvania and the Neighbouring States*, Memoirs of the Historical Society of Pennsylvania 12 (Philadelphia: The Historical Society of Pennsylvania, rev. 1876), 320.

103. See Wheeler and Eyerly, "Songs of the Spirit," for a detailed discussion of Native-language hymns in the Moravian missions.

104. A Mohican hymn attributed to Pyrlaeus and Joshua Sr. from "Pattern for a Song Booklet of the Blessed Hearts of the Brown Nations of the Mahican Delawares and Some Short Verses of the Language of the Six Nations [Mohawk]," NB.VII.R.3.91 (1746), UA Herrnhut. English translation by Masthay, *Mahican-language hymns, biblical prose, and vocabularies from Moravian sources, with 11 Mohawk hymns*. To listen to a recording of this hymn and other Mohican hymns performed by the early music choir at the Florida State University, please visit the website: ⊕ **See website chap3.1, Sound recordings: "Mohican-Moravian Singstunde."**

105. In early Pennsylvania, Native households tended to be linguistically diverse, and the use of multiple languages by families or within communities was also common outside of Moravian missions. Jane T. Merritt, "Metaphor, Meaning, and Misunderstanding: Language and Power on the Pennsylvania Frontier," in *Contact Points: American Frontiers from the Mohawk Valley to the Mississippi, 1750–1830*, ed. Andrew R. L. Cayton, Fredrika J. Teute, and the Omohundro Institute of Early American History & Culture (Chapel Hill: University of North Carolina Press, 1998), 63.

106. "Den 4ten Sept. Zum Besuch beym Sabb. Lmahl [Liebesmahl] wurden deutsche, Engl., Lateinische, Griechische, Maquaische [Mohawk], Mahikanderische, Frantzösische, Eyrische, Böhmische, Wendische, Walisische 'Holländische' Verßen gesungen. Unser dänischer Bruder Matthew Beütz, der Polacke Hancke u. Christ. Bower der Ungar blieben noch die ihrigen schuldig." English translation: Levering, *A History of Bethlehem, Pennsylvania*, 205. German

text cited in Erben, *A Harmony of the Spirits*, 239. Wendish is the Slavic language spoken in Lusetia, an area in eastern Germany.

107. *Bethlehem Diary*, May 29, 1749, MAB.

108. "Abends kamen die Sänger zusammen in unser Hause, sich im Singen zu üben. Es ist wahr, sie haben Englische Stimme." *Shekomeko Diary*, MissInd 111.1, March 7, 1745, MAB; and "Gegen Abend kam ging br. Nathanael, auf den Platz herum und sang seine Deutschen u. Ind. versel so schön u. deute. das mans fühlen Konten, das Ihn der Hld mit seiner Funckelnden Wunden innig nahe war." *Wechquetank Diary*, MissInd 122.3, June 23, 1753, MAB.

109. "Nach den Singstd. kamen die lieben Brüder von dem MühlPlaz. (alwo sie ein lediges Brüder Haus gebauet haben und da beisammen wohnen, welches auf einer Seite mit Baum Rinde, auf der andern Seite mit grünen Reißig bedeckt ist) durch den Busch herauf mit ihren Wald Hörnern und Trompeten und andrer Music und machten sich recht lustig in Gnadenhütten, mit schönen Verseln: es hörte sich gar schöne an hinter dengrossen Bergen und in dem dicken Busch, so was lieblich Tönen zu hören." *Gnadenhütten Diary*, MissInd 116.1, May 24, 1747, MAB.

110. Quoted in Wheeler and Eyerly, "Songs of the Spirit," 15.

111. Kenneth G. Hamilton, *John Ettwein and the Moravian Church during the Revolutionary Period* (Bethlehem, PA: Times Publishing Company, 1940), 268.

112. Synodal Relation 4279, March 15, 1750. See the second session of the Fifth Moravian Synod, paragraph 3.

113. *Bethlehem Diary*, September 23, 1756, MAB; cited in Larson, "Mahican and Lenape Moravians and Moravian Music," *Unitas Fratrum* 21–22 (1988): 180.

114. Lawrence W. Hartzell, "Joshua Jr.: Moravian Indian Musician," *Transactions of the Moravian Historical Society* 26 (1990): 10.

115. John Gottlieb Ernestus Heckewelder, *A Narrative of the Mission of the United Brethren among the Delaware and Mohegan Indians, from Its Commencement, in the Year 1740, to the Close of the Year 1808, Comprising All the Remarkable Incidents Which Took Place at Their Missionary Stations during That Period, Interspersed with Anecdotes, Historical Facts, Speeches of Indians, and Other Interesting Matter* (Philadelphia, 1820), 414–415.

116. Cited in Larson, "Mahican and Lenape Moravians and Moravian Music," 180.

117. *Bethlehem Diary* [date unknown], quoted in Larson, "Mahican and Lenape Moravians and Moravian Music," 178.

118. See Wheeler and Eyerly, "Singing Box 331: Re-Sounding Eighteenth-Century Mohican Hymns from the Moravian Archives," for further information on Mohican Moravian hymns and the challenges involved in reconstructing performance practices in eighteenth century Moravian communities. Also see Eyerly, "Mozart and the Moravians," for a discussion of the indigenization of Mozart's music in Inuit Moravian communities in Labrador.

HERRNHUT

Herrnhut was a typical small German town. Where once there had been a dark wood of pine and beech, a camping spot for traveling gypsies and peddlers along a lone stretch of road from Löbau to Zittau, there was now a pastoral landscape of homes and farms. Although my hosts, the Lorenz family, assured me that fairies still lingered in the nearby hills of the Oberlausitz, the forests around Herrnhut appeared thoroughly tamed. As I looked out of my window in the attic of the Lorenz family home along the Zittauer Straße, I could see the Unity Archives of the Moravian Church on the top of the next hill. In between, the spire of the Moravian church rose over the town's square, the Zinzendorfplatz. The attic was so quiet that I could hear the faraway hum of a tractor cutting hay in the August sun, and children playing down the road. In the distance, I could almost glimpse Zinzendorf's former manor house at Berthelsdorf, and the gentle rise of the Hutberg (Watch Hill) where his family and many eighteenth-century Moravians lay buried. To the south and east were beautiful views of the Czech border and the Löbauer Gebirge, and desolate views of the Polish border—a vast strip mine nicknamed "Mordor."

Each morning in the summer of 2004, Ruth Lorenz packed a bag of sandwiches and garden vegetables that I stowed in my backpack before making the one-mile trek through the town to the Unity Archives. There I spent the day in glorious silence, reading and photographing the hundreds of music manuscripts and treatises, church documents, hymn texts and poems that would form the basis of my doctoral dissertation. It was here that I learned of the Moravians' love of sound and singing, and of their belief in the power of improvised hymnody and lot-casting to provide access to the divine. But, in the midst of my musical discoveries, I also hoped to find something of my own family's history, and I was not disappointed. In the card catalog that sat in a corner of the reading room, I found the call numbers for two Lebensläufe: Johann Jacob Eyerly Sr., and Johann Jacob Eyerly Jr. The labor of reading the eighteenth-century German script gradually yielded their secrets.

Father Eyerly's Lebenslauf was fraught with heartbreak. His determination to go "to Pennsylvania to join the Brethren" meant his inability to care for a critically ill sister in Württemberg. What would happen to her if he should die in the passage of the Atlantic? Yet he wrote poignantly of the moment when his heart was truly awakened and "He [the Savior] entered my heart so powerfully that my heart and eyes overflowed with tears." Still, his faith must have been severely tested at many moments of his life. After his arrival in Bethlehem on September 14, 1753, he was appointed to serve at the mission of Gnadenhütten in the Kittatinny Mountains north of Bethlehem. Although his Lebenslauf is silent about his time in Gnadenhütten, he must subsequently have felt deeply about his experience. He was called to return to Bethlehem in August 1755. On November 24, 1755, the mission was attacked by Haudenosaunee, Delaware, and Shawnee warriors. Eleven German missionaries were killed that night—some burned, some shot or scalped and hacked to death. Susanna Nitschmann was captured and forced to march north over the following six weeks in the freezing snow to Tioga on the New York border. There, she was murdered by her captor. After her death, another captive woman was forced to sew her own clothes from Susanna's garments. So ended the first Gnadenhütten massacre. Eyerly did not talk about Gnadenhütten. His handwriting ended there. The rest of Eyerly's life was recorded by the congregation after his death more than forty years later. By then, he would have had more than a decade to consider the news of a massacre at the Gnadenhütten mission in the Ohio Country.

The Lebenslauf of his son, Johann Jacob Jr., yielded a sounded testament to his life's journey of spiritual discovery and questioning: a hymn written when he was only a teenager. In it, he sang of his "tender childhood years," when the Savior pressed him "to His heart." But, as he grew older, he grew fearful of his own sin—"I was corrupted through and through":

Da schrie ich denn um sein Erbarmen,	Then I cried out for His mercy,
daß er um seines Blutes wille,	That for the sake of His blood
aus lauter Liebe doch mich Armen,	He might envelop me, the wretched one,
in Seine Gnade möcht einhülle.	Out of pure love with His mercy.
Ich lag zu Jesu blutgen Füßen,	I lay at Jesus's bloody feet,
ich fleht und bat in solcher Noth zu ihm,	I pleaded and begged in distress to Him,
und wollt von sonst nichts wißen,	And wanted to know of nothing else,
als nur von Gnad in Jesu Tod.	Than of mercy through Jesus's death.

Da trat Er mir vor das Gesichte	Then He appeared before my face
in Seinem blutgen Marterbild.	In His bloody martyr image.
Mein ganzes Herze wurde lichte,	My whole heart was relieved,
ich ward mit Seinem Trost erfüllt.	I was filled with His solace.
All meine Angst war nun verschwunden,	All my anxiety had now disappeared,
der Trost, den ich in meiner Seel,	The comfort that I had found in my soul,
bey meines Jesu Kreutz gefunden,	By the cross of my Jesus,
war mir ein rechtes Lebens-Oel.	Was a true life's-oil for me.

As he lay on his death-bed, in the Moravian tradition of Einsingen (singing at death), he "joined in singing, quite audibly, the verses that were being sung at his bedside: 'O Lamb of God slain innocently on the beams of the cross,' and the verse 'O how comforting to me is the voice that I know; it assures me that His heart still always burns full of love, that He is forever faithful and the Savior of sinners.'" The diarist then noted, "He received the final blessing on his journey home, whereupon he gently expired in the mid-morning about 10 o'clock on 11 May, having attained 43 years 4 months and 5 days."

More than two hundred years later, the words and songs of father and son brought tears to my eyes. Their words were so poignantly written that I could almost hear their moments of doubt and faith, and the songs that they had sung.

3

SOUND & SPIRIT

Kia ghackei wehetenap　　　　　You have sacrificed yourself.
Anehenan' amuak　　　　　　　Let us be a little bee
Nitschwatsche mnated watachgaik　So that we may drink the sap
Pagachganek paquaik　　　　　　Of the bloody wounds
Watschetsch anamegiak　　　　　So that we may grow
Atanetsch geschech sanniak.[1]　　And become strong.

ON THE EVENING OF SEPTEMBER 24, 1748, IN the Moravian mission of Gnadenhütten, the German missionary Johann Christoph Pyrlaeus, known by his Mohican name of Tganniatarecheu, and the Mohican spiritual leader Tassawachamen, known by his Christian name of Joshua, sat in conversation.[2] They spoke of the sweetness of the "little Lamb," "Christ who had purchased the souls of all the world with his precious Blood." They meditated on Christ's Crucifixion and bodily suffering.[3] And then, records the diarist of the Gnadenhütten community, through the action of the Holy Spirit they sang a hymn in Mohican:

Jesu, paschgon kia　　　　　　Jesus, to you alone
Nia quege menen ntah　　　　Do I willingly give my heart.
Kia muchtsche gpenhammen　You certainly have earned it.
Gahana gmachenochganap　　Yes, you have dearly bought it
Gtauwahan machane papaquajan　With your many wounds
Nickquaak gpegachganom　　And your blood.
Nhackay waktajom　　　　　　Since I am your child,
Osatammawe.[4]　　　　　　　Forgive me.

So began an improvised Moravian *Singstunde*, sung by two men with very different backgrounds. While they shared a common belief in the unique power of the Holy Spirit to manifest in the lives of their small community through improvised song, each man may have experienced the action of that spirit in very different ways. 🌐 **See website chap3.1, Sound recordings: "Mohican-Moravian *Singstunde*."**

In the mid-eighteenth century, Moravians on both sides of the Atlantic devoted themselves to improvising hymns as a tangible representation of their commitment of body, mind, heart, and soul to the Moravian Church. To be a Moravian meant to sing like a Moravian. Like many other early Moravians who had come to the church as adults, Pyrlaeus and Joshua learned to sing and to compose hymns in the Moravian style and to participate fully in the church's vision of an acoustic Christian community. Both men were musically and spiritually gifted and listened carefully to the world around them even as young children. They had already learned to glean spiritual meaning from the soundscapes of their early home environments: a small German town dominated by the hymns and bells of its Lutheran church, and a Mohican village in the Taconic Mountains, surrounded by thick forests filled with the songs of spirits and *manitou*. What inspired Pyrlaeus and Joshua to join the Moravian Church and to "live in song"? They had each searched without success for a deeper sense of spirituality and meaning in their lives since the time they were young boys. Each man had longed to find a community that fostered both their material and spiritual needs, but instead had found a system of belief that was either failing to meet the needs of its adherents, or a pervasive sense of religious superficiality. In addition, for two careful young listeners, the Moravians' "singing utopias" were musically as well as spiritually attractive. As Joshua and Pyrlaeus rose to positions of leadership within the church hierarchy, they perpetuated Moravian traditions of singing and their associated spiritual meanings. In the process, the Moravians' "spirit songs" were communicated to and assimilated by diverse new groups of Moravians. In each mission town, new generations of Moravian Christians would actively shape their vision of a life lived in song and a community filled with the spaces of the "spirit."

While the first chapters of this book have focused on the relationship of Moravians to the natural environment of Pennsylvania, or social relationships in Moravian missions such as Bethlehem, this chapter focuses on the more intangible, spiritual spaces of Moravian Christianity. These spiritual spaces were created and manifested through sound, not through the physical materials of wood, stone, tile, and soil. For eighteenth-century Moravians such Joshua and Pyrlaeus, the vibratory power of song had the potential to connect community members directly with the Holy Spirit. There was little separation between the material and spiritual, the human and nonhuman.[5] Through song, buildings, fields, and forests became sounded, sacred spaces.[6]

* * *

Pyrlaeus and Joshua were both born in the second decade of the eighteenth century, but they came from noticeably different cultural backgrounds. They knew each other intimately—they worshipped, worked, ate, slept, sang, and composed hymns together. Both men were adult converts to Moravian Christianity, yet their stories and those of their families traverse the Atlantic, and stretch through the American colonies from New York, Connecticut, and Pennsylvania, to the Ohio Country and the Indiana Territory.

Pyrlaeus was born in Pausa, Saxony, in 1713, and showed an early aptitude for music and Christian piety.[7] His father and grandfather had been the head pastors of Pausa's principal Lutheran church, and young Pyrlaeus was expected to follow them into the pastorate. As an introspective, quiet, and spiritually inclined child, Pyrlaeus listened carefully to the soundscapes of his home, church, and village. But much of what he heard filled him with dread. In the family's cottage, Pyrlaeus's mother held regular morning and evening prayers, where the children sang hymn verses and recited scriptures. It was through singing the hymn verse, "*Ach Herr lasz mich Gnad' verlangen / Gieb mir nicht verdienten Bahn* [Ah, Lord, bestow on me Thy Mercy / Give me not what I deserve]," with his mother that he experienced his "first consciousness of Jesus's suffering."[8] Although his mother explained that Jesus's "inexpressible love" had redeemed Pyrlaeus from his sins through his death on the cross, Pyrlaeus felt irrepressible guilt for his human sinfulness. At age thirteen, Pyrlaeus was sure that God had sent him a sounded warning of the consequences of his sinful life. As he and another boy were ringing the large bell on the roof of the church building, the bell struck him and hurled him to the ground below. He was carried home in a coma, where he slowly recovered. During his time of recovery, the lonely boy began to "speak to the Savior" alone in his room, and to ask his counsel. Although he feared to hear the reprimanding voice of Jesus, the answering voice was amazingly compassionate, assuring him that he "would not be lost."[9]

At age fifteen, he left for the nearby town of Schleiz to study singing, keyboard performance, music theory, and composition, in addition to history and writing. Yet his unusual religious devotion did not endear him to his classmates: "I endured many canings and much name calling; I was called strange, and my school mates called me Pietist and made fun of me."[10] It was not until he arrived in the musical center of Leipzig to study theology at the university that his talents were appreciated. He learned voraciously from performances at the Thomas- and Nikolaikirche directed by Johann Sebastian Bach, and likely even participated in or attended rehearsals and performances by Bach's Collegium Musicum. However, unease continued to plague him. Four weeks after arriving in Leipzig, the terrible sounds of a summer thunderstorm aroused his terror of divine retribution. Lightning tore through two houses and struck the church. Pyrlaeus was on his knees: "I was sore afraid and threw myself at the feet of God."[11] In April 1735, the sudden death of a student sitting next to Pyrlaeus in his university class also left him deeply depressed.

He longed to find friends with whom he could talk freely and develop a "spiritual friendship." He longed to find some relief from his ever-present sense that divine retribution was imminent as punishment for his spiritual inadequacy. But it was not until September 1735, when he met the Moravian brother Martin Dober, who was traveling through Leipzig on his way to the theological academy at Jena, that he found a solution to his spiritual quandaries. At Dober's invitation, Pyrlaeus visited the Moravian brethren at Jena and talked with the students there. The following spring, he continued his exploration of Moravian Christianity with a journey to the community of Herrnhut in southeastern Saxony. It was there that Pyrlaeus first heard the songs and improvised hymns that dominated Herrnhut's spiritual soundscapes. Pyrlaeus listened carefully to the Moravian way of singing. He observed community members improvising songs, and singing together into the wooden floorboards of their worship halls to cleanse each other with divine vibrations. He met Zinzendorf, and began to learn how he had carefully planned this utopian vision of Christian community. He came to believe that through song, each person could sound their prayers to God, and God could respond, sounding out messages in return that were channeled through the singing body. Ultimately, it was the sounded theology of the Moravians that most struck the musically sensitive Pyrlaeus. In Herrnhut, he experienced an emotional conversion that would come to shape his future life. He quickly abandoned his plan to become a Lutheran pastor, and dedicated his life to the Moravian Church.

After a final trip to see his mother in Pausa, he worked tirelessly for the next five decades on behalf of the church in Germany, North America, and England. His first assignment in the 1730s was to serve as a school and music teacher for Moravian children in Herrnhut. He was also appointed as an Hourly Intercessor, one of the most important positions in all Moravian communities. He was responsible for praying and singing hymns on behalf of the whole community at the turning of each hour, so that a continuous connection with and supplication to God could be maintained. He also submitted his name to the missionary lot, and on August 20, 1740, the lot selected Pyrlaeus to become a missionary to Native American communities in the Pennsylvania Colony. He left Herrnhut on March 1, 1741, and arrived at the Moravians' newly purchased 500-acre tract of land at the confluence of Monocacy Creek and the Lehigh River in eastern Pennsylvania on October 26, 1741. There, he was present at Bethlehem's first Christmas service in the log cabin called the First House.

In August 1742, another new Moravian arrived in Bethlehem. Tassawachamen (baptized as Joshua) was seven years younger than Pyrlaeus. He was born around 1720 at the Mohican village of Shekomeko in the New York Colony.[12] Although there is little surviving archival evidence detailing Joshua's childhood in Shekomeko, Rachel Wheeler's speculative biography of Joshua, "An Imagined Mohican-Moravian 'Lebenslauf,' Joshua, Sr., d. 1775," provides a framework for how we might approach the reconstruction of this period of his life. According to Wheeler, as a young boy, Joshua likely learned from the men of the families in his community to fish, to hunt bear, deer, and wolves, and to build canoes. He may also have learned how to use roots and herbs for healing from his mother. His religious life would have been cultivated through great feasts celebrating the harvest or successful hunts—occasions of thanks to the Great Spirit for blessing and protection. He likely dreamed of the vision quest he would one day be allowed to undertake, to seek a lifelong and personal relationship with one of the lesser spirits, or manitou, and if so favored, to receive the gift of a vision song.[13] Like other men in his village, Joshua would have expected to find solace and the blessing of a personal spirit helper through sanctuary in the great forests that surrounded his community. Young boys such as Joshua were often sent to be alone in the forest, beginning around their tenth birthday, because only in the silent spaces of the trees could they properly hear, without hindrance by the noise of human community, the calling of the lesser manitou. In the forests, Joshua could have listened carefully to the sounded messages conveyed by birds, winds, storms, water, and animals. Each sounded message was a potential seed of a song—the most important gift of a personal spirit helper.[14]

However, it was also likely that some in Joshua's community had begun to worry that the spirits were deserting their community, and that their songs could no longer be heard. Some may have even felt that the spirits who served nearby communities of Dutch and English settlers could be more powerful. Certainly, the people living in those communities had regular access to medical care, food, and basic life needs.[15] As he reached manhood, married, and began his occupation as runner and messenger for his community, Wheeler has speculated that Joshua too would have worried: "As my wife and I began to bring children into the world, I worried much about what sort of people they would become if we continued in our current ways. Many of our old people had died in the sicknesses and many of our young men too often were under the evil *beson* of alcohol. We were not as faithful in our obligations to the spirits as we had once been, and the spirits did not always answer our petitions."[16] Several Dutch and English Christian missionaries visited his community while Joshua was a boy, but their spiritual teachings probably offered little comfort to a troubled youth who might have claimed, "They did not tell us anything we did not already know."[17] The Dutch and English missionaries relied on literacy and the written word to promulgate their teachings, preaching that the secrets to the divine had been revealed in an ancient book called the Bible. But even later in his life, Joshua would wonder why the followers of this great book did not often put its contents into practice.[18] What evidence was there that they followed their own teachings? It may also have seemed odd to think that the Great Spirit would reveal himself in a book, and not through the sounds of the wind, and in the silent spaces of the forest. True revelation for many eastern Woodlands cultures lay in the sacred heights of the mountains and in the great trees of the forests, its animals, and its insects. Spiritual power and truth was revealed in the songs communicated from spirits to humans.[19] The Dutch and English sang, but their hymns were not "spirit songs." Spirit songs attuned to the vibratory centers of the earth: sacred spaces endowed by the Great Creator with spiritual power that could be transferred from mountains, lakes, caves, waterfalls, forests, rivers, plateaus, and oceans to humans.[20]

It was not until the arrival of Moravian missionary Christian Heinrich Rauch in the summer of 1740 that Joshua may have begun to hear Christian teachings that were meaningful to him. Brother Rauch apparently did not resort to the book, but spoke words to the community that were "*felt* to be true."[21] His teachings of the Great God who had become a man, a great warrior who was killed, yet whose wounds and blood held redemptive, life-giving power likely moved Joshua's heart. This new spirit offered freedom and power for those who believed. He had allowed himself to be sacrificed to redeem the suffering of all of the Earth's peoples. This Great God, in his human form, understood the secrets of the Earth. He had taught his human disciples to seek the Great Spirit in the high places of a mountain peak where they received healing power from the divine.[22] We know that Joshua listened carefully to Rauch's messages, and that he may have also observed this new manitou at work in Shekomeko: those who were blessed in the power and blood of the Christian god were perhaps healed from alcohol, as in the case of the Shekomeko Moravian elder Shabash, or more successful at providing food for the community. Hunters may have brought home deer and bears from places that had long been deserted of game.

We may never know exactly how Joshua arrived at the decision to devote his life to the church, but we do know that in August 1742, when Zinzendorf visited Shekomeko along with several other Moravian brethren, Joshua requested to travel back with him to Bethlehem.[23] In Bethlehem, Joshua would have slept in the Gemeinhaus, and sat next to the other Moravian brethren at table.[24] He listened to spiritual teachings in the Gemeinsaal and participated in singing. At some point during his visit, Joshua felt that he was beginning to hear and feel the Spirit's presence, and his heart was moved. In Bethlehem, "the Jesus manitou" appeared to him for the first time. In response, Joshua offered his life to this new manitou, sent by the Great Spirit. He was baptized in the blood of the Savior in the newly constructed Saal of the Gemeinhaus on August 27, 1742, by Gottlob Büttner, accompanied by the singing of hymns. Joshua now began his initiation as a Moravian brother, and dedicated his life to the church.[25]

Two men, from two sides of the Atlantic, Joshua and Pyrlaeus, Tassawachamen and Tganniatarecheu. They lived for a time together in Bethlehem and Gnadenhütten in the early 1740s. 🌐 **See website chap3.2, Soundscape recording: "Gnadenhütten, Pennsylvania."** They slept together in the communal choir buildings, worked together in the workshops of the tradesmen, toiled in the agricultural fields, orchards, and pastures, and worshipped together throughout the day. But, most importantly, they sang together in the spiritual spaces that surrounded their communities. Like all Moravians, Pyrlaeus and Joshua participated in daily improvised Singstunden. These singing services were performed out-of-doors during hunting expeditions or travel between communities. They could be sung in natural areas around Bethlehem such as the terraced riverbank of the Lehigh or the orchards and gardens. And they could be privately performed in homes and choir houses as a spiritual exercise and method of communication with the Holy Spirit. Regardless of location, during a Singstunde singers were expected to channel the voice of God through their own singing body as a demonstration of their faith. Musical improvisation was a religious practice, and an individual's ability to improvise new songs was highly prized as "giving voice to the divine." Improvisers were to sing exactly what the Holy Spirit intended to communicate at that particular moment in time. In the process, they drew the intangible, spiritual world down into their communities. Improvisation not only provided the divine with a voice, but also a meaningful role in community life.[26]

Yet improvisation, and especially a communal tradition of improvisation, required skill and dedication on the part of the improvisers. Participation in the improvised Singstunden could not happen through divine inspiration only. How did Moravians such as Joshua and Pyrlaeus learn to improvise? How did they learn to manifest the spiritual world in their own community? Although Bethlehem's residents were not required to improvise new songs, but were instead tasked with spontaneously recombining memorized parts of hymns, they still needed to commit a significant amount of time to learning this process of spontaneous composition. The ritual cycle of daily life in Moravian communities provided a dedicated framework for participating in and learning how to improvise. As discussed in chapter 2, Moravians were immersed in a continuous soundscape of religious song. Singing took place every day in Moravian communities as part of a cycle of hymns and prayers that began at four o'clock in the morning and ended at ten o'clock in the evening.[27] Communal rituals circumscribed every aspect of life, melding the secular with the sacred. Waking, sleeping, eating, drinking, and working—all the activities of daily life were accompanied by hymn singing, both memorized and extemporaneous. Even pregnant women sang special hymns to help unborn children to hear the religious soundscape of the community.[28] At night, spiritual elders and watchmen walked the streets, singing hymns that protected the unconscious minds of their fellow congregants. 🌐 **See website chap3.3, Interactive map: "Spiritual Singing in Bethlehem."**

In addition to daily immersion in Moravian rituals, all communities offered "poet" and "singing" schools that incorporated vocal, poetic, and theological instruction to assist community members in learning how to sing softly and harmoniously, and to compose new hymn texts and tunes.[29] These schools also taught community members to memorize. Rather than memorizing words loosely structured around a theme—as in prose—poetic meters and rhymes helped Moravians to organize the words being memorized into larger groupings and patterns. This provided helpful models for creating new and improvised hymns. It was especially important for converts to learn how to sing quietly and harmoniously in communal contexts, and to spontaneously compose new hymn texts, based on previous tunes, as a prayerful response to the Holy Spirit. Especially in relatively isolated Moravian mission communities such as Gnadenhütten, it was the soundscape of spiritual singing that articulated a continued sense of Moravian Christian community. Moravians of all ethnic backgrounds were instructed to listen carefully for this soundscape—the "beautiful verses" and "sounds coming through the thick woods and over the high mountains."[30] These "lovely sounds" represented a sense of Christian spirituality encompassed within a simultaneously human and spiritual landscape and geography. In addition, singing navigated the boundaries of the natural, mortal world and the spiritual, intangible world in a way that the targeted phonemes of language could not do alone. The vibrations of music were free to cross boundaries to mediate, pray, and call upon the divine.

Harmonious, beautiful singing was such a notable feature of the soundscape of Moravian communal life that travelers' descriptions of singing in communities across the Atlantic World are remarkably consistent, a fact which has already been noted at the end of chapter 2. Observers as diverse as Johann Friedrich von Heinitz, Johann Friedrich Reichardt, and Johanna Schopenhauer wrote about the distinctive qualities of Moravian singing. For Reichardt, Moravian voices effected "a serenity of the body," through "pure, unaffected singing."[31] Johann Friedrich von Heinitz observed that: "They [the Moravians] sang in such a melodic and quiet manner that one thought one would hear the voices of the angels, arch-angels, and cherubim laying prostrate before the Lamb in heaven."[32] Both Reichardt and von Heinitz were cognizant of a sense of divine space represented in Moravian song. Johanna Schopenhauer, mother of philosopher Arthur, also described the "harmonious" singing of Moravians and the dedication of community members in Herrnhut to singing in a particularly soft, reverent style.[33] She noted the spiritual effect that this seemed to have on the community, and how it struck her as uniquely different than the singing of other Christian congregations. Even Johann Gottfried Herder ascribed a "truly devotional quality" to Moravian singing in his essay on Zinzendorf.[34]

Still, despite these harmonious sounds, the divine could not manifest in Moravian communities without singers who were skilled at the art of improvising hymns. Learning to improvise didn't just happen through immersion in the communal religious soundscape. In the daily "poet" and "singing" schools converts like Joshua and Pyrlaeus also learned to extemporize hymns using a memorized corpus of several thousand individual hymn verses and a selection of twenty common chorale tunes. This system of spontaneous singing took advantage of basic mnemonic techniques. The hymn verses used in the Singstunden were culled and arranged by church elders from the *Herrnhuter Gesangbuch* (the Moravian hymnal published in 1735) and its twelve subsequent appendices. They were then classified into manuscript guides called *Auszüge für Singstunden* (Guides for Singstunden). To make the hymn verses in the *Auszüge* easy to memorize, they were arranged hierarchically according to subject matter and melodic classification (fig. 3.1). The script was elegantly and beautifully written, and the layout of the page was clearly discernable at a glance, reminiscent of mnemonic techniques presented in ancient and medieval treatises on the art of memory.[35] Moravians who could not read, and therefore could not memorize from the *Auszüge*, could still take advantage of the aural teachings of those who had memorized "from the book." They could also repeat spontaneously combined hymns improvised by other residents.[36]

Moravians also learned the basic harmonic structures of around twenty chorale tunes, which could be used to sing the majority of the hymn verses in the *Auszüge*. By combining a memorized hymn verse and chorale tune, or parts of memorized verses and tunes, even the most musically and poetically ungifted Moravians were able to "draw down the Holy Spirit" by extemporaneously combining memorized materials. Improvisation of completely new material was not necessary. Zinzendorf was adept at simultaneously improvising music and text, but he did not expect this of his followers. More commonly, Moravians improvised music to an existing text, or improvised text to an existing tune. In terms of textual improvisation, daily practice in creating new hymns, using predictable patterns of versification and poetic meter, enabled Moravians to spontaneously create new text with relatively little effort.[37] With existing texts, music was improvised either by using the same melody to set hymn verses or parts of verses with the same syllabic pattern. Separate verses and their associated melodies

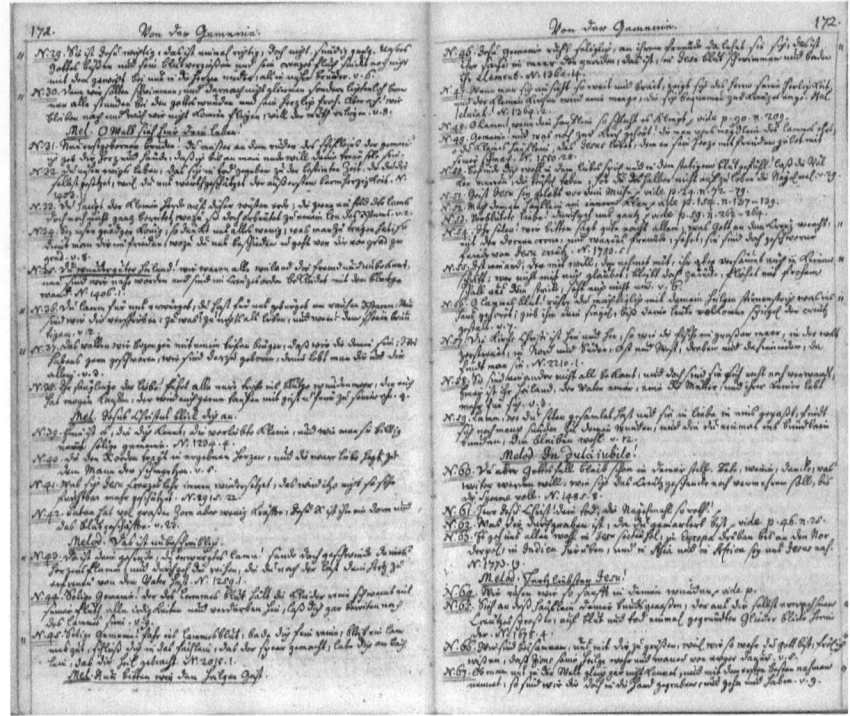

Fig. 3.1 A mid-eighteenth-century guide to creating Singstunden. Auszug für Singstunde, NB.IV.R.1.120.b, pp. 171–172, UA Herrnhut.

could also be connected together, either in the same key area or by modulating "as the Spirit dictated." An improviser could also embellish an existing tune with new melodic or harmonic material.

This method of improvisation was so successful that even young children were instructed in the art of improvisation, and were able to improvise hymns and to participate with adults in the Singstunden. Zinzendorf's son, Christian Renatus, could apparently improvise a Singstunde at quite a young age: "When my ten-year-old son plays for our private *Singstunden*, he is able to seamlessly connect one melody to another so that no one knows that the entire *Singstunde* was not expressly composed that way, for there is no hesitation, and every child sings along without looking in a book, for they know the songs by heart."[38] What was perhaps most interesting about Zinzendorf's structured method of improvisation was its inherent inclusivity. Anyone could theoretically participate in his system of religious improvisation, provided they were willing to learn its techniques. This carefully controlled method allowed him to communicate a particular style of improvisation to his followers. After learning and practicing Zinzendorf's method, most Moravians could have produced fairly consistent improvisations. While his original intention was to foster an emotional and intuitive connection with the divine by allowing improvisers to channel "God's voice" through improvisation, that voice could not be heard or communicated without a controlled and practiced system of techniques.

The success of this style of singing is aptly demonstrated by the countless stories of hymn improvisations and Singstunden that fill the pages of archival records from communities on both sides of the Atlantic. After their meeting in Bethlehem, Joshua and Pyrlaeus continued to minister and sing together in various missions, such as Gnadenhütten. The diary of that community preserves several stories of Joshua and Pyrlaeus improvising together, including the one that appears at the beginning of this chapter. Transcriptions of improvised hymns and Singstunden preserve these fleeting instances of text and music, but they are also important demonstrations of how improvisation was applied in the daily lives of Moravians. Although Joshua and Pyrlaeus began their lives in markedly different circumstances, they eventually came to share a common belief in the unique power of the Holy Spirit to manifest in their community through the sounded vibrations of music. For Pyrlaeus, improvisation of song represented a freedom of religion and a sincerity of Christian conviction that he had not found in the Lutheran Church. For Joshua and other Native Christians, the idea of divinely gifted songs may have been similar to Native musical traditions that they had learned as children.

Sacred music in Mohican communities was granted to human singers through contact between the finite world and the infinite or supernatural.[39] As a young adult, Joshua would have sought a personal and lifelong relationship with a Guardian Spirit. He would have listened for the voices of various spirits, including the souls of ancestors, animals, and plants that surrounded his community.[40] Songs of this type of visionary origin were sacred, and could not be performed. They could only be sung within the context of proper religious rituals.[41] Neither could they be attributed to individual musical agency. Singers who tried to sing vision songs of their own making could become violently ill, and were subject to communal humiliation.[42] Joshua could only have sought an authentic song of divine origin through a vision quest, alone in the forest, listening and attuning to the whispering sounds of the ancient white pines, black cherries, and Seneca hemlocks that towered over the forest floor; and the rustle of small animals—guardians of the undercanopy of blue spruce, American beeches and chestnuts, and white and scarlet oaks.

The story of a young boy seeking supernatural guidance and affiliation with the divine parallels the genesis of hymn improvisation in the Moravian Church: as a child, Zinzendorf played with the "Savior" in his room as a kind of imaginary friend, and wrote notes to him on slips of paper which he threw out of his window, so that "Jesus" would find them. He also kept bits of paper in his pockets, with Bible verses or messages, which he would draw at random to ask the Savior's answers to questions. He needed to be sure the messages came from the Savior, and not from his own wishes or thoughts, and the idea of lot casting, or randomness, occurred to him as particularly suited to this purpose. His personal childhood practices eventually shaped his vision for Herrnhut: a Christian utopia where everyone could have personal, immediate access to the divine through lot-casting and improvisation of hymns. So, while Mohican and Moravian sacred songs were musically dissimilar, the purposes of both traditions of sacred song were similar.

For both Joshua and Pyrlaeus, learning to sing and participating in communal rituals such as the Singstunden represented not only spiritual transformation but also true membership in a spatially diverse community of Christians that spanned from human to divine. Sensual experience of Christ's suffering was a necessary component of theological learning both for adults who became Moravian and for those born into the community. Joshua and Pyrlaeus learned that to be Moravian meant not just to channel the divine voice but to be transmogrified, flesh-and-blood, into Christ's own body. Every Moravian, whether German, Mohican, English, or Delaware, learned to viscerally connect with the suffering and death of Christ as if it were the suffering and death of their own body. It was through the physical and spiritual transformation of Moravian bodies that the spiritual spaces of earthly communities were revealed.

The importance of this human-to-spirit connection, and the bodily manifestation of that connection in Moravian communities, is reflected in a contemporary description of a Single Brothers' Choir Festival in 1748. On the day of the festival, the Single Brothers gathered inside the Saal of the Single Brothers' House. As they waited for the festival to begin, a solo voice sang repeatedly, *"Du weißt wohl worauf wir warten* [You know what it is we're waiting for]." Suddenly, the door to the worship hall opened, and the spiritual leadership of the Single Brethren's Choir entered, wearing white robes in imitation of Christ:

> "Today, you will be embraced by the Side Hole," they [the leaders] exclaimed. The brothers sang together of the side wound of Christ until some of them were physically unable to rise from the floor. Later that day, they returned to their communal home to find an illuminated representation of Christ, soaked with blood, covering the entrance. A brother, dressed as the Roman soldier Longinus, appeared and violently pierced the image with a spear. Blood burst from the wound and splashed onto the brothers nearby. They washed their hands in the blood. Suddenly, an image of the side hole appeared, big enough for a man to enter, and each brother bent down and entered their home, singing "deep inside, deep inside, yea, deep therein."[43]

This communal ritual, performed and experienced by the Single Brothers, elaborated upon the basic tenets of Zinzendorf's "heart theology." For Zinzendorf, the power of Christ's blood had transformed Christians physically and spiritually through a kind of holy, transformative alchemy: "His blood of reconciliation is the *proprium quarti modi* of the entire holy creature, of the entire blessed universe."[44] Not only were the Single Brothers encouraged to connect their own bodies with the divine body of Christ through ritually bathing in Christ's blood, but together the participants in the Single Brothers' Choir Festival acted out a further ritual transference of place—the Saal of their choir house was itself transmogrified from a simple room into the divine interior of Christ's Side Wound. This spatial transformation of bodies and buildings is clearly portrayed in myriad miniature devotional watercolors painted by Moravian community members on both sides of the Atlantic (fig. 3.2a–2d). In the "warm and secure nest" of the Side Wound, Moravian artists built fences and buildings, and carved tables and chairs, cups and plates, both real and imagined. They painted themselves waking, sleeping, eating, drinking, strolling, and singing within a boundary that was both mortal and spiritual, physical and aural. Those who lived in the Side Wound saw or heard, in the words of Zinzendorf, "nothing above or beyond."[45]

The Moravians' conceptions of divine transmogrification and ritual transference of place were connected to a long lineage of Christian mystics who believed in physically experiencing Christ's suffering through contact with the divine body and blood.[46] Graphic stories of the Crucifixion had assisted the Christian faithful for centuries to transfer the suffering of Christ onto their own bodies. Italian theologian Emanuele Orchi's

Wie warm sichs liegt im Seitelein/Ehre dem Seitenschrein
[How warm it is to lie in the little side/Honor the side shrine]

Im höhlgen geh ich spazieren
[In the little hole I go for a walk]

Im Seiten höhlchen schlaf ich ein
[In the little side hole I go to sleep]

Da halt ich Mahl Zeit früh und spat
[There I have meals early and late]

Fig. 3.2a–d Devotional images of the Side Wound. Miniature watercolors on card stock, mid-eighteenth century. TS.Mp.375.4.b, TS.Mp.375.4.c, TS.Mp.375.4.d, and TS.Mp.375.4.e, UA Herrnhut.

evocative and intensely descriptive sermons on the Crucifixion were just some of the published Passion narratives familiar to Zinzendorf: "You will have seen our good Jesus, out of natural fear, tear out the hair of his head. You will have seen his arms flailing, his thighs shaking, all of those milky members chilled to ice and trembling. . . . A hot river of blood courses headlong from all the torn victim's body to his feet. From the flayed Lord a pure, warm lake of blood appears. And blood again from the plummeting lashes, spattering the walls all around, hanging on the rough walls in gross, thick lumps."[47] According to cultural historian Piero Camporesi, bloody stories such as those preached by Orchi facilitated contact with the sensual self:

The ambiguity of the metaphors, and the erotic symbolism of these visionary experiences, far from canceling the sensual language of the flesh, actually reinvigorate and excite it, expressing as they do the omnipresence of bodiliness and the senses, through which even religious allegories must necessarily pass. We all possess an anatomical, physiological sieve, without which it seems that discourse cannot begin, cannot be ordered with any rationality, cannot render itself intelligible. At the origin of mystical language is a sensual alphabet, which strives, by its innermost calling, to be delivered from the senses and from the body, not by canceling them, but by sublimating them, by transferring them to God, by immersing them in a laver of "thirsting, slaking ... concupiscence."[48]

For many Christian mystics such as Orchi, this sublimation of the body manifested itself in the mystical transference of human flesh into the tormented body of Christ, the resurrected body that breathes new life into its creatures. Ingestion of the body and blood of Christ in Holy Communion, like the aqua vitae of legend, effected the magical transference of divine life to the believer.[49] Ingestion of the divine promised hope of redemption from sin and eventual release from the bonds of a corrupted physical body, and the imparting of a new spiritual body.[50] Those who bathed in and ingested the blood were transformed through an almost alchemical process. They shed their former bodies and assumed new, clean ones, in preparation for the passage into eternal life, as expressed in the text of a hymn from the *Herrnhuter Gesangbuch*:

WEnn ich Ihn essen kan	When I can eat Him,
so ist mirs am gesündsten,	It is healthiest for me,
und wenn mein lieber mann	And when my beloved man
sein öl läßt in mich dünsten;	Lets his oil steam in me;
weil aber diese gnad	However, because this grace
in einem sacrament,	[Is] in a sacrament,
das man nicht immer hat,	That one does not always have,
dem leib wird zugewendt.	Attention is turned toward the body.
Ich hör Ihn ins gespräch	I hear Him get into a conversation
mit dem versucher kommen;	With the tempter;
mein JEsus ist gar träg	My Jesus is very languid,
vom leibe abgekommen,	Out of His body,
Das sprechen wird Ihm schwer:	Speaking becomes difficult for Him:
wenn Satan auf Ihn sticht,	When Satan pricks Him,
bet't er so sprüchel her,	He prays little phrases,
wie ers zusammen kriegt.[51]	Whenever He can piece them together.

Fig. 3.3 Zinzendorf and the Saviour in Six Allegorical Scenes [Zinzendorf und der Heiland in sechs allegorischen Vorstellungen], c. 1750. Watercolor on paper, artist unknown. TS.Mp.376.12, UA Herrnhut.

This mystical transformation of human flesh is clearly depicted in an anonymous Moravian devotional painting entitled *"Zinzendorf und der Heiland in sechs allegorischen Vorstellungen* [Zinzendorf and the Saviour in Six Allegorical Scenes] (fig. 3.3)."[52] In the six different spiritual spaces represented in the painting, Zinzendorf bathes in the blood that spurts from Christ's side wound, and adores the sweat drops of Christ's agony that fall to earth.[53] Every sense is alive for Zinzendorf: he touches, tastes, smells, hears, and sees Christ, drawing salvation through physical contact. Bits of poetry surround the six allegorical scenes in the painting—small love songs sung to the Savior by Zinzendorf. The painting depicts them lying together in the grave's bed, transformed by the *mors mystica* (mystical death), as Zinzendorf sings:[54]

Ave, mein lieber Mann!	Hail, my dear man!
Ave, vor deinen Plan!	Hail, to your plan!
Ave, für deinen Fleiß!	Hail, for your diligence!
Ave, für deinen Schweiß!	Hail, for your sweat!

Ave, fürs Todes-eis!	Hail, for death's ice!
Ave, du Mund so blaß,	Hail, you mouth so bare,
Ave du Wangen naß,	Hail you cheeks so wet,
Ave du Blick so graß,	Hail you glance so great,
Dornichtes Scheidelein!	Thorn-crowned little head!
Wundes, wundes, wundes Häutelein!	Wounded, wounded, wounded little skin!
Ave, ave, ave Seitelein![55]	Hail, hail, hail little side!

Zinzendorf identifies with the Crucified Christ through the most personal means possible—his senses. Sensual contact causes Christ to live. Blood courses through veins, hair stirs on arms, and Jesus presses his face into the dirt of Gethsemane in fear of his approaching torture and death. Zinzendorf is there not just as a witness, but to experience the suffering as if it were his own body. According to Zinzendorf, if believers could be crucified and buried along with Christ, and sing with him in the ancient, sacred spaces where his body walked the earth—under the olive trees of Gethsemane, on the stark hillside of Golgotha, and inside of the cold, stone-walled garden tomb—they would feel his salvation "in their hearts."

Graphic Crucifixion narratives and paintings such as this elicited sensual encounters and spatial connections with the divine. Dozens of extant mid-eighteenth-century devotional paintings from Herrnhut, Bethlehem, and other Moravian communities across the Atlantic World portray Moravians caressing Christ's body on the cross, and bathing in the blood that flows from his wounds. They depict Moravians lying inside of Christ's grave, and feeling the scars of the Crucifixion with their own hands, as represented by a painting from Herrnhut in which Moravian sisters touch and kiss Christ's wounds as he lays in the tomb (fig. 3.4).[56] It is this sensual contact with the body of Christ that leads them toward an understanding of the theological doctrines of Christ's death and Resurrection. Rational proof is not necessary—they simply understand. Every part of their body, every cell, comprehends. As members of Christ's body, they experience the world through his senses. They breathe when he breathes, sing when he sings, and die when he dies.

Although Moravians cleverly employed paintings, narratives, and rituals to transform themselves spiritually, they also believed in the power of sound to change the nature of both the body and the spirit through a kind of vibrational alchemy. As a young child, Zinzendorf had internalized his grandmother's respect for the teachings of the Christian mystic Jakob Böhme.[57] For Böhme, music tapped into the spiritual vibrations of God's

Fig. 3.4 Moravian sisters kiss and touch Christ's body in the garden tomb, c. 1750. Watercolor on paper, artist unknown. TS Mp.376.10, UA Herrnhut. Text translation: "Do not ask me how I do it; I have enough when I can be more than one day in the grave where I can see, kiss, have, and embrace Jesus's dear corpse."

voice. He believed that Christians should sing in harmony with the "universal vibrations," allowing their "hearts to sound" of that which was beyond words and reason.[58] Zinzendorf was very attracted to Böhme's ideas of "vibrational alchemy" because they matched his own understandings of the potentially transformative effects of Christian song. According to linguistic scholar Patrick Erben: "In following alchemical concepts of transforming substances from a lower to a higher level of purity, Moravian hymnists hoped to enlist their divinely inspired compositions to lift themselves and the community onto a higher spiritual level."[59] Alchemy in general, and certainly for Zinzendorf, was considered a model of universal purification and renovation. The Christian alchemy of both Böhme and Zinzendorf aimed for an earthly and cosmic renovation, which was reached through the sublimating powers of music. Hymn singing, in particular, could modify and transform human beings by connecting body and mind. This alchemical or mystical transformation into a higher spiritual form allowed human beings to articulate themselves in a higher spiritual language that was capable of expressing fully the divine meanings of the universe.[60]

These divine meanings were especially communicated through the improvised hymns sung during communal rituals such as the Singstunden. Believers improvised songs that sounded their longing to be caressed and cradled inside Christ's body, pierced and gashed by thorns and nails, their mouths overflowing with blood. They prayed to become "little wound bees which burrow into the side hole of Christ."[61] They sang together very softly, prostrate upon the wooden floors of the choir houses, meditating upon graphic representations of the suffering Christ. Moravians believed that these controlled ecstatic experiences allowed their bodies to resonate and attune to God's word. Moravian litanies, hymns, and rituals emphasized the sacred in the everyday, connecting the physical bodies of worshippers with the divine body of God, in the physical form of Jesus's blood.[62] Out of the blood of Christ was born the Moravian *Blutgemeine* (blood community), a physical and communal praxis of embodied theology that included both human worshippers and the divine presence of the Holy Spirit.[63] The blood of the wounds (*Wundenblut*) transformed Moravians through the "*feurigen Leuchtkraft des Blutes Christi* (fiery light-craft of Christ's blood)." According to Zinzendorf, this transformation was physically visible through the light of Christ's wounds which shone out of the eyes of the transformed and immediately distinguished them from other Christians: "All day long we gaze into Jesus's wounds . . . He always stays in our countenances with His wounds-tears, in the light of His nailmarks, where we customarily sit . . . that remains constantly in front of our eyes, that distinguishes us from all other people, the countenance glows before our brows, it looks out of our eyes."[64]

This "daily gaze" into Jesus's wounds assumed its most powerful form in the Singstunden. To truly participate in a Singstunde, an improviser needed to silence their own thoughts and feelings in favor of the "divine voice." True channeling of the divine represented transformation and sublimation of the worshipper's own body, in exchange for access to inner thoughts and feelings—God's voice.[65] Just as at Passover the ancient Hebrews marked their houses with the blood of a sacrificial animal against the nightly terrors of the Angel of Death, Moravians, too, believed in metaphorically marking themselves with the sacrificial *Lammesblut* (Lamb's blood). Those so marked were visible from the outside as singing members of their community, and from the inside as transformed members who sang in harmony with the vibrations of the divine voice. The Moravian way of singing marked those chosen and sealed by the blood of the Lamb. As they listened to the voices of those around them, this mark was as audible as if their foreheads were literally painted with a cross of blood. In the words of one hymn: "Their mouths were filled with blood, and they sang together in joyful union with the heavenly spheres."[66] In so doing, the core tenets of their theology were transmitted, consumed, and ingested through the sensual vibrations of the singing body. Just as Moravians believed that the blood of Christ had transformed believers, the physical nature of singing, in communion with others, produced sensations and vibrations that cleansed the body. Rituals of communal hymn singing channeled the vibrations of the "divine voice" into the body and marked it as a member of Christ's Blutgemeine. The human body itself became a spiritual space.[67]

Perhaps the most important test of the vibratory powers of song came at the point of death.[68] As Moravians made the final passage between the human and spirit world, referred as a homecoming (*Heimgang*), their mortal bodies were rebirthed as spiritual bodies.[69] Since the vibrations of singing had the capacity to easily transcend these boundaries, it was natural for community elders such as Zinzendorf to advocate for the importance of song at the end of life.[70] This practice was referred to as *Einsingen*, or singing at the deathbed of the dying.[71] Despite the pain or delirium that could precede death, Moravians often sang. If they could not sing for themselves, the community sang for them. The *Gnadenhütten Diary* preserves an account of the final hours of Esther, a Delaware Moravian, who died there in childbirth in 1754. As she lay "in her greatest agony," incapable of speech or song, the attendant women sang "wound verses" over her, as she pointed on her body to the wounds of Christ. Even though Esther, in extreme pain, was physically incapable of song, the community sang for her, and she participated as she could with gestures.[72] Singing was a healing ritual that helped the dying to ease the transformative passage from this world to the next, and most Lebensläufe recorded the last songs of the dying. These song transcriptions fulfilled an important purpose: they were a definitive demonstration that the person's commitment to their religious faith had endured until their final moments. They assured surviving family and the Moravian brethren and sisters that the dying person had successfully navigated one of life's most difficult passages.

They were transformed—blood, sinew, and bone—as Christ had been rebirthed through the pain and suffering of his crucified body into a spiritual being. For all Moravians, Native or European, it was song, and especially the communal experience of song, that allowed them to successfully cross the boundary of life and death and to harmonize with the heavenly singers they expected to join whose singing permeated the spiritual spaces beyond the human world.[73]

* * *

At the end of his life, Zinzendorf made his passage from the world in song. As he lay dying on the evening of May 9, 1760, from an unknown illness, more than one hundred people sang with him in preparation for his passage from the mortal world.[74] The diarist for the Herrnhut community would later record that Zinzendorf exclaimed to those present that it was "time to go home," and that "nothing stood in between" him and God. He mused on the events of his own life—a transatlantic tale that had influenced thousands of people of different ethnicities and backgrounds. He wondered how God had allowed him to do this work, and exclaimed that his life's purpose was truly fulfilled. He specifically requested that the children of Herrnhut come to his bedside, so that they could witness one of most miraculous transformations of human life. He spoke to them of "finding rest in the wounds" that he could already see before him with the sight of the dying. Mirroring the story of Christ's Crucifixion, the Herrnhut diarist recorded that Zinzendorf, too, closed his eyes for the final time at that most magical and sanctified hour of the day, the ninth hour when Christ himself had submitted his body to death.

We do not know how Pyrlaeus died, except that he was in Herrnhut at the time of his passing. The final entries in his Lebenslauf ceased around the time of the death of his wife, Susanna Benezet, on October 8, 1779. Distraught, Pyrlaeus did not even record the last songs of his wife, instead exclaiming: "I am still here in this valley of sorrows, but my soul awaits the call of my Savior that it may follow her into eternal rest."[75] Before the death of Susanna, he had left behind three young children in Pennsylvania that he never saw again, and three more in England that he met only briefly again in the year before Susanna's death. From the last months of 1779 until his death in 1785, Pyrlaeus was alone. Although we cannot know for certain, we might imagine that he may have found some measure of comfort in singing during his last hours, just as he had done since his childhood.

Joshua, like Zinzendorf and so many other Moravians, also "went home to the Savior" in song. After living in several different mission communities in eastern Pennsylvania, he and his family settled in the early 1770s in the Ohio Country. Following the deadly conflicts of the Seven Years' War and Pontiac's Rebellion, Joshua and other Delaware and Mohican Moravians could no longer live safely in Pennsylvania. In 1772, they accepted an offer of sanctuary from Netawatwees, the head of Delaware settlements in the Ohio and Tuscarawas River Valleys. After a difficult and dangerous journey on foot across the Allegheny Mountains, Joshua and other Moravians built three new mission communities along the Tuscarawas River: Gnadenhütten, Schönbrunn, and Salem. Joshua lived and served as a spiritual elder in Gnadenhütten for three years until his death in 1775. As he lay suffering and unable to sing on the evening of July 31, 1775, one of Joshua's closest friends, the missionary Johann Schmick, sang hymn verses on his behalf. Joshua remembered his Moravian brothers and sisters in Bethlehem, and asked Schmick to send them his final greetings. At two o'clock on the morning of August 1, as Schmick sang, Joshua died.

Joshua did not live to see the terrible suffering that would soon afflict the Delaware and Mohican Moravians in Ohio, including his own son and his grandchildren. Not even their removal from Pennsylvania could save them from the encroaching conflicts of the American Revolution and its aftermath. After attending school at the Antes farm and the Native school at Gnadenhütten, the younger Joshua often led worship services, helped with translations, including an important hymnal in the Delaware language, and served as a diplomatic interpreter.[76] But, like many northeastern Native Americans of his generation, Joshua suffered tremendous upheaval. He was born in 1741, the year of Bethlehem's founding, and moved at least a dozen times from his home in the New York Colony to the Indiana Territory, where he lived at the Moravian mission along the White River at the time of his violent death in 1806. Along the way, he buried his wife and ten children.[77] According to the diaries of the White River mission, the cumulative effect of these tragedies took their toll on Joshua, and his Christian faith waivered. He turned to alcohol, and began to boast that "if, after the heathen manner, he wanted to make use of the dream of his youth, he could also do evil, for in his vision a bird had appeared unto him and said: 'I am a man-eater, and if you wish to feed me, you need but point out to me some one, and then I will put him out of the way.'"[78] When emissaries of the Shawnee Prophet (Tenskwatawa) came to the White River mission seeking allies to join their nativist movement, Joshua refused. He was accused by Delaware allies of the prophet of harboring a medicine bundle that he used to do evil and was sentenced to be burned as a witch. During his last hours on March 17, 1806, Joshua apparently spoke and sang in a language his tormentors could not understand. His final act during his execution was to sing, presumably appealing in German to the Spirit whom he hoped might redeem his suffering.[79]

Notes

1. Mohican hymn attributed to Joshua, Bathsheba, and Pyrlaeus, and sung to the melody, "O Opfer Lämmelein." From "Pattern for a Song Booklet of the Blessed Hearts of the Brown Nations of the Mahican Delawares and Some Short Verses of the Language of the Six Nations [Mohawk]," NB.VII.R.3.91 (1746), UA Herrnhut. English translation by Masthay, *Mahican-language hymns, biblical prose, and vocabularies from Moravian sources, with 11 Mohawk hymns.*

2. After publicly affirming their faith, baptized Moravians were given new Christian names as a sonic marker of their membership in the church and the worldwide community of Christians. Children born to baptized parents were baptized as infants and given a Christian name. The practice of receiving a new name was familiar to Native people who converted to Moravian Christianity, because Eastern Woodlands peoples sometimes assumed new names at important points in their lives. The practice of renaming at baptism meshed with Native tradition in this regard. Moravian missionaries who made particular connections with Native communities were also renamed, and given a new Native-language name, as was the case with Zinzendorf (Johanan), Pyrlaeus (Tganniatarecheu), Zeisberger (Ganousserascheri), and Conrad Weiser (Tarachiawagon). See Kyle Fisher, "After Gnadenhütten: The Moravian Indian Mission in the Old Northwest," 17–19, and 35. Michael McNally has also discussed the process of baptismal naming as a form of linguistic colonialism: "Naming the Legacy," in *Native Americans, Christianity, and the Reshaping of the American Religious Landscape*, ed. Joel W. Martin and Mark A. Nicholas (Chapel Hill: University of North Carolina Press, 2010), 292.

3. *Gnadenhütten Diary*, MissInd 116.4, September 24, 1748, MAB.

4. This hymn verse was recorded in "Verses for the Use of the Indians in Pisgachtigok (Connecticut)," MissInd 331.2 and 331.3, MAB. In this manuscript, it is identified as having been sung to the chorale tune "Herr Jesu Christ! dein Tod," also known as "In dulci jubilo." From "Pattern for a Song Booklet of the Blessed Hearts of the Brown Nations of the Mahican Delawares and Some Short Verses of the Language of the Six Nations [Mohawk]," NB.VII.R.3.91 (1746), UA Herrnhut. English translation by Masthay, *Mahican-language hymns, biblical prose, and vocabularies from Moravian sources.*

5. Peter Charles Hoffer, Olivia Bloechl, Richard Cullen Rath, and Leigh Eric Schmidt have written on the relationship between everyday life in the American colonies and the belief of both European colonists and Native Americans in the power and omnipresence of the supernatural within the physical world. Manifestations of angels, demons, ghosts, witches, sea monsters, and the invisible spirits of disease and pestilence were regularly

experienced as a common part of daily life: "[In the] American colonies, the invisible world wore a different cast than it does today. It was just as close—indeed, it overlapped the visible world—but its presence was discerned in a different fashion and its impact felt in different ways. . . . Then, everyone . . . conceded the potency of the unseen. The invisible crossed into the visible, and the unseen had power over everyday life." Hoffer, *Sensory Worlds of Early America*, 77–78; also see Bloechl, *Native American Song at the Frontiers of Early Modern Music*; Rath, *How Early America Sounded*; and Schmidt, *Hearing Things: Religion, Illusion, and the American Enlightenment* (Cambridge, MA: Harvard University Press, 2000).

6. As Belden Lane has argued in *Landscapes of the Sacred*, the imaginative aspects of space and landscape in religious life cannot be underestimated. It is through the religious imagination that practitioners create places or landscapes that embody experiences of the divine. In the history of Christian thought, in general, landscapes were often spiritually conceived. These imaginative places represented and concretized experiences of faith, creating "theologies of place." God could, in fact, be imagined as a place. Belden Lane, *Landscapes of the Sacred: Geography and Narrative in American Spirituality*, Isaac Hecker Studies in Religion and American Culture (New York: Paulist Press, 1988), 238–245. For Moravians, hymns and songs that encouraged worshippers to "live inside the body [of Christ]," or to become "little wound bees that burrow in the Side Wound [of Christ]," constituted an important spatial dimension of Moravian religious experience.

7. Pausa is a village in the Vogtlandkreis district of Saxony, near Plauen and Schleiz. Johann Christoph Pyrlaeus, *Lebenslauf*, XII 1785, MAB.

8. Johann Christoph Pyrlaeus, *Lebenslauf*, XII 1785, MAB; translated in Albert G. Rau, "The Autobiography of Johann Christoph Pyrlaeus," *Transactions of the Moravian Historical Society* 12, no. 1 (1938): 18.

9. Rau, "The Autobiography of Johann Christoph Pyrlaeus," 19.

10. Rau, "The Autobiography of Johann Christoph Pyrlaeus," 19.

11. Rau, "The Autobiography of Johann Christoph Pyrlaeus," 19.

12. As discussed in the Introduction, Shekomeko was a Mohican village on the banks of Shekomeko Creek, now located near Pine Plains, NY, in Dutchess County. The Moravian missionary Christian Heinrich Rauch founded a mission there in 1740, a year before the construction of the First House in Bethlehem. Members of the Mohican community in Shekomeko, including Joshua, became the first Native American Moravian congregation. For a detailed discussion of the subsequent history of this Mohican Moravian community, including a more detailed narrative of Joshua and his family, see Wheeler, *To Live Upon Hope*. For a historically informed and well-researched account of Joshua's life, modeled on the tradition of the Moravian Lebensläufe, see Rachel Wheeler, "An Imagined Mohican-Moravian 'Lebenslauf,' Joshua, Sr., d. 1775," *Journal of Moravian History* 11 (2011): 29–44. I would like to thank Rachel Wheeler for providing much of the information on Joshua's life that informs this section of the chapter. There is also a brief biography of Joshua's son, Joshua Jr. (1741–1806), in Hartzell, "Joshua Jr., Moravian Indian Musician."

13. "It was common when a person passed to adulthood to seek the favor of a particular spirit. In exchange for their devotion, the spirit promised to care for the individual. . . . Knowledge of how to engage the lesser spirits was mostly passed through families and directly from the spirits." Wheeler, "An Imagined Mohican-Moravian 'Lebenslauf,' Joshua, Sr., d. 1775," 33.

14. See Bobby Lake-Thom, *Spirits of the Earth: A Guide to Native American Nature Symbols, Stories, and Ceremonies* (New York: Plume, 1997), 26–27, for descriptions of animal and spirit helpers.

15. Wheeler, "An Imagined Mohican-Moravian 'Lebenslauf,'" Joshua, Sr., d. 1775," 33.
16. Wheeler, "An Imagined Mohican-Moravian 'Lebenslauf,'" Joshua, Sr., d. 1775," 33–34.
17. Wheeler, "An Imagined Mohican-Moravian 'Lebenslauf,'" Joshua, Sr., d. 1775," 34.
18. Later in his life, Joshua told one visitor to Gnadenhütten, Pennsylvania, that he wondered why the white people, who since childhood were able to "read in the Bible about the savior that died for us on the cross, and shed his blood, and wants to make us holy and yet they remain dead and cold in their hearts," while the Indian people, "do not know much, yet we love the Savior." When asked by the same visitor if he could read, Joshua responded, "Yes, not much, but I know five important letters and my brothers in Gnadenhütten know these letters too. I read these letters day and night, and when I go in the woods to hunt, and I shoot deer, so I read these letters, not with my head, but with my heart." Joshua then showed the visitor his picture of Jesus, pointed to the wounds and said, "See, these are my five letters that I love to read. They give me strength and power in my heart. I think about them all day long, wherever I am." "Br. Josua besuchte der Br. Martin u. gab ihm Nachricht, so wol von ihm als den andern Brüder, die mit ihm auf der Jagd gewesen waren daß sie alle Abend hätten ihre Singstunde mit einander gehalten, u. er hätte den Brrn auch manchmal was vom Hld gesagt. Sie wären stille u. gehörsam gewesen, hätten auch vieles erzählt, wie verschiedene von ihnen einen Zeit her schlecht gelebt hätten, u sich gar manche Seligkeit, die sie hätten geniessen können verdorben." *Gnadenhütten Diary*, January 26, 1754, MissInd 118.1, MAB.
19. Lake-Thom, *Spirits of the Earth*, 178–188.
20. Lake-Thom, *Spirits of the Earth*, 182. Also see Elizabeth Tooker, ed., *Native North American Spirituality of the Eastern Woodlands: Sacred Myths, Dreams, Visions, Speeches, Healing Formulas, Rituals, and Ceremonials*, Classics of Western Spirituality (New York: Paulist Press, 1979).
21. Wheeler, "An Imagined Mohican-Moravian 'Lebenslauf,'" Joshua, Sr., d. 1775," 34.
22. Joshua, and other Native Christians, would likely have understood the story of the Transfiguration, as transmitted in Matthew 17, in terms of its relationship to the natural environment of the mountaintop on which the event transpired: "After six days Jesus took with him Peter, James, and John the brother of James, and led them up a high mountain by themselves. There he was transfigured before them. His face shone like the sun, and his clothes became as white as the light. Just then there appeared before them Moses and Elijah, talking with Jesus. Peter said to Jesus, 'Lord, it is good for us to be here. If you wish, I will put up three shelters—one for you, one for Moses and one for Elijah.' While he was still speaking, a bright cloud covered them, and a voice from the cloud said, 'This is my Son, whom I love; with him I am well pleased. Listen to him!' When the disciples heard this, they fell facedown to the ground, terrified. But Jesus came and touched them. 'Get up,' he said. 'Don't be afraid.' When they looked up, they saw no one except Jesus." Matthew 17:1–8 NIV (New International Version). Also see Lake-Thom, *Spirits of the Earth*, 184.
23. For an account of this trip, see the first chapter of this book, and Zinzendorf, "Narrative of a Journey to Shecomeko, in August of 1742," in *Memorials of the Moravian Church*, especially 56–57.
24. During Joshua's first visit to Bethlehem, he most likely stayed in the Gemeinhaus, because the Single Brothers' House was not completed.
25. Levering, *A History of Bethlehem*, 143.
26. The improvisation of hymns was a musical parallel to the process of communicating with Christ through the lot. For a discussion of lots and their relationship to the improvisation of hymns, see the chapter "Faith, Reason, and Lots," in Eyerly, "'Singing from

the Heart'"; and Eyerly, "*Der Wille Gottes*: Musical Improvisation in Eighteenth-Century Moravian Communities."

27. "From the *Singstunden* on, people sang throughout the day.... Thus, they lived in song [Von den Singstunden aus sang man durch den Tag.... So lebte man im Lied]." Wollstadt, *Geordnetes Dienen in der christlichen Gemeinde*, 84–85.

28. For children born into Moravian communities, instruction in hymn singing began *in utero*: mothers were to sing a special collection of hymns intended for the unborn. Once born, children were divided into groups by age and gender and received further daily instruction in theology and hymnody through these groups. See Craig Atwood, "The Union of Masculine and Feminine in Zinzendorf's Piety," in *Masculinity, Senses, Spirit*, ed. Katherine Faull, 29–30; and Bettermann, *Theologie und Sprache bei Zinzendorf*.

29. See Reichel, *Dichtungstheorie und Sprache bei Zinzendorf*: 12. Anhang zum Herrnhuter Gesangbuch.

30. "Nach der Singstd. kamen die lieben Brüder von dem MühlPlaz. (alwo sie ein lediges Brüder Haus gebauet haben und da beisammen wohnen, welches auf einer Seite mit Baum Rinde, auf der andern Seite mit grünen Reißig bedeckt ist) durch den Busch herauf mit ihren Wald Hörnern und Trompeten und andrer Music und machten sich recht lustig in Gnadenhütten, mit schönen Verseln: es hörte sich gar schöne an hinter den grossen Bergen und in dem dicken Busch, so was lieblich tönen zu hören." *Gnadenhütten Diary*, May 24, 1747, MissInd 116.1, MAB.

31. "Ein alter ehrwürdiger Mann aus dieser Gemeine führte mich auch in ihr Bethaus, um da ihrem Gottesdienste mit beyzuwohnen. Die edelste Simplicität des Gebäudes, die vollkommenste Stille der Zuhörer, auf deren Gesichtern sich die Allgegenwart Gottes, die zärtlichste Liebe zu ihm und dabey eine gewisse Ruhe mahlte, die unser Körper nur hat, wenn die Seele in der äussersten Arbeit ist; zu diesen noch die einfachste Musik, und ein reiner ungekünstelter Gesang flößten mir eine gewisse andächtige und selige Empfindung ein, die ich in unsern gewöhnlichen Kirchen noch nie empfunden habe. Und dennoch kam meine Andacht der ihrigen nicht gleich." Johann Friedrich Reichardt, *Briefe eines aufmerksamen Reisenden die Musik betreffend* I (Hildesheim: G. Olms, 1977), 47–49.

32. "Nun fing man diese heilige Handlung mit mehreren Gesängen an, die sehr langsam, leise und melodisch gesungen wurden.... Nach dieser Rede legten sich alle Brüder und Schwestern mit dem Gesicht auf die Erde, um das unbefleckte Lamm anzubeten und blieben in dieser Stellung so lange, wie das 'Te deum laudamus' dauerte, das sie zusammen in 2 Chören sangen, d.h. die Brüder fingen eine Strophe an und die Schwestern beendigten sie. Sie sangen dieses 'Te deum' in einer so melodischen und leisen Weise, daß man die Stimmen der Engel, der Erzengel und Cherubinen zu hören glaubte, die sich vor dem Angesichte des Lammes im Himmel niedergelegt haben. Die Stellung der Brüder und Schwestern, die alle auf dem Boden lagen, das Angesicht auf der Erde, und dieser melodische Gesang haben mich derart entzückt und sich mir bis in die Tiefen meines Herzens eingeprägt, so daß ich meine Freudentränen nicht zurückhalten konnte, weil ich noch eine Kirche auf der Erde sah von jesustreuen Christen, die ihm die Ehre geben, die ihm gegeben werden muß." Von Heinitz, quoted in Helmut Hickel, *Das Abendmahl zu Zinzendorfs Zeiten* (Hamburg: L. Appel, 1956). See also Vogt, "Festive Stillness: The Sound of Moravian Music in the Perception of Outside Visitors."

33. "Von dem Gottesdienst der Gemeine würde ich gar nichts gesehen haben, wenn es nicht zum Glück einem alten ehrlichen Herrnhuter eingefallen wäre, sich gerade heute begraben zu lassen. Die Thaten des guten Mannes beschränkten sich während seines ganzen sechs und siebenzigjährigen Lebens auf nichts, als auf die Verfertigung von Siegellack

und englischem Pflaster, deshalb sahen wir aber doch die ganze Gemeine ihm zur letzten Ehre zwischen den weißen kahlen vier Wänden des hohen Betsaales versammelt. Der eisgraue Pfarrer setzte sich ganz bequemlich in einen mächtigen Großvaterstuhl; sein ziemlich unverständlicher, aber gewiß gut gemeinter Vortrag der Lebensgeschichte des Verstorbenen machte indeß auf mich keinen sonderlichen Eindruck, einen desto tieferen der leise harmonische Gesang der Gemeine. Dieser ist das Rührendste, Herzergreifendste, was ich jemals gehört habe, jeder Ton spricht mächtig das Gefühl der reinsten Andacht, der demüthigsten Ergebung und Gottesverehrung aus. So hat noch keine Kirchenmusik mein heiligstes Gefühl erregt, wie dieser einfache Gesang, und wenn sie noch so herrlich vom hohen Dome wiederhallte." Johanna Schopenhauer, *Ausflucht an den Rhein und dessen nächste Umgebung: im Sommer des ersten friedlichen Jahres* (Leipzig: Brockhaus, 1830), 254. Quoted in Ekkehard Langner, "Eine Ortsgemeine um 1800—Die Herrnhuter in Neuwied in Reiseberichten der Zeit," *Unitas Fratrum* 4 (1978): 61. See also Vogt, "Festive Stillness: The Sound of Moravian Music in the Perception of Outside Visitors," 20.

34. In his essay on Zinzendorf, Johann Gottfried Herder also noted the communal nature of Moravian hymnody: "In the more selected songs, when they seem to be the direct language of the heart, when many people and everyone moves and sways in One harmony, then the song is rightly the key to a community, that should be 'a collection of souls.'" Johann Gottfried von Herder, "Zinzendorf," in *Johann Gottfried von Herder's sämmtliche Werke*, ed. Johann Georg Müller (Tübingen: Cotta, 1828), XXIV, 37.

35. See Quintillian, *Institutio oratoria*, 11.2:22733; and Fortunatianus, *Artis rhetoricae libri*, III, in Halm, ed., *Rhetores latini minores*, 129. Translated by Mary J. Carruthers, *The Book of Memory: A Study in Medieval Culture*, 86. See also Carruthers, *The Book of Memory* (Cambridge: Cambridge University Press, rev. 2014), 88–121.

36. For a detailed discussion of the Moravian method of improvisation, and its relation to literacy and memorization from written sources, see Eyerly, "*Der Wille Gottes*: Musical Improvisation in Eighteenth-Century Moravian Communities."

37. The following passages, from a sermon given by Zinzendorf on the occasion of the introduction of the *Londoner Gesangbuch* to the Moravian congregation in London in 1754, describe the mnemonic properties of meter and rhyme: "The practice of cloathing divine Thoughts in Metre, is perhaps as universal as Speech itself; and has two Grounds for it. First, that when our Affections are strongly moved, which surely Religion may be allow'd to do, singing or a sort of Modulation of the Voice is what the Heart naturally chooses to vent itself by. Secondly, that the comprizing of important Truths or Counsels in Verse, is a Help to their being remember'd, and a Kind of *Memoria technica*.... Rhymes are a very good Thing,... they render the Elocution more melodious and are a present help to Memory. For it would be impossible for our little Children to hold these Speeches of an Hour long out of their Hearts, if the Verses were not rhymed. They help memory, make a deeper Impression upon their little minds; and what is so learnt by heart is not to be so easily forgotten. That is the true Clash with reason, why we cannot enter into the renewed Way of Versification without rhyming." Sermon given by Zinzendorf on 2 September, 1754, in London. Johannes de Watteville copied these passages into the blank pages of his copy of the *Londoner Gesangbuch* (1754). This copy is preserved in the Unity Archives, Herrnhut.

38. Alice May Caldwell, "Music of the Moravian *Liturgische Gesänge* (1791–1823): From Oral to Written Tradition" (PhD diss, New York University, 1987), 82.

39. Robert Harold Adams, *Songs of Our Grandfathers: Music of the Unami Delaware Indians* (Dewey, OK: Touching Leaves Indian Crafts, 1977), 15.

40. Adams, *Songs of Our Grandfathers: Music of the Unami Delaware Indians*, 16.
41. Adams, *Songs of Our Grandfathers: Music of the Unami Delaware Indians*, 17.
42. Adams, *Songs of Our Grandfathers: Music of the Unami Delaware Indians*, 17.
43. For a description of the festival, see the *Jüngerhausdiarium* (1748), and R.4.C.III.7.b, UA Herrnhut. Also see Peucker, "'Inspired by the Flames of Love': Homosexuality, Mysticism, and Moravian Brothers around 1750," *Journal of the History of Sexuality* 15, no. 1 (2006): 30–64; and A. P. Hecker, *Gespräch eines Evangelisch-Lutherischen Predigers* (Berlin: Buchhandlung der Realschule, 1751), 47–50.
44. Zinzendorf, "Wundenlitanei Homilien," in "Reden während der Sichtungzeit in der Wetterau und in Holland," *Hauptschriften* 3, no. 1, 8. Quoted in Atwood, *Community of the Cross*, 101.
45. Zinzendorf, *Nine Public Lectures on Important Subjects in Religion, Preached in Fetter Lane Chapel in London in the Year 1746*.
46. See Lane, *Landscapes of the Sacred*, 244.
47. Orchi, "La passione per il Venerdì Santo," in *Prediche quaresimali*, 409–10. Quoted in Camporesi, *Juice of Life: The Symbolic and Magic Significance of Blood* (New York: Continuum, 1995), 55–56.
48. Camporesi, *Juice of Life*, 70.
49. The rite of Holy Communion was an important communal ritual for Moravians, as well as for many other Christian communities. It was an act of physical transcendence, described by Camporesi as "the darksome, complex rite in which the inexplicable transubstantiation of wine into blood is effected by the ritual invocation." Camporesi, *Juice of Life*, 59.
50. "From the fountain of suffering, from the fountain of blood from the heart ground of the little Lamb, there is always something supernatural, that is connected with the same Creator's Spirit and grace, that brings life into all conversation and association, in the love among one another, in all life from Jesus' blood." Zinzendorf, "Gemeinreden," in *Hauptschriften* 4, no. 44, 237. Quoted in Atwood, *Community of the Cross*, 100–101.
51. *Synode Protokoll* (1745), 142. Later published in the *Herrnhuter Gesangbuch* as hymn 2085, verses 1 and 14.
52. "Zinzendorf und der Heiland in sechs allegorischen Vorstellungen [Zinzendorf and the Saviour in Six Allegorical Scenes]." Devotional painting (1748), artist unknown, TS.Mp.76.12, UA Herrnhut.
53. According to Moravian theologian Craig Atwood, even the sweat of Christ's body was considered to be sacramental: "Every drop of the crucified Christ's sweat was a special blessing. In this way, it functioned essentially like holy water." Atwood, *Community of the Cross*, 183.
54. Of parallel importance to the "blood and wounds" theology of the Moravians is the tradition of bridal mysticism which flourished in the 1740s in such communities as Herrnhut and Herrnhaag. See Peucker, "The Songs of the Sifting. Understanding the Role of Bridal Mysticism in Moravian Piety during the Late 1740s," *Journal of Moravian History* 3 (2007): 51–87; and *A Time of Sifting*.
55. This hymn text, which appears in the picture just above Christ's grave, was composed by Zinzendorf on December 4, 1747, and later published in the *Herrnhuter Gesangbuch* as hymn 2325.
56. Artist unknown, Andachtsbilder Mappe 9. TS Mp.375.9, UA Herrnhut.

57. Burkhard Dohm, "Des Blutes Licht-Tinctur: Alchimistische Koncepte in Herrnhutischer Poesie," in *Künste und Natur in Diskursen der Frühen Neuzeit*, ed. Hartmut Laufhütte et al., Wolfenbütteler Arbeiten zur Barockforschung (Harrassowitz Verlag, 2000), 1173. Zinzendorf was most certainly introduced to the philosophies of Böhme through his grandmother, Henriette Katherina von Gersdorf. See Erich Beyreuther, *Der junge Zinzendorf*, vol. 2. (Marburg an der Lahn: Verlag der Francke-Buch-handlung, 1957), 51. See also Dohm, *Poetische Alchemie: Öffnung zur Sinnlichkeit in der Hohelied- und Bibeldichtung von der protestanischen Barockmystik bis zum Pietismus*, Studien Zur Deutschen Literatur 154 (Tübingen: Max Niemeyer Verlag, 2000).

58. "Contemplating a natural world pervaded literally and metaphorically by the sounds, smells, and images of the Holy Spirit, mystics [such as Böhme] remained hopeful that human beings, too, could regain a glimpse of that spirit by penetrating the enduring links between divinely created systems such as star constellations, music, mathematics, and even language." Erben, *Harmony of the Spirits*, 204.

59. Erben, *Harmony of the Spirits*, 235–236.

60. Erben, *Harmony of the Spirits*, 235–236.

61. This terminology is found in many Moravian hymns, including hymns 1267 and 1316 in the *Herrnhuter Gesangbuch*. Paul Peucker's recent book, *A Time of Sifting*, discusses the theological and social meanings of the "blood and wounds" and "mystical marriage" terminology found in Moravian hymns from the 1740s and 50s on both side of the Atlantic.

62. "Nearly every aspect of life in Bethlehem was incorporated into communal rituals in order to bring the secular into the sacred sphere by connecting daily life to the life and death of Jesus (Atwood, *Community of the* Cross, 157)."

63. For more information on the Moravian *Blutgemeine*, see Atwood, "Zinzendorf's 'Litany of the Wounds,'" *The Lutheran Quarterly* 11 (1997): 189–214.

64. "Wir sehen den ganzen tag in die Wunden JEsu (sic) hinein . . . Er bleibt uns immer im Gesicht mit seinem Wunden-Ritzen, in seiner Nägel-Maale Licht, da pflegen wir zu sitzen . . . das bleibt uns beständig vor Augen, das distinguirt uns von allen andern Menschen, das Gesicht leuchtet vor unsrer Stirn, es blickt uns aus unsern Augen heraus." Zinzendorf, *Die drey und dreyßigste Homilie*, 358. Quoted in Dohm, "Des Blutes Licht-Tinctur," 1175.

65. Katherine Faull, "Faith and Imagination: Nikolaus Ludwig von Zinzendorf's Anti-Enlightenment Philosophy of Self," in *Anthropology and the German Enlightenment: Perspectives on Humanity*, ed. Katherine Faull (Lewisburg, PA: Bucknell University Press, 1995), 24–27; and Katherine Faull, "Imagining and Learning: Utopian Visions in Early Moravian Communities," paper presented at Moravian College, Bethlehem, PA, 21 April, 2006. Faull applies Foucault's philosophies of the self to the Lebensläufe, however the same ideas also apply to the Singstunden. See Leigh Eric Schmidt, "Sound Christians and Religious Hearing in Enlightenment America," in Smith, ed., *Hearing History*; and Schmidt, *Hearing Things: Religion, Illusion, and the American Enlightenment*.

66. HG 2144.

67. The Moravians were certainly not the only spiritual community or religious movement to recognize the power of song to create space and place. Song has often been employed as an anchoring point for a spiritual geography that includes both human and divine elements. As Gary Tomlinson has observed in his work on the religious songs of the

Inca's: "Human song signals a location in place and time.... [it] joins people in expressive communion ... [and] facilitate[s] contact with the divine itself.... The sacred is, at root, an impulse to find oneself in a dwarfing cosmos, and song, marking humans' places in the world, comes to be the medium of choice for sacred utterance in countless societies.... The act of singing one's place in song ranges widely. It reaches out through the mundane landscape traversed in human lives to encompass the heavens ... and invisible realms." Tomlinson, *The Singing of the New World*, 124.

68. For an excellent overview of Moravian deathways, as well as the deathways of other European and Native American communities in the American colonies, see Erik R. Seeman, *Death in the New World: Cross-Cultural Encounters, 1492–1800*, Early American Studies (Philadelphia: University of Pennsylvania Press, 2010).

69. Moravians tended to describe death as a "verwandelung des Fleisches [transmutation of the flesh], after 1 Cor. 15:44 (NIV): "we are sown a natural body, and raised a spiritual body."

70. Bettermann, *Theologie und Sprache bei Zinzendorf*, 19.

71. Peucker, *Herrnhuter Wörterbuch: Kleines Lexicon von brüderischen Begriffen* (Herrnhut: Unitätsarchiv Herrnhut, 2000), 23.

72. "Diesen morgen um 2 Uhr wurden Geschw Macks zu der kranken Esther geruffen, weil es scheine, sie würde heimgehen. Wir sungen ihr einige Versgen, sie zeigte uns auch gleich die Wunden, die Hld. hätte in seinen Hände, sie erhohlte sich auch bald wieder, u. weil ihre Freunde alle da waren, fing sie an u. sagte: lieben Freunde! Ihr seht das ich krank bin, u. werde bald zum Hld gehen ich freue mich darauf von Herzen u. danck dem Hld, daß er mich wieder hat nach Gnadenhütten gebracht, u. hat mich sel. gemacht, das war seine Absicht, warum er mich hier hergebracht, u. warum ich auch her gegangen bin, um Ihm meines ganzes Herz zu geben. In ihrer harten u. scheren Kranckheit legte es sich bey allen vorkommenden Umständen deutl. dar, das ihr Herz [verliebt] war in die Marter [?] des Hlds. Ja wenn sie in den grösten Schmerzen lag, u. man sung ihr Wunden Versgen, wies sie uns gleich wo Hld die Wunden hätte." *Gnadenhütten Diary*, MissInd 118.1, March 5, 1754, and Schmick March 5 and March 6, 1754, MAB.

73. Zinzendorf's contemporary, Immanuel Swedenborg, also wrote of the relationship between the singing of the angels and the songs of mortal Christians in his *Spiritual Diary* of 1748: "Through the alchemy of hymn singing, the angels 'had been forming a golden crown with diadems about the head of our Saviour.'" Quoted in Schmidt, *Hearing Things: Religion, Illusion, and the American Enlightenment*, 218. According to Schmidt, "What most caught Swedenborg's attention was ... the wonders of how all the angels sang together.... Spontaneous in their affections, yet marked by unanimity, those who performed these canticles were utterly devoid of self-love. Sometimes the delights of their harmonies were so overwhelming as to be entrancing.... In this representation of the heavenly choirs, Swedenborg offered a model of the self's engulfed subjection to the divine whole, and an exaltation of Christian mutuality over rational autonomy (218–219)."

74. The story of Zinzendorf's death was recorded in copies of the *Gemeinnachrichten* (*Jüngerhausdiarium*) for May 1760, MAB and UA Herrnhut.

75. Rau, "The Autobiography of Johann Christoph Pyrlaeus," 25.

76. Hartzell, "Joshua Jr.: Moravian Indian Musician," 1–19. For a detailed discussion of Joshua's life, see Wheeler, *To Live Upon Hope*, chapter 12, "The Cooper and the Sachem."

77. Two of those murdered at the second Gnadenhütten massacre were the daughters of Joshua Jr., Joshua Sr.'s granddaughters. It has also been speculated that another victim of the

massacre was Joshua's spinet, which would have burned during the destruction of the village. Hartzell, "Joshua Jr.: Moravian Indian Musician," 12.

78. Lawrence Henry Gipson, ed., *Moravian Mission on the White River: Diaries and Letters, May 5, 1799, to November 12, 1806*, Indiana Historical Collections 23 (Indianapolis, IN: Indiana Historical Bureau, 1938), 621.

79. Gipson, *Moravian Mission on the White River*, 417–418, 561; and Heckewelder, *A Narrative of the Mission of the United Brethren*, 413. "Joshua's last musical performance seems to have been hymn singing during his final confinement. As has been the case over and over again, the singing of hymns during one's last moments on earth provided a source of comfort for Moravians, so it was for Joshua." Hartzell, "Joshua Jr.: Moravian Indian Musician," 13. According to missionary Johann Heckewelder: "Joshua, before, and after being placed on the burning pile, prayed most fervently to God his Saviour, and continued either praying or singing praises to the Lord, until his strength was exhausted, and death closed his career." I would like to thank Rachel Wheeler for sharing her unpublished research on Joshua and his time in Indiana, including information on his death.

MORAVIAN RUN

In the summer of 2015, while packing up a lifetime of objects and furniture in preparation for closing up my parents' farm in Cooper Township, Pennsylvania, I came across my mother's old yearbook, *The Cotohisc*. It was vol. 10, no. 9, dated May 28, 1954—a celebratory marker of the school year's end at the nearby Cooper Township High School. In a section entitled "A Short History of Cooper Township," a particular passage struck me:

> Long after the Revolutionary War the area which was to become Cooper Township was still a vast forest filled with deer, bear, wolves, panthers, and many other wild animals. A trail between the Indian villages at Lock Haven and Clearfield crossed the Moshannon near Peale and passed through the township by way of what was to become Kylertown. In 1758 Frederick Post, a worker among the Indians, traveled this trail with two Indians on his way to Fort Duquesne. They camped overnight on an island in the Moshannon near Peale. The island is now known as Post Island. In 1772 a band of about a hundred Moravians, traveling from Bradford County to Ohio under the Leadership of John Ettwein, passed over this trail. Some of the party died on the journey and were buried near Peale by the small stream which is now called Moravian Run.

Interspersed with the familiar names of Peale, Moravian Run, and Post's Island, I immediately recognized the names of Johannes Ettwein (1721–1802) and Christian Frederick Post (1710–1785), missionaries with the Moravian Church in the eighteenth century. Was *The Cotohisc* correct? Were the island and stream near my parents' farm connected with the Moravian Church? Could I unravel the mysteries of Moravian Run and my grandfather's beloved fishing spot on Post's Island? Thanks to the previous fourteen years of examining Moravian archival records on both sides of the Atlantic, I knew exactly where to search for the answers.

I began with the works of Paul Wallace, the historian who had translated Johann Jacob Eyerly's travel diary. Wallace had located that diary while completing research for his two best-known books: *Indians in Pennsylvania* (1981) and *Indian Paths of Pennsylvania* (1965). According to Wallace, there was a trail that passed through the area of Cooper Township. It threaded its way over the Allegheny Front along the West Branch of the Susquehanna River, which led from Shamokin and Madame Montour's settlement of Otstonwakin to the Great Island at Lock Haven and the "Bald Eagle's Nest." From there, it crossed the mountains to a sleeping place called "Snow Shoe" and eventually to Moshannon Creek. At Moshannon Creek, the path split into two branches. The southern branch crossed the creek and continued to where it swung around the bend of a mountain to the place called "Post's Island." From there, it ran through the area of the abandoned mining village of Peale, across Moravian Run, and through the modern village of Grassflat to the towns of Kylertown and Clearfield (Chinklacamoose).

The Great Shamokin Path was one of the most important avenues of connection for Native communities on both sides of the Alleghenies for hundreds, perhaps even thousands, of years. It was along this path that both Christian Frederick Post and Johannes Ettwein had traveled in 1758 and 1772. In their travel journals, I discovered numerous references to the forests, mountains, and streams that were now a part of "Cooper Township" and the landscapes that surrounded my parents' farm. Post's journey, along with western Delaware representatives Pisquetomen and Keekyuscung, had taken him to the town of Kuskusky (New Castle) with wampum belts from the Pennsylvania government in Philadelphia and the eastern Delaware tribes headed by Teedyuscung. These belts were meant to ensure peace on the western border of colonial settlements following the Penn's Creek and Gnadenhütten, Pennsylvania, massacres. Post was chosen as the "white" emissary due to his ability to speak multiple Native languages as well as English and German. His linguistic abilities had been nurtured by his first two wives, Rachel (Wampano) and Agnes (Delaware), and his multiple mission assignments. Born in Prussia, Post had served the Moravian Church in Pennsylvania, Ohio, New York, Connecticut, Labrador, and Nicaragua. During his journey along the Great Shamokin Path, Post had camped for the night at an island in the Moshannon Creek, in an area he described as "wild, broken, and mountainous." Somewhere between the island that now bore his name and Chinklacamoose, his diary recorded "wolves making a terrible music."

Johannes Ettwein's journey was undertaken along with church elders such as Joshua Sr. and Johann Roth, to remove the Native Moravian congregation at Friedenshütten on the north branch of the Susquehanna near Wyalusing from the increasingly tense standoff between European and Native settlements in the areas north of the Wyoming Valley. The church had sought permission from Delaware communities in Ohio to remove the congregation beyond the boundaries of colonial settlement. There, it was hoped, they would at last be safe after having been decimated by disease and religious and ethnic hatred in Shekomeko, Bethlehem, Gnadenhütten, Nain, Philadelphia, and two different settlements named Friedenshütten. For more than thirty years, this Native Moravian congregation, which included Joshua Sr. and Joshua Jr. and many of the people whose stories, words, and songs have filled this book, had moved to various places in eastern Pennsylvania. In Ohio, they hoped at last to find a home.

But the journey from Friedenshütten to Ohio over the Great Shamokin Path was filled with death and starvation. ⊕ **See website chap4.3, Interactive map: "The Journey of the Native Gemeine from Friedenshütten II to Friedenstadt."** They were tormented by violent summer thunderstorms, rocky streams, mud slides, and bugs that bit like "fire." The crossing of the Allegheny Front was almost unbearable for the women, children, elderly, and sick. Somewhere near Snow Shoe, Johann Roth's travel diary began to record the story of a young boy baptized with the name Nathan: "3rd July, 1772. A sick boy, when visited by me, expressed his desire to be washed with the blood of Jesus. 'In my heart I pray God that he may grant me this,' he said with a soft heart. Hence his ardent desire was fulfilled when this afternoon on his sick-bed, in holy baptism, he received the name of Nathan." Ettwein, too, was struck by Nathan's story and included it in his own diary: "A poor little cripple, aged ten years, a son of the late Jonas, whom his mother had carried all the way in a basket from one station to another, was very weak today, and expressed the wish to be baptized. Bro. Roth administered the sacrament and named him Nathan." Despite the joy of the travelers at the "great abundance of fish in the rivers and brooks" and a "peculiar turtle, the size of a goose, with a long neck, pointed head, eyes like a dove, scales on its back," and a covering that was soft, leathery, and liver colored, the next few days of travel were especially treacherous. On July 8, Ettwein noted: "Advanced six miles to the West Moshannek over precipitous and ugly mountains, and through two nasty rocky streams." By the next day, their progress was completely halted: "July 9. Advanced but two miles to a run in the swamp. July 10. Lay in camp, as some of our horses had strayed."

As they lay stuck in that "swamp" on the night of July 10, beyond Post's Island and near a small stream that ran into the Moshannon from the rocky hillside above, Nathan quietly died: "In the night from the 10th to the 11th, Nathan, mentioned above as having received baptism, died in his sleep. For quite some time he had been ill with an infected foot, and in consequence of it, he had been reduced to mere skin and bones, and had to be carried in a basket; from one day to the next his troubles were ended by the dear Savior's calling him home into the Happy Kingdom; he was the son of the late Brother Jonas. His face in death bore a serene and happy expression; in lack of a coffin, the body was wrapped in tree bark and was buried." His grave was also marked by Ettwein according to the custom of recording important events in trailside tree carvings and paintings: "July 11th. We found Nathan released from all suffering—his death had been unobserved. His emaciated remains were interred along side of the path, and I cut his name into a tree that overshadowed his lonely grave." Although neither diary mentions the singing of hymns, it is almost certain that Nathan's burial was attended by the singing of many hymns in honor of his "journey home." The next morning, Nathan's mother, Joshua Sr., Joshua Jr., Roth, Ettwein, and the congregation traveled on, leaving Nathan in that "lonely" place at the confluence of Moshannon Creek and Moravian Run. 🌐 **See website chap1.3, Interactive sound map: "The Great Shamokin Path (Moravian Run)."**

Two centuries later, Nathan's story lived on without a stone or a coffin. The residents of Cooper Township still remembered his grave, even if they had forgotten his Christian name. The passing of the Moravian congregation and the delegation of Christian Frederick Post were imprinted in the names of islands and streams. *The Cotohisc* was correct. These places—Peale, Moravian Run, Post's Island—were all connected with the Moravian Church and with the very group of people whose stories I had committed to record in this book. Without any knowledge or forethought, I had somehow come full circle from those 250-year-old stories of Moravian Christians to the very forests and streams I had loved as a child. As I packed up my parents' lifetime of possessions, I could see the Moshannon Mountains and the edge of the Allegheny Front from the kitchen window. How many times over the past forty years had I unknowingly looked out of that same window to the place of little Nathan's grave and the winding contours of the Great Shamokin Path, now vanished into the undergrowth of the forest or submerged under the pavement of US Route 53? I had a strange sense that this was but one part of a longer, more intricately connected tale of landscape, hardship, singing, home, and family—a story that had been sounded and would continue to be sounded in this place for generations to come.

4

1782

> For there they that carried us away captive required of us a song;
> and they that wasted us required of us mirth, saying,
> Sing us one of the songs of Zion.
> How shall we sing the Lord's song in a strange land?
>
> —Psalm 137: 3–4

IN THE EARLY 1760S, IN RESPONSE TO FINANCIAL pressure from the Moravian Church in Europe and the incursion of the Seven Years' War on Bethlehem, church elders voted to dissolve the communal Economy. Bethlehem's thriving industries lost money, and the resultant economic trauma curtailed the expansion of further mission settlements, as well as the church's ability to financially support existing outlying missions. Elders also initiated a lease program that allowed Moravian families to rent land from the church, and to build single-family homes and farms on church-owned property. This shift in the spatial distribution of Bethlehem's people had a profound impact on the soundscapes that had characterized religious and social life during the first twenty years of the settlement. But it was not only these internal changes that altered the soundscapes of Bethlehem. Surrounded by increasingly numerous settlements in the Delaware and Susquehanna River Valleys, the spiritual soundscapes of Bethlehem were shifting in favor of the bustle of independent homes and businesses, farmsteads, and government-controlled navigable roadways that accompanied the westward trajectory of Pennsylvania's colonial expansion. Forests were clear cut, and towering stands of oaks, beeches, and chestnuts were shipped en masse to wood-starved Europe. Pennsylvania's landscape was changing rapidly, and with it the natural soundscapes that had characterized life for early settlers and Native Americans.

* * *

In 1783, the German botanist and zoologist Johann David Schöpf took the King's Road that ran from Philadelphia to the ford of the Lehigh River near Bethlehem's Crown Inn. Ever keen to observe and collect plants, seeds, and bird and animal specimens, his travel notes are remarkably specific. On his way from Philadelphia, he noted forests that were "in large part composed of the several kinds of North American oaks, the sassafras, tulip-tree, sour gum, chestnut, birch, wild-ash, and others." Still he lamented the lack of birds and "indigenous quadrupeds." There were very few flowers and "no great variety of plants." He despaired of finding seeds and other useful cultivatable plants: "Nothing thus far which as a product of the country might be highly recommended for adoption in other lands." The soil, he noted, was only of "moderate goodness." And the few houses he met along the road were simply "mean cabins."

However, as soon as he ascended the Lehigh Mountains and came upon a view of the valley of the Monocacy Creek and the Lehigh River and Bethlehem's pastoral fields and buildings, he immediately noted the features of a place worth writing about. The Lehigh, he extolled, had an almost "magical beauty," with "soft, clear, pure water flowing over a rocky bottom." On its banks grew "almost all the finest North American shrubs and trees, . . . the calamus, the rhododendron, cephalanthus, sassafras, azalea, tulip-tree, magnolia, and many others which we desire consumedly as guests in our gardens." These riparian trees and bushes had an elegant grace, wrote Schöpf, not found in their "wild" cousins. They were desirable as "guests" in the garden—useful in their way to human cultivation.[1]

Where once Thomas Pownall had seen a "long rolling Sea of Woods" in his approach to Bethlehem in the summer of 1754, now visitors such as Schöpf admired a sublime, civilized, beautiful, and cultivated space.[2] Many different visitors passed through Bethlehem in the later eighteenth century: congressmen, patriots and revolutionaries, military convoys, Native and settler refugees, and scientists studying the American wilderness. Like the travelers who had once described the sounds of Moravian singing, observers of Bethlehem's new landscapes were similarly consistent in noting a "pleasant, ordered, pious" community. Revolutionary leader John Adams wrote in a letter to his wife, Abigail, that the Lehigh Valley was "a Country better cultivated and more agreably diversified with Prospects of orchards and Fields, Groves and Meadows, Hills and Valleys, than any We had seen."[3] He admired Bethlehem's gardens and fields, and the "Rowes of Cherry Trees, with spacious orchards of Apple Trees on each Side of the Cherry Walk."[4] Bethlehem's orderliness was especially astonishing as it appeared suddenly "rising up, one above another, lofty buildings in this presumptive wilderness" after "so long a road through such wild regions. . . . with tedious sameness of bush and forest." Young New York socialite Judith Sargent Murray also noted pleasing and romantic views of the valley: "Upon an eminence is *Bethlehem*, the cultivated scene is displayed before us—a chain of verdant hills encircle it, and this little Eden is embosomed in the midst. . . . a terrestial Paradise."[5] It was this cultivated view of the Lehigh Valley and Bethlehem that also delighted the Marquis de Chastellux in the early 1780s:

> [When] issuing out of the woods at the close of the evening in the month of May, [I] found myself on a beautiful extensive plain, with the vast eastern branch of the Delaware on the right, richly interspersed with wooded islands, and at the distance of a mile in front the town of Bethlehem, rearing its large stone edifices out of a forest, situated on a majestic, but gradually rising eminence, the background formed the setting sun. So novel and unexpected a transition filled the mind with a thousand singular and sublime ideas and made an impression on me never to be effaced.[6]

Within this cultivated and charming space, Schöpf noted that Bethlehem was comprised of sixty buildings made of limestone, three or four stories in height, including the impressive structures of the Single Brothers' and Single Sisters' Houses. The entire town showed "the mark of order and constant industry," oil and grist mills, tannery and dye works all run with German efficiency. Even the beer was excellent and "all the European potherbs flourish exceedingly at Bethlehem, under the good care of exact and indefatigable gardeners."[7]

Around six hundred people lived in and around Bethlehem, Schöpf observed. Most were German and a few English. Almost everyone could speak both languages, including the "purest and best German." He was astonished at this multi-lingual capability. But, had Schöpf visited Bethlehem in 1745, he could have heard Bethlehem's residents singing in thirteen different languages.[8] He could have conversed with Delaware, Mohican, Afro-Caribbean, Hungarian, Polish, and Wampano residents. But this was 1783. Bethlehem's Economy had ended twenty-one years earlier. Native members of the Gemeine had been driven from Bethlehem in 1763. They were now in Ohio and Michigan, and it was one year since the second Gnadenhütten Massacre. What was left of Bethlehem was single-family homes and farms, well-run agricultural and industrial enterprises, and ordered well-planned spaces. Indeed, noted Schöpf, "in a shorter time than any other people, they [the Moravians] have changed numerous wildernesses to flourishing spots."[9]

* * *

The American colonies may have opened up new opportunities for European immigrants, but the rapid influx of people, technologies, plants, animals, and microbes had quickly transformed the environmental space of Penn's Woods. Where once Pennsylvania's Native American communities had perceived a sacred landscape of woods interspersed with human settlements, Europeans imagined (and created) a cultivated scenery of fields, pastures, villages, cities, and farmsteads, marked on maps and divided by their county and township names. The landscape of Native country had been flexible—communities moved from place to place; fields turned to scrub, and were recultivated or returned to forest. But, with astonishing rapidity, these Native places were given new names and new political, social, geographic, and sonic markers by Pennsylvania's settlers: Shamokin became Sunbury, Welagameka became Nazareth, Menagaschuenk became Bethlehem, and the sacred lands at the Forks of the Delaware River became Easton.

🌐 **See website chap4.1, Interactive map: "The Pennsylvania Frontier."**

The new European-style landscape of Pennsylvania was fixed by invisible lines represented by laws, treaties, and deeds. Where Native gardens had mixed a variety of crops, English and German fields were sown with a single crop, uniform in appearance, odor, and texture.[10] The beauty of Pennsylvania's natural resources and forests now lay in their usefulness as a source of extractable biomass and personal wealth. Even in the early days of Penn's Woods, William Penn had been well aware of the attractive natural resources of the Pennsylvania landscape. This had in fact been the most important marketing tool for attracting settlers from Europe: "The Country is in Soyle good, aire sereen . . . & sweet from the Cedar, Pine & Sarsefrax, with a wild mertile that all send forth a most fragrant smell. . . . Turkys of the wood, I had of 40 & 50 pound weight."[11] The pamphlets that he published and distributed in England emphasized these resources and their potential uses:

> Something of the Place. For Timber & other Wood, there is Variety for the use of man, as Oak, Chesnut, Wallnut, Popler, Cedar, Beech, &c. For Fowl, Fish

and Wild Deer, they are reported to be plentiful in those parts, and English Provision grows there, and is to be had at reasonable Rates: The Commodities that the Countrey is thought to be capable of, are Silk, Flax, Hemp, Wine, Sider, Wood, Madder, Liquorish, Tobacco, Potashes and Iron; and it doth actually produce Hides, Tallow, Pipestaves, Beef, Pork, Sheep, Wool, Corn as Wheat, Barly, Rye, and also Furrs, as your Beaver, Peltree, Mincks, Racoons, Martins, and such like; store of which is to be found among the Indians, that are profitable Commodities in Europe.[12]

Penn's marketing had worked. By the late eighteenth century, Pennsylvania was quickly becoming a commodified landscape. But it was not just the commodification of natural resources that increasingly produced misunderstandings and dissension between Native and settler communities. Ideas about property and use of land were ultimately the most divisive force. Settlers, accustomed to the densely packed landscapes of Europe, were intent on putting this vast land resource toward generating prosperity and income for the thousands of immigrants who now crossed the Atlantic. They were accustomed to a hierarchical idea of land: whoever held the "title" owned the profits of that parcel of earth. It was the working of the land—the using of it—that ensured ownership. In fact, "wastefulness" of the land was justification for the termination of property rights.[13] Native Americans did, of course, also farm and practice agriculture, but European observers often downplayed these methods of farming because they were unfamiliar, and therefore did not constitute true "working" of the land. Native ideas about land use also included communal ownership of rights to farm and hunt, and therefore did not accord with European-style laws of single-person ownership for perpetuity.[14] In addition, according to many colonial officials, such as Thomas Pownall, differing theories on control and use of land were fundamentally genetic and could not be modified:

> The holy books, after having given a philosophical account, cloathed in drama, of the origin of things, seem to confine their real narrative to the history of the white family, to that race of people who have been LAND-WORKERS from the beginning; who, where ever they have spread themselves over the face of this globe, have carried with them the art of cultivating vines, and fruit trees—and the cultivation of bread corn: who, wherever they have extended themselves, have become *settlers*; and have constantly carried with them the sheep, goat, oxen and horse, domiciliated and specially applied to the uses and labour of a settlement.... The red family, were originally, wherever found, and are yet in most parts, *Wanderers*. The Tartars are in one part wandering herdsmen, and in other parts hunters and fishermen. The American Inhabitants (Indians, as we call them, from the word Anjô, or Ynguo, signifying a man in their language) are the same race of people.[15]

For early botanists, collectors, farmers, settlers, and government officials such as Pownall, these differences were inherent and fixed, and not subject to change or negotiation. Native people, for Pownall, were "*not land-workers, but hunters; not settlers, but wanderers.* They would therefore, consequently, never have, as in fact they *never had, any idea of property in land*: of that property, which arises from a man's mixing his labour with it." European settlers, too, were under the harness of their own existing cultural ways. They "still continue land-workers; and have made settlements in the parts of America which they occupy; and have transported thither bread-corn, sheep, oxen, horses, and other usually domestic animals."[16]

As the eighteenth century progressed, these differences were viewed as increasingly threatening, from the standpoint of both sides. For essayist J. Hector St. John de Crèvecoeur, the character of Americans was intimately related to the natural environment, and the national character therefore depended upon the very quality of the soil tilled by American farmers.[17] The spell of the woods and "Indians who had the ways of the woods about them" were to be eradicated. Men are like plants, wrote Crèvecoeur, and their goodness proceeds from the soil on which they are located. Frontiersmen rooted in Native soil could transform into lawless, idle, rude, and savage beings. Settlers on the American frontier risked becoming merely carnivorous animals, tearing at the flesh of wild animals. Settler farmers on the edges of European colonization risked falling under the spell of the woods, and could be bewitched by the intrusion of wild plants and animals into their gardens and fields. And, should they turn primarily to hunting, instead of farming, there was no redemption.[18] Crèvecoeur decried: "After this explanation of the effects which follow by living in the woods, shall we yet vainly flatter ourselves with the hope of converting Indians? We should rather begin with converting our back settlers."[19]

Later in his essay *Letters from an American Farmer*, Crèvecoeur seemed to relent. Native people did cultivate and farm, he admitted. And many thousands of Europeans had chosen to live in proximity to Native communities and even to adopt some of their ways.[20] In this seeming turn from his earlier pronouncements, though, Crèvecoeur evidenced the double standard of his time. While the American "wilderness" was seen as dangerous and hindered the progress of settlement and industry, there was no doubt that evocative frontier figures such as Daniel Boone who roamed the wilderness and lived among Native communities also took hold of the settler imagination.[21]

As James Merrell and Daniel Richter have argued, whether one looked east from Indian Country or west from the colonies, Pennsylvania's natural environment was fundamentally the same—millions of acres of trees.[22] But while there might have been some eastward- and westward-facing congruences between Native and settler viewpoints on the natural environment of Pennsylvania, there was never a suggestion in the European worldview, or even a desire, of a world filled with trees.[23] For European settlers, the woods needed to go. Armed with axes and torches, they carried out the biblical injunction to subdue the earth: "Transplanted Europeans hacked away at the forest, dreaming of the day when the trees would be gone, the land cleared, the very climate and air forever changed."[24] Given enough time and enough people, the landscape could and did shift dramatically under the European view of land use and management.[25]

Indeed, Pennsylvania's forests were quickly disappearing as the border between European colonial settlements and Native country shifted relentlessly westward. In the 1750s and 1760s, escalating conflicts between Native Americans, European settlers, and the European colonial empires of France and England during the Seven Years' War brought rapid industrialization to Pennsylvania: roads were quickly built for transport of arms, people, and goods; new forts and settlements were installed. In consequence, Bethlehem was also changing.

The End of the Economy

In 1762, Bethlehem's elders established a new "Brotherly Agreement," and Bethlehem shifted dramatically away from a communal economy to a cash-based commercial system in the style of nearby English and German settlements.[26] Although the church continued to own the majority of Bethlehem's land holdings, individual parcels could now be leased by church members for private homes or businesses. This organization remained in place for the next eighty-two years until Bethlehem made its final transition, opening land purchases and business opportunities to non-Moravians in 1844.

As Katherine Carté Engel has detailed in her study of financial life in Bethlehem, the end of the Economy was both gradual and sudden, and simultaneously precipitated by internal and external pressures.[27] The geographically ambitious and successful Moravian missionary projects of the 1730s and 1740s had been financed in large part by the fortunes of the Zinzendorf family and other wealthy Moravians and members of the

European nobility. Due to this considerable network of wealth and social connections, the Moravian Church had experienced unprecedented success in mission work. Their ability to operate within an existing British and Danish Atlantic economic and political system had also permitted the creation of mission communities in almost every location occupied by British and Danish colonial forces. Yet their immersion in this network also meant a connection with ongoing tensions between the European powers of Britain, France, Spain, and the Holy Roman and Hapsburg Empires. With the outbreak of the Seven Years' War in 1754, the Moravians were immersed in a global conflict that spanned continents, from the Ohio Valley to Calcutta. According to Engel:

> The same sort of ties that knit together Moravians in Bethlehem, London, Amsterdam, and Herrnhut bound together people of all walks of life and multiplied in countless ways to create a single Atlantic World, at once extraordinarily diverse and deeply interconnected. Imperial conflicts that began in London and Paris started military campaigns in New England. . . . Shocks on one continent were thus quickly felt in another, and the Moravians, who had settled communities in nearly every corner of the Atlantic world, were insulated from very little.[28]

The Seven Years' War pitted France and England against each other in a struggle to protect land holdings in North America. The French established a network of forts in Ohio and the western side of the Appalachians behind the boundary of English settlements, hemming the British on the eastern side of the mountains. The Six Nations also sought to preserve their political ties and land holdings, sometimes at the expense of the tribal groups they had traditionally protected: the Delaware, Shawnee, Nanticoke, and Susquehannock. As Rachel Wheeler has argued, it was the complicated nature of tribal affiliations combined with alliances with the British or French that resulted in the attack on the mission at Gnadenhütten, which was destroyed in 1755 by a French-funded alliance of Delaware and Shawnee warriors in the first Gnadenhütten Massacre.[29] With that attack, the western boundary of the war rolled back south and east to the Lehigh Valley, and Bethlehem was suddenly on the edge of the North American front. Benjamin Franklin and colonial military leaders passed through on forays. The ruined mission at Gnadenhütten was sold to the government and converted into Fort Allen by Franklin. The protected space of Bethlehem's newly erected palisades and watchtowers became a haven for settlers from the north and west who fled their homes and farms, causing Bethlehem's population to swell temporarily to more than one thousand people. The Moravians now preserved a narrow zone of safety between the hostile powers of England and France and their Native allies. As Bethlehem's elders struggled to feed, house, and protect visitors and refugees and their own Gemeine, they sought financial support and direction from the church headquarters in Europe. But Herrnhut was itself caught in the European front of the war. Frederick the Great had invaded Saxony and Hapsburg Empress Maria Theresa's attempts to recapture this territory caused the two armies to march back and forth across the area of Herrnhut for two years. The Moravians in Herrnhut were starving, and the Bethlehem community was obliged to send money to Europe to help their fellow church members. The final blow to Bethlehem's financial security came with the sudden death of Zinzendorf in 1760. In the wake of his passing, the debt he had accrued during his lifetime became due and his creditors demanded to be paid.

By 1762, Bethlehem was in crisis. As war continued to ravage the countrysides of Pennsylvania and Germany, there was barely enough money to sustain the surviving missions at Nain, Wechquetank, and Pachgatgoch, and certainly not enough money to consider further expansion. The communal Economy in Herrnhut and other European communities had already ended after failure to generate enough income to pay Zinzendorf's debts. Under pressure to conform to the changing economic and social circumstances of the worldwide church, Bethlehem's elders at last voted to end the communal system in their own settlement. Missionaries continued to receive a salary from the church, and the Single Sisters, Single Brothers, and Widows continued to occupy their houses, especially those who did not have families.[30] But other members were suddenly tasked with paying for their own homes and rented land, as well as food, clothing, and other daily necessities.

For many members of the Gemeine, this was a traumatic shift. The Eyerly family in Nazareth struggled to find enough money to support their family, as related in the Lebenslauf of Christina Elisabeth Eyerly, written by her children after her death: "they [Johann Jacob Sr. and Christina Elisabeth] worked in the *Oeconomie* until this arrangement was dissolved. It was then up to them to look after their own needs. At the beginning however, they experienced great impoverishment. But since they trusted in the help of the Lord in a childlike way, He did not abandon them and they were not ashamed." Although most industries remained under church control, church leaders established a system of salaries for trades and other economic work. This allowed Johann Jacob Sr. to earn a salary for serving as the blacksmith in Nazareth, and to build his own house in a new section of the town that had recently been made available for individual rental lots.[31] Christina Elisabeth's Lebenslauf proudly states: "In 1782 she moved into a newly built house in New Nazareth with her dear husband."[32] Other women, such as the former eldress of the Single Sisters' Choir, Anna Johanna Seidel, also found the transition difficult: "In 1762, the first task was the turn around of the communal economy, which was a difficult job that caused my dear husband [Nathaniel Seidel] and me many sleepless nights. But the Savior stood by us with His grace even under these difficult circumstances."[33]

The end of the Economy also produced radical changes in the practice of singing in Bethlehem. With the shift from communal housing to single-family homes, the daily cycle of singing and worship that had sustained the liturgical life of the Gemeine ended for those families who resided in private houses. Heads of households were now responsible for prayer and singing in their own homes, but with the new economic demands of sustaining a household, it was not possible to sustain the intensity and frequency of worship and singing. This led to a direct decline in singing hymns by memory and the ability to improvise new hymns. Improvisation had been directly tied to the frequency of worship services and occasions for singing, as well as the daily singing and poet schools run by the Gemeine. There was still instruction available in music and singing, but, like the worship services, it was not possible to have such a frequent and rigorous schedule of "practicings" and training in learning to improvise and compose new hymns. Musicologist Alice Caldwell has documented this shift from the 1750s onward by studying printed versions of Singstunden, called *Liturgische Gesänge*, that were used in Bethlehem after the end of the Economy.[34] These publications filled the gap left by the decline in oral tradition and memorization that had sustained the ability of Moravians to improvise. By 1823, when revised editions of the original later eighteenth-century *Liturgische Gesänge* appeared, the tradition of memorized singing was almost entirely forgotten. Caldwell has stressed that these nineteenth-century musical settings are independent compositions that do not show roots in the oral tradition. However, these later books of *Liturgische Gesänge*, accompanied with the eighteenth-century publications, provide an interesting case study of the transition from an oral to a written musical practice in the Moravian Church.

The new spatial distribution of Bethlehem's residents also had important implications for their ability to comprehend a communal identity articulated through sound. Bethlehem's tightly compacted spatial design and communal housing had allowed shared modes of hearing and listening that resulted in socially articulated communal and religious soundscapes. Now, many community members were spatially dispersed and living in individual homes, resulting in little commonality of sounded experience except for community worship services. Moravians also gradually adopted worship modes from other Protestant denominations, reading texts and tunes and singing from hymnals in the style of neighboring Lutherans, Methodists, and Presbyterians. Post-Economy Bethlehem was no longer a sounded Gemeine.

The End of the Missions

The frontier wars of the mid-eighteenth century had important implications for the spatial distributions of missions and Native Christians within the Moravian Gemeine that had little to do with changes in Bethlehem's Economy, but which nevertheless impacted its singing practices and soundscapes. With the end of the Seven Years' War in 1763, and the signing of the Treaty of Paris, colonial boundaries were redrawn, and the French ceded almost all of their North American lands to the British. But it was not just political boundaries that were remapped by the war. Mental geographies of race were also transformed. Before the war, Native and settler communities in eastern Pennsylvania had lived for the most part in common geographic spaces. But war atrocities on both sides had now caused each group to see their interests as fundamentally distinct and mutually exclusive, making physical copresence difficult.[35] The year 1763, like 1762, was a crucial year in Moravian history. This was the last year that there was a Native Moravian community in the area of Bethlehem or Native Americans living in Bethlehem's Gemeine. In the turbulent two years following the destruction of Gnadenhütten in 1755, a decision had been made in Herrnhut, and verified by the lot, to build a new Native Moravian community called Nain in direct proximity to Bethlehem along the Monocacy Creek. This new community was to house people who fled from Gnadenhütten. However, 1757 was not 1742. As Katherine Faull has stated: "What might have been perceived as a possibility in the still relatively peaceful earlier era of white settlement, where the Native population was regularly included in the conferences and treaties of the Colonial government on almost an equal footing, in accordance with the utopian vision of the pacifist Quaker William Penn, now flowed almost perversely against the tide of racial separation, suspicion and fear that marked the rivers and valleys of Pennsylvania."[36] In addition, the building of Nain showed little awareness by church elders in Herrnhut of the cultural needs of Native Americans to have adequate hunting grounds near their communities, which they had had in Gnadenhütten. Nain was planned out with kitchen gardens, space for domestic animals, and separate family homes. But there was no space for inhabitants to be able to hunt in the densely packed European American settlements that now ringed Bethlehem, traversed by newly constructed roads and turnpikes. Nevertheless, the lot and the majority of Native Moravians themselves agreed to build Nain, after receiving permission from the Pennsylvania colonial government, the Six Nations, and the Delaware leader Teedyuscung.

From 1757 to 1763, Nain residents enjoyed an active liturgical and social life, as discussed in chapter 2. But in 1763, just months after the end of the Seven Years' War, the Ottawa leader Pontiac attacked British forts and European settlements in Ohio and Pennsylvania. This resulted in tensions with the European neighbors of Nain. There was no desire on the part of these neighbors to have Native people living in their midst, even those who had been stalwart "friends" of the Bethlehem Gemeine. Tensions became so bad that Nain residents were eventually afraid to leave their village. In July 1763, the Nain Moravians petitioned the Pennsylvania governor for protection, affirming their allegiance to Pennsylvania and their faith in the Christian God, for they "loved our dear Savior." The Bethlehem Moravians also issued a document to assist European settlers in recognizing the differences between "friendly" and "hostile" Native Americans. This document was entitled: "Marks whereby Christian Indians may be distinguished from Wild Indians," and was written by Timothy Horsfield, the keeper of the Strangers' Store and Justice of the Peace for the local county:

> They [Native Christians] are always clothed. They are never painted, and wear no feathers, but hats or caps. They let their hair grow naturally. They carry their guns on shoulders, with the shaft upwards. When meeting a settler, they will call to him, salute him, and coming near, will carry their guns either reversed or on the shoulder. Lastly, they intend, when they go out hunting, to get a pass of Mr. Timothy Horsfield, if he be at home, or else of their ministers, Mr. John Jacob Schmick, at Nain, or Mr. Bernhard Adam Grube, at Wechquetank.[37]

The Moravian leadership was hopeful that open communication with settlers living around Nain would allow them to see the Native Moravians as fellow Christians, and not as a threat. But racial misunderstanding and hatred had been going on since the founding of Nain, according to a letter written by Horsfield in 1756. The letter addressed the grievances of a Native Moravian who was assaulted in the woods and implored help from the Pennsylvania government: "I beg leave to mention to your honour [Governor William Denny], that a few Days Since as one of our Indians was in the Woods a small distance from Bethlehem, with his gun, hopeing to meet with a Deer, on his return home he met with two men, who (as he Informs) he saluted by takeing off his Hat; he had not gone far before he heard a gun fired, and the Bullet whistled near by him, which terrified him very much, and running thro' the thick Bushes . . . came home much frighted."[38]

The escalating violence addressed in Horsfield's letter continued to grow worse in the years between 1756 and 1763. Even though some of Bethlehem's "neighbors" may have realized that their anger was misplaced on the very small group of Native Christians who were actually "friends," they were unwilling to consider living near them. The situation finally became untenable in October 1763, when an Irish tavernkeeper named John Stenton was killed by an attack at his home near Bethlehem. A Native Moravian man named Christian Renatus was falsely accused of his murder, and the residents of Nain were held under a virtual house arrest.[39] They were not allowed to keep their hunting weapons and were subject to daily monitoring.[40] On November 8, 1763, the 127 Native Christians of the Moravian *Gemeine* from the communities of Nain, Bethlehem, Nazareth, and Wechquetank were forcibly moved to Philadelphia "for their protection." There, they were settled first in an army barracks, and when their safety could not be guaranteed, they were moved south of the city to a former quarantine house on Province Island in the Delaware River. 🌐 **See website chap4.2, Interactive sound map: "Journeys of the Native *Gemeine*, 1763–1772."**

It was a heart-breaking journey for the Moravians. The Native women who had lived in the Single Sisters' choir in Bethlehem were removed from the only home they had known for many years. Maria, a Mohican from Shekomeko, had lived in Bethlehem for fifteen years. Despite the terrible conditions of the journey to Philadelphia and on Province Island, as well as the homesickness felt by many, Maria and other members of the Native congregation held regular morning and evening worship services in exile. The communal diary of the Native *Gemeine* (see fig. 4.1) records numerous occasions where singing and worship continued to provide comfort in the face of anguish and suffering:

> December 10. In the evening, there was a blessed *Singstunde*. December 24: We had a blessed Christmas Eve. First, a pleasing Love Feast was held with the

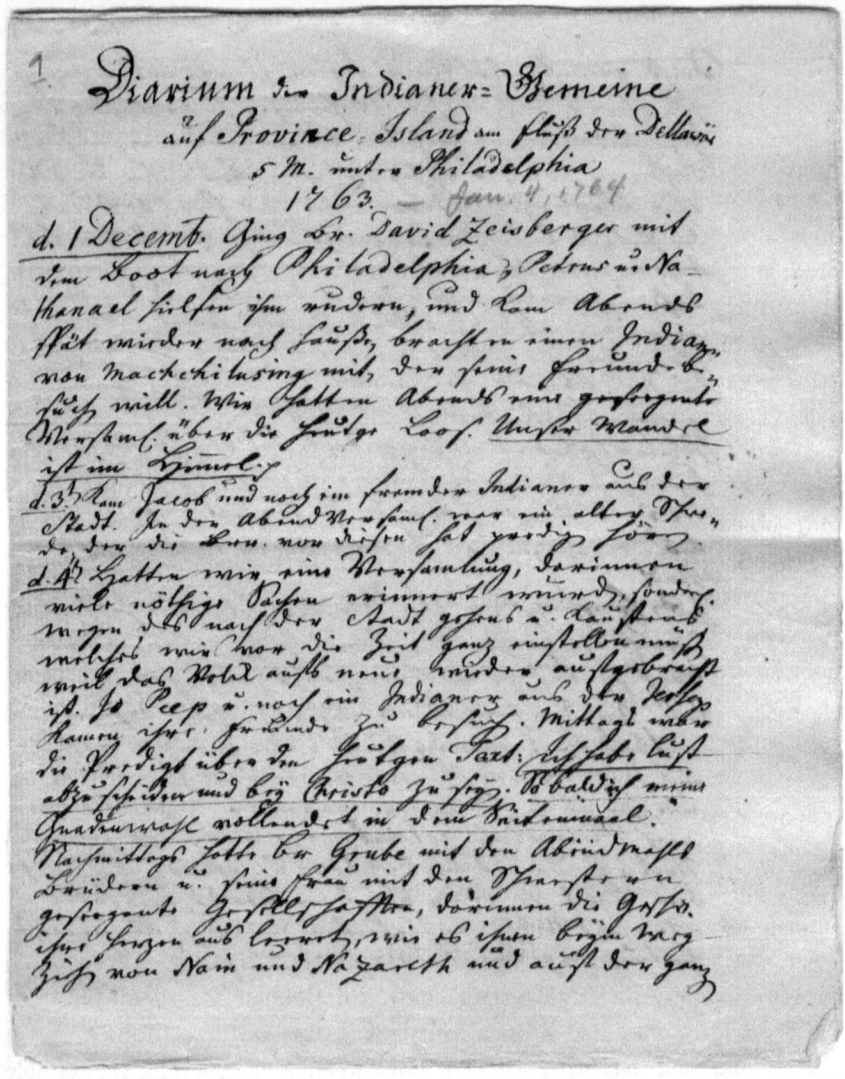

Fig. 4.1 First page of the diary of the Indian Gemeine on Province Island, Philadelphia, compiled by Bernhard Adam Grube, December 1, 1763 to January 4, 1764. MissInd 127.1, MAB.

children, in which Br. Schmick told them with blessed feeling of the Savior's birth. They were very glad upon hearing this and sang several pretty verses to the little Jesus child. The young Josua [Joshua Jr.] played the *Spinet*, and Elias played the *Citter* [cittern]. Afterwards, the adults had their *Agape* with biscuits and tea. Br. Grube talked on the Daily Text. Finally we kneeled before our most dear little Jesus, thanked him from our hearts for his painful birth and incarnation, and recommended us all to his faithful heart.[41]

According to the congregational diary, the Native brothers and sisters thought often of Bethlehem: "The Indian Brethren gave him [David Zeisberger] many heartfelt greetings for the *Gemeine* in Bethlehem; their hearts are always tender when they think of Bethlehem."[42] But despite the prayers of the congregation, violence continued to escalate throughout southeastern Pennsylvania. In December 1763, angered by the Pennsylvania government's protection of the Native Moravians, a mob of Scots-Irish settlers who called themselves the Paxton Boys after the settlement of Paxtang along the Susquehanna River, murdered a group of Susquehannock at their settlement on Conestoga Manor, a land grant given to them by the Penn family.[43] This act of genocide effectively ended the lineage of the Susquehannock tribe. The Moravian congregation was hurriedly evacuated by boat to an island further south on December 31, in fear that the mob would come to Philadelphia. But the Paxton Boys did not come, and the congregation returned to Province Island.

During the tense weeks that followed, the congregational diarist recorded numerous occasions in which curious Philadelphians, or the soldiers who guarded Province Island, attended worship with the Moravians:

> January 3: We had a nice quiet day. Only four Quakers to visit. We spoke to a few more brothers and sisters in preparation for the *Abendmahl*. The Savior was also merciful and allowed us to have the *Abendmahl* today. After the evening service, which Br. Schmick held, was the Love Feast of the Communicants, then the *Absolution* and the blessed taste of the body and blood of our Lord. O how the dear hearts, who had almost despaired of enjoying of the great thing, praised [the Lord]. We therefore thanked [Him] with tears for this great blessing, He knows best when we will have this blessed day again. The people who were sent here from Philadelphia to guard us made it nice, [they] were quiet and watchful during the service, and were amazed that the Indians have so many services here.[44]

However, in February 1764, five hundred Paxton supporters marched on Philadelphia. They were only held off from descending on the Moravians due to the efforts of Benjamin Franklin. During the ensuing negotiations, they presented a pamphlet of their distresses and concerns to the Pennsylvania government. This document, entitled "Declaration of the Distressed and Bleeding Frontier Inhabitants," accused the government of ignoring settlers but "protecting Indians at the public Expence." Franklin printed and responded with his own pamphlet, accusing the Paxton Boys in return of wicked and sinful behavior: "with the Scriptures in their Hands and Mouths, they can set at nought that express Command, *Thou shalt do no Murder*; and justify their Wickedness, by the command given *Joshua* to destroy the heathen. Horrid Perversion of Scripture and of Religion."[45] In fear for the safety of the Moravians, and also in concern for the general welfare of Philadelphia, the government tried to transfer the Moravian refugees to New York and New Jersey, but the Moravians were rejected by both governments and were returned to Philadelphia, where they remained confined in the barracks until 1765. The former pestilence house where they lived was damp and cold, and likely still embedded with microbes from its time as a place of quarantine. During that time, fifty-six members of the Gemeine died of diseases and ill health.[46]

Despite the year of forced confinement, and the death of almost half of the congregation, the Moravians continued to hold worship services and to sing together. There must have been many doubts and difficult conversations during this time, and likely expressions of anger toward both the Moravian Church and the Pennsylvania government. However, the congregational diary did not record those conversations. What the various members assigned to write the diary did record was the frequency of worship, prayer, and singing. The diarists also recorded various musical performances. It was during this time that Joshua Jr. played the spinet for the governor of Pennsylvania and other Philadelphians who came to visit the Moravians, an event that would become a legendary part of early Pennsylvania history.[47] Diarists also remarked that visitors noted the beauty of the singing: "January 16: Most of the Indians went to the water to gather oysters, and they brought a good portion of them back to the house. In the evening, we held a very blessed service, many white people from the city as well as Highlanders were there and were amazed by the Indians and their beautiful singing. Some people now receive a different idea of our Indians. A soldier said to the people [that] God would wish that all white people were as good Christians as these Indians are." But as the months of 1764 wore on, the diary was increasingly filled not just with songs, but with the terrible heartbreak of losing family and friends:

> June 30: We explained various matters to our people in a special service, particularly about going home [dying], because various of them have let themselves come into reasoning about it. Anton said at last: It is true what our wise brothers say, I know their hearts, and whoever says or thinks something against them, he speaks or thinks against the Savior, and I will have no part of it. Gottlob, Nathanael's son, got the pox. Because his parents have not themselves had it, the last two to be baptized, old Br. Nicodemus and Peter, volunteered to care for Gottlob. They were placed in a special room on the south side. In

the evening, Br. Grube closed this week with a *Singstunde*. In this month, we have particularly experienced the visits of the Savior, in that he took ten little brown sheep into his arms and had four people buried in his death. May he be thanked 1,000 times by our four poor hearts.

Despite all that they had experienced, the Moravian Sisters still longed for Bethlehem. When the congregation was finally released from confinement by the government after agreeing to move away from the area of Bethlehem into the "wilderness" on the north branch of the Susquehanna below the New York border, many still longed for "home" and for the friends they had made in the Gemeine: "December 10: Several sisters came, crying and lamenting their concerns, particularly about going to the wilderness, which would be unendurable for them if no Brethren were to go with them, since their heart's desire was to ever remain in connection with the Brethren." But the Sisters had little choice and on March 19, 1765, several elders, including Joshua Sr. penned a farewell letter to the Pennsylvania Governor:

> Address of the Christian Indians to Governor John Penn
>
> To the Honourable John Penn Esqr.
>
> Lieutenant Governor & Commander in Chief of the Province of Pennsylvania etc.
>
> The Address of the Indians at the Barracks near the City of Philada.
>
> Being about to depart in two Days hence with our Wives and Children from the Philada. Barracks where we have Sojourned above a year back in & intending to go back into the Woods of Wachelusing on Susquehanna & to settle there. We think it is our firm Duty to take a friendly leave from You by presenting our hearty Thanks for Your great Goodness to us. We do not come with a String or Belt of Wampum agreeable to the Custom among Indians but and as we cannot speak Your Tongue we must endeavour to express our grateful Hearts in by this Writing and hoping You will take it accept of it in Your usual Benevolence from Your poor Indians. We all acknowledge Your great Kindness to us in the late War, we were then in Danger of our Lives from the white people and You have taken us in Your Protection so that we could live in peace & quietness in the Barracks[.] You have as a good father provided Food & Raiment for us our Wives & Children[.] You have given medicines & nursing for our Sick and have buried our dead, and we have been rejoiced to hear that Your tender care for us extends still further that You will give us some Meal till our Indian Corn shall ripen[.] We have indeed great occasion for such[.] Your Goodness as are all [illeg.] are poor and have many superannuated impotent persons amongst [us.] For all this we offer You our grateful Hearts and moreover We thank You for the Liberty we have enjoyed during these last diff times of difficulty to have our Ministers with us & the to daily

to attend divine Service. By these means we have been kept in the Way of our Salvation & have heard the good Words of our God & Creator that we shall love him & all mankind and be in friendship with the English and we further rejoice & thank You that Mr. Schmick one of our Ministers & David Zeisberger a Brother of Bethl., [illeg.] shall go to live with us on Susquehannah to influence us poor Indians in the knowledge of Truth of the Gospel[.] We have great occasion for such daily influence as there is many of our Indians who are not as they ought be & some of them know nothing at all of our Creator.

Your Benevolence & protection towards us are great & consider in our Eyes & have made an Impression in our hearts that never can wear out and we will always relate all this & testify to the Indians on Susquehanna & testify & declare to them that we are & will for ever remain true friends to the English.

We have another Request to make of You & hope which is that You will give us some Powder & Shot to make use of on our tedious & difficult Journey in Killing some Game for our Wives & Children. This is all we have to say for the present, & we wish that the Almighty may bless You. These words come from us who have submitted this address & from all the men Indian Men Women & Children now at the Barracks and we are Your true & faithful friends.

Johannes Papunhang

Josua Sen.

Anton

Sam. Eawens[48]

On March 22, the Gemeine departed for Nain and Bethlehem. There they stayed for a week in what remained of their former homes. During this time, Nain's buildings were sold to interested families in Bethlehem on March 30.[49] On the following day, the community held a Lovefeast, and then the Native Moravians left for the northern branch of the Susquehanna and the Wyalusing Valley north of Wyoming. 🌐 **See website chap4.2, Interactive sound map: "Journeys of the Native *Gemeine*, 1763–1772."** The five-week trip was arduous and filled with starvation, sickness, death, and despair. Still, the Moravians continued to sing. The familiar patterns and traditions of hymn singing likely provided comfort and continuity during a time of extreme mental and physical stress. On the morning of April 3, as they set out for northern Pennsylvania, they blessed and consecrated their journey with a hymn: "May your holy blood rain over us; there could be nothing better to bless us on our journey [from the hymn: *Dein schweiß und blut laß ueber uns regnen, uns kan auf erden nichts besser segnen*]."[50] On Easter morning, April 7, the congregation remembered the Gemeine at Bethlehem, who would even then be celebrating the Easter service in the

Gottesacker. They also remembered those who had passed away in Philadelphia and were never buried in the sacred space of Bethlehem or Nain, but in a rented "potter's field":

> We remembered our fifty-six Indian brothers and sisters who were buried at the potter's field in Philadelphia and in spirit we were also in the cemetery in Bethlehem. Yet here we were sitting in the wilderness where it had been raining very hard all day and how happy and thankful each person was that we had water to drink and with which to cook, which had accumulated in the puddles. What blessed and comforted us above all was feeling the closeness of our risen Savior. Even in the forest, this gave us such blessed joy on feast days that we could forget all the difficulties.[51]

The following day, the Gemeine ended their journey with "a blessed little singing service [*Singstündgen*]."[52] Throughout the trip, as they prepared to climb a mountain, or struggled to cross a rain-swollen creek, or enjoyed an unexpected feast of one old bear and three young bears, they sang.[53] By Thursday, May 2, the travelers had eaten their last portion of flour. Distressed, the children were crying from hunger, and the men searched frantically for wild potatoes in the pouring rain and tapped chestnut trees to drain syrup for the children to lick. Still, that day ended with a "moving *Singstunde*."[54]

After more than five weeks in the forest, they arrived in Wyalusing, below the town of Machilusing, and sent a messenger to the Haudenosaunee at Onondaga requesting permission to build a Christian village and to continue to have a German missionary from Bethlehem. Within months, the congregation had built a new mission—a second Friedenshütten—with forty wood-frame shingled houses, and 250 acres of corn (see fig. 4.2). From 1765 to 1771, there was hunting, planting, harvesting, fishing, and regular worship services: Singstunden, morning services, baptisms, communions, and burials.[55] But in 1768, the Haudenosaunee ceded their remaining lands in eastern Pennsylvania to Britain in the treaty of Fort Stanwix. Friedenshütten lay on the British side of the new border. Once again, the Native Gemeine left their homes. Netawatwees, a Delaware chief in Ohio's Tuscarawas Valley, had granted the Moravians a tract of land near his village of Goschachgünk (Coshocton). It was then that the congregation undertook the journey during which little Nathan and many others lost their lives in the difficult crossing of the Allegheny Mountains. By 1773, all of the Native Moravians had relocated to Ohio, bringing an end to the Moravian missions in Pennsylvania. 🌐 **See website chap4.3, Interactive map: "The Journey of the Native *Gemeine* from Friedenshütten II to Friedenstadt."**

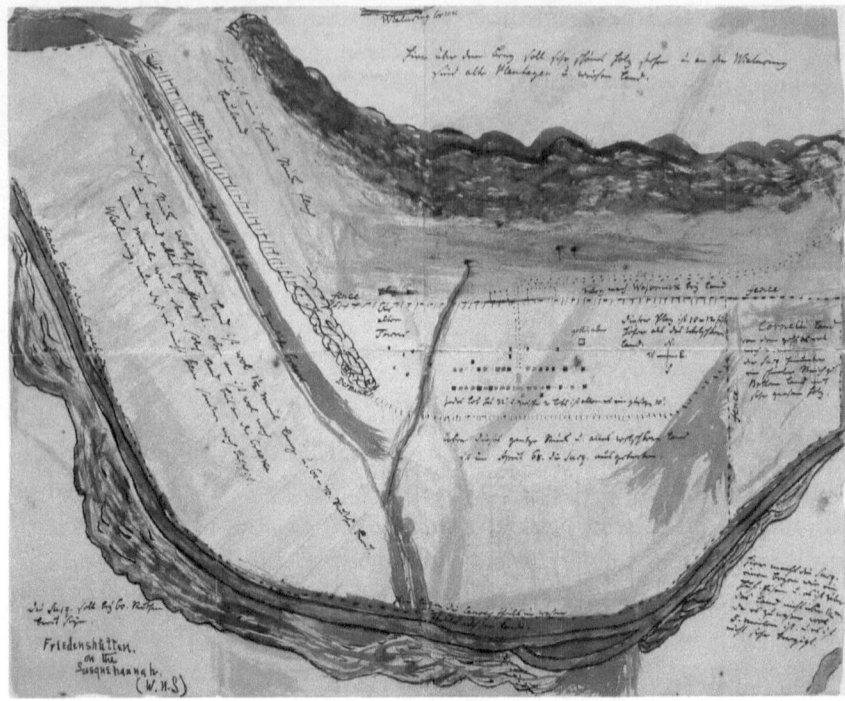

Fig. 4.2 Johannes Ettwein's sketch for a map of Friedenshütten on the Upper Susquehanna River, 1768. Watercolor on paper. DP f.161.1, f.226 (formerly PP EJ 1372), MAB.

Gnadenhütten

The first few years of the new Gnadenhütten, Schönbrunn, and Salem missions were successful and the villages grew and prospered. But within three years, the Revolutionary War, a new global conflict between Britain, the American colonies, France, Spain, and the Netherlands, would once again place the Moravians on the edge of the western boundary of colonial settlements (see figs. 4.3 and 4.4 for early maps of the Moravians' land holdings in the Ohio Country). 🌐 See website chap4.4, Static map: "The Ohio Country, 1782" and website chap4.5, Historic document and map collection: "Ohio." This geographic position would have lasting consequences for the Moravian missions, just as it had in Pennsylvania. The commanders of the English forces at Fort Detroit strategically sought to engage Native American communities in Ohio in an alliance that would create a western front behind the American forces stationed at Fort Pitt. As a result, the Tuscarawas Valley became a crossroads for Native, British, and American war parties, militias,

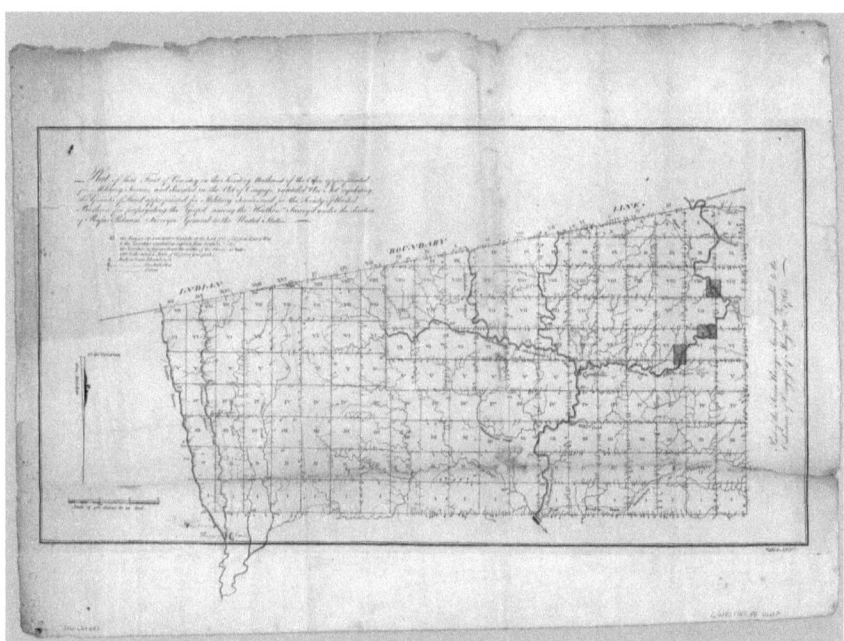

Fig. 4.3 "Part of the seven ranges survey'd agreeable to the ordinance of congress of May 20th, 1785." Published in Philadelphia, 1785. Map of the Ohio Territory showing the boundary of Indian Country and the land granted by the United States Congress to the Moravian Church for the missions at Salem, Gnadenhütten, and Schönbrunn. Surveyed and rendered by Rufus Putnam, Geographer and Surveyor of the United States, 1789. Library of Congress. Public Domain.

and military expeditions. Schönbrunn and Gnadenhütten were briefly abandoned by the Moravians in 1777 when they sought refuge with Netawatwees at Goschachgünk. But when false rumors circulated in 1778 that the Americans had suffered a crushing defeat, and were now sending a military force against the Delaware Nation in retaliation, Netawatwees relinquished his protection of the "American-loving" German missionaries who served the Moravian communities. The entire Native American population of Ohio was moving quickly toward an opinion that Americans of any ethnicity or background were not trustworthy, and they strongly urged the Moravians to side with the British. The Gemeine was forced to return to Gnadenhütten, Schönbrunn, and Salem, where they hoped they would be left in peace.

In the fall of 1781, Major Arent Schuyler DePeyster, the English commander at Fort Detroit, ordered his Wyandot allies to seize the Native

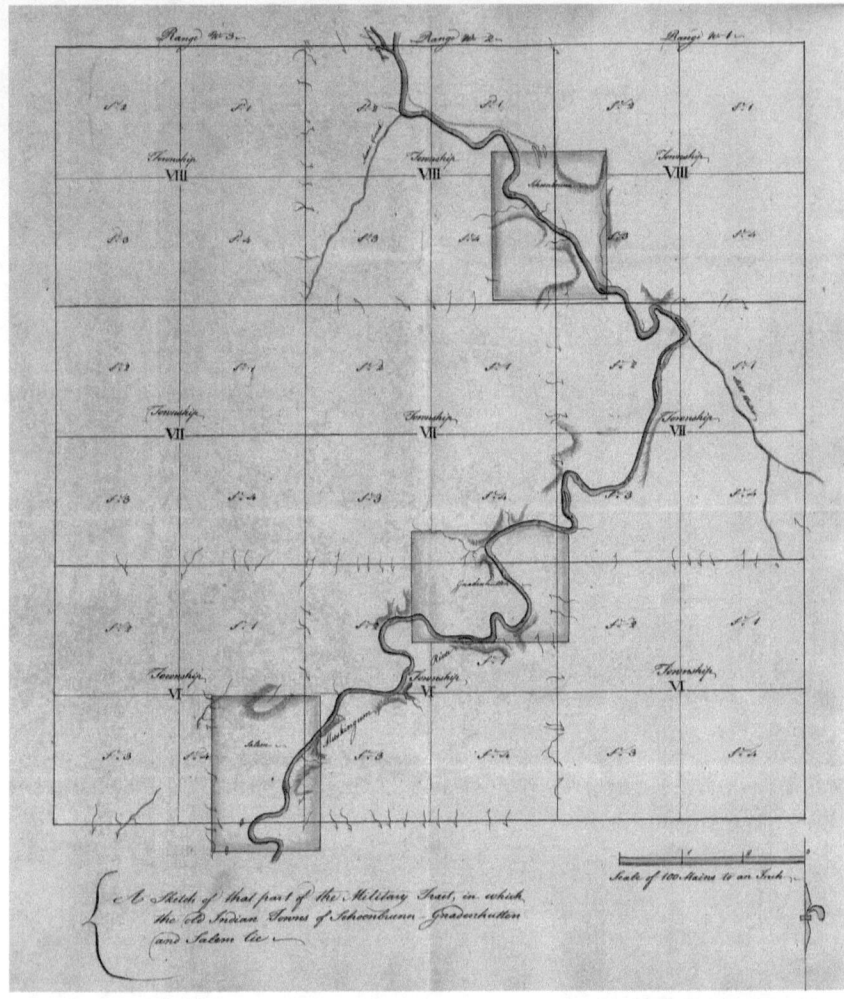

Fig. 4.4 Zoning map depicting land granted to the Moravian missions at Schönbrunn, Gnadenhütten, and Salem, taken from the 1785 map of the Old Northwest Territory, adapted from an original survey by Rufus Putnam, ca. 1785. Ink and watercolor on paper. DP 226.2, MAB.

Moravians and their "teachers" and bring them by force to Wyandot settlements on the Sandusky River in northwestern Ohio. He suspected that the German missionaries were sending letters to Fort Pitt that contained information about English and English-allied Native forces. He was correct about the letters, although the Moravians did not intend treachery. David Zeisberger and the other missionaries in Ohio were very aware of the position of Bethlehem, Nazareth, and Lititz within the American territories, and they were anxious to preserve peaceful relations with both sides. When the Wyandot forces arrived at Gnadenhütten, Schönbrunn, and Salem, the Moravian Gemeine refused to abandon their villages. It was late fall, and they were worried about spending the winter in Sandusky, far away from their corn supply. However, they were forced to leave their stores of corn and march for twenty days through Ohio in the cold to a place prepared for them: Captive's Town. Upon their arrival, four German missionaries and three Native elders were taken to Detroit for trial. They were eventually exonerated of trading secrets to the Americans and released. But in their absence, tragedy struck.

Many members of the Moravian congregation, after facing starvation in Sandusky, had returned to the Tuscarawas Valley to gather what corn remained in their fields. There, in early March 1782, they were seized by a militia of roughly 160 western Pennsylvanians out of Washington County. Just as had happened at Nain, and with the Paxton Boys during the Moravians' confinement in Philadelphia, these western Pennsylvania settlers accused the American military of not being able to protect their homes and farms from Native American attacks. Three separate and deadly raids in the winter of 1782 had caused many members of the Washington County militia to decide it was time to take matters into their own hands.[56] They were not interested in finding the actual perpetrators who had killed their family members and friends. The militia was mustered on March 1 and assembled on March 4 at Mingo Bottom on the western side of the Ohio River. From there, they traveled quickly over the Moravian Trail to Gnadenhütten, arriving on March 7.

What happened next was subsequently told and recorded from many different viewpoints. Some of the most reliable accounts are preserved in the Moravian records, especially accounts recorded by David Zeisberger after interviewing the two teenage boys, Jacob and Thomas, who escaped from the militia. Many members of the militia would never admit to having been to Gnadenhütten and Salem, let along discuss what happened there, so the perspectives of the other side are sparse. However, the surviving records do allow for a reconstruction of the events of March 7 and 8. As the militia approached Gnadenhütten, they surprised a young Moravian named Joseph Schebosch in the forest.[57] Joseph was immediately killed and scalped and his body thrown in the bushes. The Pennsylvanians then surrounded the Moravians in the cornfields outside of the town and invited them to enter the village. They were promised safety. The Moravians knew immediately that these were Americans, and they believed that they had been saved from their captivity by the English and Wyandot. Now they could be taken to Pittsburgh where they would be housed and fed, just as their community had been received in Philadelphia twenty years earlier. So, they went willingly with the men into the village.

In the meantime, some of the militia went south along the river to Salem and encouraged the residents of Salem to come to Gnadenhütten, which they did.[58] However, the promises of food and shelter were quickly dispelled. After the arrival of the congregation from Salem, both groups were seized and forced into separate houses: "They trusted and went, but were all bound, the men being put into one house, the women into another." Isolated into these houses, the men, women, and children "began to sing hymns and spoke words of encouragement and consolation to each other."[59] Once the Moravians had been locked up, the members of the militia voted whether they should take them to Pittsburgh or kill them and burn the village. Whether through group pressure or the steadfast desire for revenge, all but eighteen members voted to kill the Moravians and destroy Gnadenhütten. When the congregation was informed of this decision, they requested time to prepare for death: "Shut up in their two prisons, the converts began to sing and pray, to exhort and comfort one another. . . . Abraham, surnamed the Mohican, took the lead. . . . 'I will hold fast to Him until I die. I believe that He will not cast me off, but pardon all my sins.' As the hours wore away, and the night deepened, and the end drew near, triumphant anticipations of heaven mingled with their hymns and prayers. Converted heathen taught their Christian slayers what it means to die as more than conquerors."[60]

In the morning, the Moravians continued to sing, as the militia chose two mission buildings for "killing houses"—one for the men and one for women and children. Elderly Abraham was the first to be taken, and Judith, a widow. Others followed in pairs. They were struck on the head with mallets, tomahawks, and other gathered implements from the mission buildings, including Joshua Jr.'s cooper's mallet, then scalped and thrown into a pile in each building. Only two people escaped. Thomas was hit on the head only once and then scalped. Initially, he lost consciousness. When he awoke, he was deep in the pile of bodies. As he stirred, he noticed Anton also moving, but a member of the militia entered the building and killed Anton. So Thomas lay motionless under the bodies until he heard the militia leave the village. Then, in the darkness and in terrible pain, he ran and hid in the woods. Jacob, another young boy, had also managed to hide in the cellar beneath the floorboards. As night fell, he forced his way through a small window and also fled into the woods, where he found the distressed and bleeding Thomas. Together, they escaped across country to Sandusky, where they brought the news of the massacre to the community. Zeisberger's diary records their story:

> Our Indians were mostly on the plantations and saw the militia come, but no one thought of fleeing, for they suspected no ill. The militia came to them and bade them come into the town, telling them no harm should befall them. They trusted and went, but were all bound, the men being put into one house, the women into another.... Then they began to sing hymns and spoke words of encouragement to another until they were all slain, and Abraham was the first to be led out, but the others were killed in the house. The sisters also afterwards met the same fate, who also sang hymns together.... Two well-grown boys, who saw the whole thing and escaped, gave this information. One of these [Thomas] lay under the heaps of slain and was scalped, but finally came to himself and found opportunity to escape. The same did Jacob, Rachel's son, who was wonderfully rescued. For they came close upon him suddenly outside the town, so that he thought they must have seen him, but he crept into a thicket and escaped their hands.... The boy who was scalped and got away, said the blood flowed in streams in the house, which was set on fire.... They burned the dead bodies, together with the houses.[61]

The boys later related their experiences of hiding in the woods, seeing the houses burn, and hearing the shouts and exclamations of the American attackers "noisily exulting in their gruesome accomplishment."[62] One of the boys could understand English. He remembered everything he had heard. As one man had picked up the Joshua's mallet, he exclaimed, "How exactly this will answer for the business." Then he crushed the skulls of Abraham and thirteen others before remarking, "My arm fails me; you go on in the same way. I think I have done pretty well!" In the end, the militia had ninety-six scalps: twenty-eight men, twenty-nine women, and thirty-nine children who had been bound with ropes, stunned by mallet blows, scalped, and burned even as they sang Christian hymns in supplication to the Savior they hoped would end their suffering.

So ended one of the greatest tragedies of the eighteenth century—the Gnadenhütten Massacre, "a stain on the frontier character that time cannot wash away," in the later words of President Theodore Roosevelt. Joshua Jr., whose two daughters had been murdered at Gnadenhütten, could never bring himself to understand the acts committed against the Moravian congregation that day on March 8, 1782. According to the missionary Johann Heckewelder, who had served at Gnadenhütten: "The murder of his two beloved and only daughters, between the ages of fourteen and eighteen years, by [Pennsylvania militia officer David] Williamson's party, at Gnadenhutten, on the Muskingum, in 1782, was a hard stroke for him to bear. Often, very often has he been seen shedding tears, on this account.... He could not conceal his astonishment, that a people, who called themselves Christians, and read the scriptures, which he supposed all white people did, could commit such acts of barbarity."[63] For the next twenty-four years until his death in 1806, Joshua would wonder and he would mourn.

David Zeisberger attempted to comfort himself that the Native congregation now rested in the arms of the Savior away from any further harm. The last statement in Zeisberger's diary for March 23, 1782, read: "This news sank deep into our hearts, so that these our brethren, who as martyrs, had all at once gone to the Saviour, were always day and night before our eyes and in our thoughts and we could not forget them, but this in some measure comforted us that they passed to the Saviour's arms and bosom in such resigned disposition of heart where they will forever rest, protected from the sins and all the wants of the world."[64] What Zeisberger does not seem to acknowledge is that the Native congregation suffered directly as a result of their association with the Moravian Church. In consequence of the Moravian desire to create new, multicultural Christian communities, the Native Gemeine perished. Although Zeisberger does not appear to question his actions as a missionary, did Zeisberger and other European missionaries, despite their intentions, hasten or even participate in the destruction of the people they hoped to save? Had the Moravians' insistence on creating shared communities ultimately rendered Native Moravians particularly vulnerable to acts of violence?

Questions surrounding the singing of the Native congregation are also particularly urgent. What had the act of singing represented to Abraham and Judith and those who died at Gnadenhütten? What messages had they chosen to voice in their last hours? And why had the Pennsylvania militia refused to recognize their songs as the sincere spiritual expressions of fellow Christians? The singing of the Gnadenhütten community was not coincidental, nor unexpected, and reveals important insights about the beliefs and practices of Native Christians who affiliated with the Moravian Church. Intercultural practices of singing were at the very center of interactions between Moravian missionaries and Native communities in the eighteenth century. Hymns were central not only to European Moravians' missionary philosophies but also to the lived Christian practice of Native Moravians. For the Moravian congregation at Gnadenhütten, hymns were both beson and appeal to spiritual powers both old and new. For Northeastern Woodlands cultures, song had long been central to deathways. Proper grieving of loved ones and the honoring of ancestors depended on traditions of sacred song that healed survivors and connected with the spiritual world surrounding human communities. Like the Moravian practice of Einsingen, these traditions of songs for the dead and dying possessed powers that bridged mortal and spiritual realms. The role of song at the end of life was also important to personal and community identity during times of war. Haudenosaunee and Algonquian peoples often adopted war captives to reinforce the strength of the community. Strength could also be gained by sacrificing a captive through fire and torture. Men were trained from youth to withstand pain, and not to cry out when tortured, but to sing in the face of death. According to Zeisberger's own *History of Native American Cultures in North America*, "Captives often endure torture with the greatest fortitude, sing of their heroic deeds accomplished in war, and do not let their captors notice fear or terror."[65] Succumbing to pain and fear was devastating—representing a loss of identity for both the person and their home community. To sing was to defy one's captors and to maintain honor and integrity even in the midst of extreme violence and pain.

So on that cold March morning in 1782, the Native congregation at Gnadenhütten sang because song was essential to daily life, and it was essential to death. They sang because song was a powerful medicine to right wrongs. They sang in defiance of the Pennsylvania militiamen who brought such tremendous destruction and horror upon their community. Listening carefully to the events of that day allows us to understand something fundamental about the identity of the Gnadenhütten Gemeine. In every sense, this was a community that was Native *and* Christian. Yet this was a perilous identity to hold in the 1780s on the Ohio frontier. It was the very hybridity of the congregation's German-Native songways that allowed the Christian members of the American militia to dismiss the Moravians' prayers and songs as foreign, alien, and hostile, rather than the demonstrable signs of a fellow Christian community. They were a sonic symbol of the community's tenuous position between the worlds of Native Americans and European settlers. Viewed by many non-Moravian groups as not truly Christian because of their ethnic heritage, their membership in a German church, rather than an English Protestant or even Catholic denomination, had also pushed them to the very margins of the British- and now American-controlled Atlantic World. Other Native American communities viewed them as traitors to the burgeoning nativist movements that dominated most Native communities in the wake of the Seven Years' War and Pontiac's Rebellion. As "outsiders" in every sense of the word, they were particularly vulnerable to acts of violence.

The grim realities of the massacre at Gnadenhütten demand a careful examination of the dichotomies between the Moravian mission agenda and the ultimate fate of Native Moravians. Although singing had for a brief span of time in the early history of the Moravian missions created a congruent space for exchange of spiritual and cultural ideas, ultimately it was those very hymns that created the most tragic and deepest misunderstandings. Instead of a diverse community of Christians capable of singing together in joyful praise before the throne of God, there was now only a small remnant of a once-vibrant Native Christian community. While European Moravian settlements in Pennsylvania continued to gradually merge into the mainstream of American small-town life, the remaining members of the Native Gemeine removed farther westward. After a brief stay along the Huron River at New Gnadenhütten, they moved several more times, before finally settling along the Thames River near Ontario, Canada, at Fairfield and New Fairfield.

The multicultural "singing utopias" envisioned by the Moravians were now completely incongruent with the escalating and often racially motivated violence that continued on well past 1782 and into the nineteenth century with the forced removal of Eastern Woodlands people into the Midwest and northward into Canada. As Daniel Richter and Katherine Faull have so poignantly articulated, the Moravian experiment did not fit within the developing binary structures of racial politics or the expansionist thrusts that were the hallmarks of subsequent American history. Competition for natural resources and control of land and differing cultural practices had engendered difficulties that seemed increasingly insurmountable. And in the wake of successive waves of war, acts of extreme violence had engendered nationalist sentiment on both sides. Still, in despite of the horrors of Gnadenhütten and the naivete or complicity of the Moravian Church, the history of the Moravian missions may still teach us about another vision of Pennsylvania, and indeed America, that might have been.

Notes

1. Johann David Schöpf, *Travels in the Confederation, 1783–1784: From the German of Johann David Schoepf; Translated and Edited by Alfred J. Morrison* (New York: B. Franklin, 1968), quotes from 134–159.
2. Pownall, *A Topographical Descriptions of the Dominions of the United States of America*, 100.
3. Letter from John Adams to Abigail Adams, February 7, 1777, Adams Family Papers: An Electronic Archive of the Massachusetts Historical Society: accessed January 24, 2017, www.masshist.org/digitaladams/.
4. Letter from John Adams to Abigail Adams, September 23, 1777, Adams Family Papers: An Electronic Archive of the Massachusetts Historical Society: accessed January 24, 2017, www.masshist.org/digitaladams/.
5. Lady A. Constantia (Judith Sargent Murray), "Description of a Journey to Bethlehem, Pennsylvania," *The New York Magazine, or Literary Repository (1790–1797)* (August, 1790).
6. Marquis de Chastellux, *Travels in North America in the Years 1780, 1781, and 1782*, vol. 2, ed. Howard C. Rice, Institute of Early American History and Culture (Chapel Hill: University of North Carolina Press, 2011), 321–334.
7. Schöpf, *Travels in the Confederation, 1783–1784*, 138.
8. "Den 4ten Sept. Zum Besuch beym Sabb. Lmahl [Liebesmahl] wurden deutsche, Engl., Lateinische, Griechische, Maquaische [Mohawk], Mahikanderische, Frantzösische, Eyrische, Böhmische, Wendische, Walisische 'Holländische' Verßen gesungen. Unser dänischer Bruder Matthew Beütz, der Polacke Hancke u. Christ. Bower der Ungar blieben noch die ihrigen schuldig." English translation: Levering, *A History of Bethlehem, Pennsylvania*, 205. German text cited in Erben, *A Harmony of the Spirits*, 239.
9. Schöpf, *Travels in the Confederation, 1783–1784*, 148.
10. Hoffer, *Sensory Worlds of Early America*, 76. As William Cronon has argued, the "frontier process" had drastic consequences for the ecosystems and landscapes of North America (William Cronon, *Changes in the Land: Indians, Colonists, and the Ecology of New England* (New York: Hill and Wang, 1983), vii). However, it is important to remember that not all of the environmental changes that took place after European colonization were caused by settlers. But, as Cronon asserts, colonists and Native people practiced lifeways that had different impacts on the landscape, and it is possible to conjecture that precolonial agricultural and cultural practices were perhaps better at maintaining an environmental equilibrium (*Changes in the Land*, 9–12).
11. William Penn to the Earl of Sunderland, July, 1683, in *Remember William Penn*, 90–91.
12. William Penn, *A brief account of the Province of Pennsylvania in America, lately granted under the great seal of England to William Penn, &c*, Library of Congress Digital Archives, accessed July 17, 2016, www.loc.gov/resource/rbpe.14000010a.
13. Even today, Pennsylvania state law dictates that land owners who do not use their land for twenty-one years can be taken to court and their rights-of-ownership terminated under the provision of "adverse possession."
14. Mackintosh, "New Sweden, Natives, and Nature," in *Friends and Enemies in Penn's Woods*, ed. Pencak and Richter, 9–11.
15. Pownall, *Administration of the Colonies*, 223–224.

16. Pownall, *Administration of the Colonies*, 223. "America, in its natural state, is one great forest of woods and lakes; stocked not with sheep, oxen, or horses; not with animals of labour, and such as may be domesticated; but with wild beasts, game and fish; vegetating, not with bread-corn, but with a species of pulse, which we call maize; of which there is great doubt whether it be indigenous or not.—All therefore that this country afforded for food or raimant, must be hunted for. The inhabitants of consequence would naturally be, as in fact they were, *not land-workers, but hunters; not settlers, but wanderers*. They would therefore, consequently, never have, as in fact they *never had, any idea of property in land*: of that property, which arises from a man's mixing his labour with it (224)"; and "The race of white people migrating from Europe, still continue land-workers; and have made settlements in the parts of America which they occupy; and have transported thither bread-corn, sheep, oxen, horses, and other usually domestic animals, that are domiciliate with these settlers (226)."

17. Richard Slotkin, *Regeneration Through Violence: The Mythology of the American Frontier, 1600–1860* (Middletown, CT: Wesleyan University Press, 1973), 260.

18. J. Hector St. John de Crèvecoeur, *Letters from an American Farmer and Other Essays*, ed. Dennis D. Moore, The John Harvard Library (Cambridge, MA: Belknap Press of Harvard University Press, 2013), 33.

19. J. Hector St. John de Crèvecoeur, *Letters from an American Farmer and Other Essays*, 38. Also see Slotkin, *Regeneration Through Violence: The Mythology of the American Frontier*, 261–262.

20. Slotkin, *Regeneration Through Violence: The Mythology of the American Frontier*, 265.

21. Slotkin, *Regeneration Through Violence: The Mythology of the American Frontier*, 267.

22. Merrell, *Into the American Woods: Negotiators on the Pennsylvania Frontier*, 27; also see Daniel K. Richter, *Facing East From Indian Country: A Native History of Early America* (Cambridge, MA: Harvard University Press, 2001).

23. Merrell, *Into the American Woods: Negotiators on the Pennsylvania Frontier*, 27.

24. Merrell, *Into the American Woods: Negotiators on the Pennsylvania Frontier*, 27.

25. Yi-fu Tuan, *Space and Place: The Perspective of Experience* (Minneapolis: University of Minnesota Press, 1977), 166.

26. "The 1762 Brotherly Agreement," Bethlehem Digital History Project: accessed September 14, 2016, http://bdhp.moravian.edu/community_records/regulations/brotherly_agreement/batranslation.html.

27. For detailed information on the ending of the Economy, see Engel, *Religion and Profit*, chapters 5–8; Engel, "The Strangers' Store," 112–113; and Engel, "Br. Joseph's Sermon on the Oeconomie," in *The Distinctiveness of Moravian Culture*.

28. Engel, *Religion and Profit*, 135–136.

29. Wheeler and Hahn-Bruckart, "On an Eighteenth-Century Trail of Tears: The Travel Diary of Johann Jacob Schmick of the Moravian Indian Congregation's Journey to the Susquehanna, 1765," 48.

30. Engel, "Br. Joseph's Sermon on the Oeconomie," in *The Distinctiveness of Moravian Culture*, 122.

31. The Eyerly home is still extant, and is located on Main Street in Nazareth.

32. "Einige Personalia der Christina Eÿerlin geb. Schwarzin," UA Herrnhut.

33. Faull, *Moravian Women's Memoirs*, 127. Also see, "Projects for the intended change of the Bethlehem Economy, re: Single Sisters, 1761," Bethlehem Digital History Project, accessed April 5, 2017, http://bdhp.moravian.edu/community_records/meeting_minutes/singlesisters/singlesistrans.html.

34. Caldwell, "Music of the Moravian *Liturgische Gesänge* (1791–1823)."
35. See Peter Rhoads Silver, *Our Savage Neighbors: How Indian War Transformed Early America* (New York: W. W. Norton, 2009).
36. Faull, "The Nain Indian House," accessed May 13, 2017, http://storiesofthesusquehanna.blogs.bucknell.edu/2014/01/24/the-nain-indian-house/.
37. Leibert, "Wechquetank," 71–72.
38. Martin, *Historical Sketch of Bethlehem in Pennsylvania*, 17.
39. Engel, *Religion and Profit*, 182–183; Wheeler and Hahn, "On an Eighteenth-Century Trail of Tears," 49; and Merritt, *At the Crossroads*, 272–281.
40. See Wheeler and Hahn, "On an Eighteenth-Century Trail of Tears," n. 10.
41. "The Diary of the Indian *Gemeine* on Province Island, in the Delaware River, 5 miles under Philadelphia, 1763–January 4, 1764," Bethlehem Digital History Project: accessed April 5, 2017, http://bdhp.moravian.edu/community_records/christianindians/provincediary/1764province.html.
42. "The Diary of the Indian *Gemeine* on Province Island, in the Delaware River, 5 miles under Philadelphia, 1763–January 4, 1764," Bethlehem Digital History Project: accessed April 5, 2017, http://bdhp.moravian.edu/community_records/christianindians/provincediary/1764province.html.
43. Merrell, *Into the American Woods: Negotiators on the Pennsylvania Frontier*, 16.
44. "The Diary of the Indian *Gemeine* on Province Island, in the Delaware River, 5 miles under Philadelphia, 1763–January 4, 1764," Bethlehem Digital History Project: accessed April 5, 2017, http://bdhp.moravian.edu/community_records/christianindians/provincediary/1764province.html. All quotations come from this document.
45. Wheeler and Hahn, "On an Eighteenth-Century Trail of Tears," 50–51.
46. See Merritt, *At the Crossroads*, 320, for the Lebenslauf of Judith, a Nain resident who died in Philadelphia.
47. Wheeler and Hahn, "On an Eighteenth-Century Trail of Tears," 52; and Hartzell, "Joshua, Jr.: Moravian Indian Musician," 3.
48. Bethlehem Digital History Project: http://bdhp.moravian.edu/personal_papers/letters/indians/1765indianaddress.html (accessed 5 April, 2017).
49. One of the houses from Nain is still extant in Bethlehem, and is located on Heckewelder Place.
50. HG 1956, verse 18. Wheeler and Hahn, "On an Eighteenth-Century Trail of Tears," 56.
51. Wheeler and Hahn, "On an Eighteenth-Century Trail of Tears," 61.
52. Wheeler and Hahn, "On an Eighteenth-Century Trail of Tears," 61.
53. Wheeler and Hahn, "On an Eighteenth-Century Trail of Tears," 62, 65, 66–69.
54. Wheeler and Hahn, "On an Eighteenth-Century Trail of Tears," 74.
55. Wheeler and Hahn, "On an Eighteenth-Century Trail of Tears," 54.
56. Olmstead, *David Zeisberger: a Life Among the Indians*, 331.
57. Joseph was the son of the Quaker convert John Bull (Schebosh) and his Native American wife, Christiana.
58. The people at Schönbrunn were saved because a messenger sent by Zeisberger to inquire about their safety found the body of Joseph Schebosh and the tracks of militia. He warned the congregation at Schönbrunn and they were able to escape before the militia entered the village.
59. Olmstead, *David Zeisberger: a Life Among the Indians*, 333.
60. "The Tragedy of Gnadenhütten," 48.

61. Olmstead, *David Zeisberger: a Life Among the Indians*, 333.
62. "The Tragedy of Gnadenhütten," 49–50.
63. From a summary of Joshua's life written by missionary Johann Heckewelder, quoted in Libin, "What Instrument Did Joshua Make?" in *Proceedings of the Seventh Bethlehem Conference on Moravian Music* (Winston-Salem, NC: Moravian Music Foundation, 2007).
64. Olmstead, *David Zeisberger: A Life Among the Indians*, 334.
65. Zeisberger, *History of the North American Indians*, 107.

EPILOGUE

Petquotting

On July 10, 1789, several Moravian Brethren, including David Zeisberger, wrote a letter to the new president of the United States, George Washington. They cordially congratulated him on his appointment and expressed their steadfast belief that his administration would be "salutary and a blessing to that nation whose unanimous voice has called you to preside over it." With the letter they enclosed a small book on the history of the Moravian missions, and a request: "Permit us at the same time to recommend in a particular manner the Brethren's mission among the Indians in the Territory of the United States, which is at present at Petquotting on Lake Erie and in a very dangerous situation to your kind notice and protection, and to lay before you the ardent wish and anxious desire we have of seeing the light of the glorious Gospel spread more and more over this Country."[1] This mission along the shore of Lake Erie, they hoped, would be supported by Washington's administration as an advantageous situation for both the Moravian Church and the fledgling United States government, given the need to control access to the Lakeshore Path, an important conduit between the Six Nations and Delaware communities in Ohio. However, the letter subtly reminded Washington of something else he would certainly remember. A terrible tragedy that had happened under his watch as commander of the Revolutionary Army just seven years prior—Gnadenhütten. Although Washington and the Congress had ordered an investigation, none of the perpetrators had ever been brought to justice. Not one person could be found who would admit to having been to the Moravian towns in the Tuscarawas Valley. And the stalwart and even violent allegiances that bound the inhabitants of settler communities along the Pennsylvania–Ohio border forestalled any further attempts to break the wall of silence.

But no one in 1789 had forgotten. The most notorious massacre of the Revolutionary War continued to be written about and retold in pamphlets and stories well into the nineteenth century. In 1792, two thousand acres were gifted by the United States Congress to the Moravian Church in partial compensation for the loss of their communities in Ohio. And in the summer of 1794, Johann Jacob Eyerly walked across Pennsylvania to survey those lands for the church. This keen observer of Pennsylvania's natural and sounded resources could not help but have noticed the irony of his journey. In the name of the church, his survey would claim the lands of Native people who swore kinship with those who had been killed at Gnadenhütten. Out of their suffering, the church would profit in the end. Native places now became Moravian spaces. Eyerly and those of his generation who faced the first years of that new America witnessed and perpetuated new land deals and treaties, new maps and new frontiers that would gradually claim Native territories. Their mission to Christianize also gradually became a mission to claim new lands on behalf of the church. Bethlehem and other Moravian towns such as Lititz, Emmaus, and Nazareth prospered and continued to grow, but the Native Moravian communities at Nain, Friedenshütten, and Gnadenhütten were abandoned and built over by settler homes and businesses. As the Native Moravian congregation settled farther west in communities such as Petquotting in Ohio, Michigan, Indiana, Ontario, and Wisconsin, their descendants were displaced from their traditional lands. Into their place came my ancestors, establishing farms and businesses and burying their dead in new "homelands."

With new frontiers, new maps, new stories, and new songs carved by my ancestors and other early settlers in Pennsylvania had come new waves of people claiming a place within the landscape of Penn's Woods. Once, as a child, I was taken on an archaeological field trip. We were bused to a local farm field just a few miles from our elementary school. There, an archaeologist parent volunteer gave us little trowels and dirt sifters. My classmates' digging unearthed several flint arrowheads, carved from the local shale. My trowel unearthed a bead. It was blue, with a beautiful transparency that sparkled in the sunlight. I wondered about the people who had left it there. How had it come to be here in the earth in this farmer's field? The parent volunteer explained that this field was the site of an "old Indian village." Here, along the creek, an important water source for the community, there had once been homes and other sorts of agricultural fields. Lives were lived, stories were told, songs were sung. Now the village left its traces on the land in bits of material culture that were dug up and plowed up in the fields along the creek side. It occurred to me then that we were newcomers to this land, my classmates and I. Settlers living in new spaces. People occupying new villages built over the material remains of former generations who lived at this place near the confluence of the Moshannon Creek and the western branch of the Susquehanna River. Like all children, I was curious to know more about the people who had come before me.

Writing this book has allowed me to do just that. It has taken me out onto the land and into the places of its songs and stories. And it has challenged me to consider the many histories of Pennsylvania—my home. In August 2017, I visited a Mohican descendant community at the Stockbridge-Munsee Reservation in Wisconsin to sing hymns in Mohican from the Moravian Church's archives in Bethlehem. After the service and a potluck meal at the Lutheran Church of the Wilderness, a church council member and his family conversed with me about the ways that their community still remembered and longed for the lands between the Hudson and the Housatonic Rivers, the villages of Shekomeko, Wechquadnach, Potatik, and others. Maheekunik, The Land Where the Waters Are Never Still. That conversation caused me to consider the ways that my perspective and relationship to my ancestral history as a descendant of a European Moravian differed from those who traced their ancestry to members of the Native congregation. We both gained something valuable from singing songs once written by our ancestors, but the songs we sang that day represented the potential revival of a community disrupted and a language no longer spoken. We were certainly both driven by a curiosity and desire to know ourselves better by understanding where we came from, but my desire for knowledge was not a struggle for cultural survival and sovereignty.

There are many paths that lead toward and away from the places we call home. In the 1880s, my mother's Swedish ancestors came to the area of the Moshannon and the Moravian to mine coal in the bituminous tunnel mine at Peale. They dreamed and saved to buy farmland. The forests around Peale were cleared for homes and businesses. Rivers were channeled into alternate courses and mountainsides were leveled with blast powder. The acoustic ecologies of strip and tunnel mines, factories, and railroads rang for miles around the western branch of the Susquehanna. But within forty years, these new soundscapes faltered and Peale returned to the forest. The instability of central Pennsylvania's coal industry left my family again in the poverty they had left Sweden to escape. Many moved farther west. My family remained, living and struggling on the farm that embodied their hopes and dreams. They loved the land with a fierceness born of poverty and adversity, a love that was bequeathed to the next generations. Yet, our more ancient ancestors had once inhabited other homelands. How deeply connected were we to this land? Objectively, historically, we were transitory. But this knowledge was not easily reconciled with my love of the places I had inhabited as a child.

In the summer of 2018, I returned to the Moravian and the Moshannon. My mother had passed away, and I had sold our family's farm in the spring of 2017. For the past twenty years, my life had taken me on paths that led far away from Pennsylvania. Painfully, I had to acknowledge that those paths would never lead me back. I struggled to reconcile my love of that place with the knowledge that I had no right now to traverse that land. Those rights belonged to a new family. But standing on the edge of those pastures and fields, and listening to the rushing rapids of the Moshannon, a class II stretch of whitewater, I began to hear different stories, larger stories—stories of Pennsylvania's past and of its present. Stories of the places my family had inhabited, loved by ancient families and new. Pains of losses that would never quite heal, written into the contours of the land. And memories of the many people in the past and present who shared a heritage in those stories. In the distance, I could see the windy hilltop where my own mother lay recently buried, and the tree-covered slope of Moravian Run where little Nathan had been laid to rest two centuries ago. The sounds of rushing waters and harsh winds coming off the Allegheny Front merged with the quieter sounds of the nearby meadows and forests. Songs of the present and of the past.

Note

1. Letter from the Society of the United Brethren for Propagating the Gospel to Gen. George Washington, President of the United States of America, 10 July, 1789. MissInd 314.4, MAB.

GLOSSARY: A MORAVIAN VOCABULARY

Arbeitslied work song

Aaywãakun place

Asuwakàn song

aus dem Herzen gesungen sung from the heart

Auszug für Singstunden guide for Singstunden

Beson spiritual medicine

Blutgemeine blood community

Dichterschulung poet school

Einsingen singing at death

Fremde strangers

Freunde friends

Gemeine congregation, community

Hausgemeine home community

Heimgang [lit. "going home"] death

Herrnhuter Gesangbuch The main hymnal used by eighteenth-century Moravian congregations

Kmeende congregation, community

Ksheexaayik Nuxkohmaawãakun sacred or pure song (hymn)

Lebenslauf spiritual (auto)biography

Lekhikàn map, book

Liederpredigt hymn sermon

Liebesmahl [lit. "lovefeast"] Communal Singstunde accompanied by a simple meal

Lied song

Liturg [lit. "liturgist"] leader of a Singstunde service

Los, das lot or lottery

Losung Commonly translated into English as "watchword." Daily scriptural passages, chosen by lot at the beginning of each year by church elders, and used by Moravians for daily spiritual guidance or meditation.

manitou spiritual being

Nuxkohmaawãakun song

Oeconomie [lit. "Economy"] Moravian communal economic system

Ootaanaak town, village, community

Osowheekun map, book

Pilgergemeine pilgrim's or missionary community

Salàmpwënikàn bell

Seiten Höhlgen [lit. "little Side Hole"] spear wound in Jesus's side

Singstunde [lit. "singing meeting"] Moravian worship service centered on a given idea or theme, involving the singing of memorized hymn verses or the improvisation of new hymns.

Tawtaw bell

Utènink town, village, community

BIBLIOGRAPHY

Primary Sources

"Abraham Steiner's Account of his Journey with Johann Heckewelder from Bethlehem to Pettquotting on the Huron River near Lake Erie, and Return, 1789," MAB.
Auszug für Singstunde. NB.IV.R1 120b, UA Herrnhut.
The Bethlehem Diary, vols. I and II. Bethlehem, PA: Archives of the Moravian Church, 1971.
Büttner, Gottlob. *Diary* (Shekomeko). MissInd 112, MAB.
———. *Journal.* MissInd 112, MAB.
Choral-Buch enthaltend alle zu dem Gesangbuche der Evangelischen Brüder-Gemeinen vom Jahre 1778 gehörige Melodien. Leipzig: Breitkopfischen Buchbruderen, 1784.
A Collection of Hymns, with Several Translations from the Hymn-Book of the Moravian Brethren. Second edition. London: At the Bible and Sun, Little-Wild-Street, near Lincoln's-Inn-Fields, 1743.
Correspondence of George Fabricius to Augustus Spangenberg. November 8, 1755. MissInd 118.6, no. 23, MAB.
Correspondence of Johann Schmick to Augustus Spangenberg. July 20, 1755. MissInd 118.6, MAB.
Eyerly, Johann Jacob, Jr. "Ein Bericht von den Reise den Brüden Jacob Eyerly iun. und Johann Heckewälder zur ausmessung der Landes am Lake Erie, welcher von den General Assembly in Pennsylvanien den Societät den Brüden zur Ausbreitung des Evangelii unter den Heyden geschenkt worden, im May und Juny 1794." *Personalia Heckewälder* I, MAB.
Gemeinnachrichten. May 1760. MAB and UA Herrnhut.
Gnadenhütten Diary. MissInd 116–18, MAB.
Goshen Diary. MissInd 171, MAB.
Gregor, Christian. *Choral-Buch: A facsimile of the first edition of 1784, compiled and edited by Christian Gregor.* Edited by James Boeringer. Winston-Salem, NC: Moravian Music Foundation Press, c1984.
Hagen, Johannes. *Diary.* MissInd 112, MAB.
Heiden Collegia, MissInd 217.12b, MAB.
Herrnhuter Gesangbuch: Christliches Gesang-Buch der Evangelischen Brüder-Gemeinen von 1735. Erich Beyreuther, Gerhard Meyer and Gudrun Meyer-Hickel, eds. Hildesheim: G. Olms, 1981.
Hirten Lieder von Bethlehem, Zum Gebrauch Vor Alles Was Arm Ist, Was Klein Und Gering Ist. 4971. Sower, Christopher, 1695–1758, printer, 1742.
Jüngerhaus Diarium. UA Herrnhut.
Pachgatgoch Diary. MissInd 114, MAB.
"Pattern for a Song Booklet of the Blessed Hearts of the Brown Nations of the Mahican Delawares and Some Short Verses of the Language of the Six Nations [Mohawk]." NB.VII.R.3.91 (1746), UA Herrnhut.

Probe zu Einem Gesang-Büchel, UA Herrnhut. Copy in MissInd 331.2, MAB.
Pyrlaeus, Johann Christoph. *Lebenslauf*, XII 1785, MAB.
Report of the Brown Brethren in Gnadenhütten, 1753. April 9, 1753. MissInd 319.4.17, no. 31, MAB.
Shekomeko Diary. MissInd 111, MAB.
A Short Account Which Br. Rauch Gave at the Lovefeast before the Lords Supper, August 23, 1746. MissInd 319.5.3, MAB.
Slover, John, H. H. Brackenridge, and Knight, eds. *Narratives of a Late Expedition against the Indians; with an Account of the Barbarous Execution of Col. Crawford; and the Wonderful Escape of Dr. Knight and John Slover from Captivity, in 1782*. Philadelphia: Printed by Francis Bailey, in Market Street, 1783.
Spangenberg, August Gottlieb. *An Account of the Manner in Which the Protestant Church of the Unitas Fratrum, or United Brethren, Preach the Gospel, and Carry on Their Missions among the Heathen. Translated from the German of the Rev. August Gottlieb Spangenberg*. London: printed and sold by H. Trapp, No. 1. Pater-Noster Row, for the Brethren's Society for the Furtherance of the Gospel; also sold at all the Brethren's Settlements and Chapels in Great-Britain and Ireland, 1788.
———. *Hymns Composed for the Use of the Brethren. By the Right Reverend, and Most Illustrious C. Z.* London: Published for the benefit of all mankind. In the year, 1749.
———. *The Life of Nicolas Lewis, Count of Zinzendorf and Pottendorf. Written in German by A. G. Spangenberg: Translated by L. T. Nyberg*. Bath: printed for T. Mills, Bookseller, and S. Hazard, Printer; sold at the Brethrens Chapels, and by the booksellers of Great Britain, Ireland, and America, 1773.
Synodal Relation 4279. March 15, 1750, MAB.
Synode Protokoll, 1745. UA Herrnhut.
"Verlass der vier Synoden der evangelischen Brüder-Unität, von den Jahren 1764, 1769, 1775, und 1782." MAB.
"Verses for the use of the Indians in Pisgachtigok (Connecticut)," MissInd 331.2 and 331.3, MAB.
Wequetank Diary. MissInd 122, MAB.

Secondary Sources

A True History of the Massacre of Ninety-Six Christian Indians, at Gnadenhuetten, Ohio, March 8th, 1782. New Philadelphia, Ohio: 1847.
Adams, Robert Harold. *Songs of Our Grandfathers: Music of the Unami Delaware Indians*. Dewey, OK: Touching Leaves Indian Crafts, 1977.
Alves, Daniel, and Ana Isabel Queiroz. "Exploring Literary Landscapes: From Texts to Spatiotemporal Analysis through Collaborative Work and GIS." *International Journal of Humanities and Arts Computing* 9, no. 1 (March 1, 2015): 57–73.
Anderson, Benedict R. *Imagined Communities: Reflections on the Origin and Spread of Nationalism*. New York: Verso, rev. ed. 2006.
Anderson, Elizabeth, Avril Maddrell, Catherine Mary McLoughlin, and Alana Vincent, eds. *Memory, Mourning, Landscape*. New York: Rodopi, 2010.

Anderson, Fred. *Crucible of War: The Seven Years' War and the Fate of Empire in British North America, 1754-1766*. New York: Vintage Books, 2001.
Anderson, Isobel. "Soundmapping Beyond The Grid: Alternative Cartographies of Sound." *Journal of Sonic Studies* 11 (January 14, 2016).
Anderson, Kay, ed. *Handbook of Cultural Geography*. London: Sage, 2003.
Andrews, Edward E. *Native Apostles: Black and Indian Missionaries in the British Atlantic World*. Cambridge, MA: Harvard University Press, 2013.
Artiss, Tom. "Music and Change in Nain, Nunatsiavut: More White Does Not Always Mean Less Inuit." *Études/Inuit/Studies* 38, no. 1/2 (2014): 33-52.
Atkinson, Niall. *The Noisy Renaissance: Sound, Architecture, and Florentine Urban Life*. University Park, PA: The Pennsylvania State University Press, 2016.
Atran, Scott, and Douglas L. Medin. *The Native Mind and the Cultural Construction of Nature*. Cambridge, MA: MIT Press, 2008.
Atwood, Craig D. *Community of the Cross: Moravian Piety in Colonial Bethlehem*. Max Kade German-American Research Institute Series. University Park, PA: Pennsylvania State University Press, 2004.
———. "Little Side Holes: Moravian Devotional Cards of the Mid-Eighteenth Century." *Journal of Moravian History* 6 (2009): 61-75.
———. "Sleeping in the Arms of Christ: Sanctifying Sexuality in the Eighteenth-Century Moravian Church." *Journal of the History of Sexuality* 8, no. 1 (1997): 25-51.
———. "The Union of Masculine and Feminine in Zinzendorf's Piety." In *Masculinity, Senses, Spirit*, edited by Katherine Faull. Lanham: Rowman & Littlefield, 2011.
———. "Understanding Zinzendorf's Blood and Wounds Theology." *Journal of Moravian History* 1 (2006): 31-47.
———. "Zinzendorf's Litany of the Wounds." *The Lutheran Quarterly* 11 (1997): 189-214.
Axtell, James. *After Columbus: Essays in the Ethnohistory of Colonial North America*. New York: Oxford University Press, 1988.
———. "Babel of Tongues: Communication with the Indians in Eastern North America." In *The Language Encounter in the Americas, 1492-1800: A Collection of Essays*, edited by Edward G. Gray and Norman Fiering, 16-18. European Expansion and Global Interaction. New York: Berghahn Books, 2000.
———. *Beyond 1492: Encounters in Colonial North America*. New York: Oxford University Press, 1992.
———. *The Indian Peoples of Eastern America: A Documentary History of the Sexes*. New York: Oxford University Press, 2009.
———. *The European and the Indian: Essays in the Ethnohistory of Colonial North America*. New York: Oxford University Press, 1981.
———. *The Invasion Within: The Contest of Cultures in Colonial North America*. Cultural Origins of North America 1. New York: Oxford University Press, 1985.
Bakeless, John. *Daniel Boone, Master of the Wilderness*. Lincoln: University of Nebraska Press, 1989.
Baker, Geoffrey. *Imposing Harmony: Music and Society in Colonial Cuzco*. Durham, NC: Duke University Press, 2008.
———. "Indigenous Musicians in the Urban 'Parroquias de Indios' of Colonial Cuzco, Peru." *Il Saggiatore Musicale* 9, no. 1/2 (2002): 39-79.
———. "Music at Corpus Christi in Colonial Cuzco." *Early Music* 32, no. 3 (2004): 355-367.

———. "Music in the Convents and Monasteries of Colonial Cuzco." *Latin American Music Review/Revista de Música Latinoamericana* 24, no. 1 (2003): 1–41.

Balay, Olivier. "Discrete Mapping of Urban Soundscapes." *Soundscape: The Journal of Acoustic Ecology* 1 (2004): 1–6.

Bandt, Ros. *Hearing Places: Sound, Place, Time and Culture*. Newcastle, UK: Cambridge Scholars, 2009.

Barker, Adam J. "Locating Settler Colonialism." *Journal of Colonialism and Colonial History* 13, no. 3 (December 2012).

Barr, Daniel P. "Did Pennsylvania Have a Middle Ground? Examining Indian-White Relations on the Eighteenth-Century Pennsylvania Frontier." *The Pennsylvania Magazine of History and Biography* 136, no. 4 (2012): 337–363.

Barr, Juliana, Edward Countryman, and William P. Clements, eds. *Contested Spaces of Early America*. Early American Studies. Philadelphia: University of Pennsylvania Press, 2014.

Barthes, Roland. *The Responsibility of Forms: Critical Essays on Music, Art, and Representation*. 1st ed. New York: Hill and Wang, 1985.

Bartram, John. *Travels in Pensilvania and Canada*. March of America Facsimile Series, No. 41. Ann Arbor: University Microfilms, 1966.

Basso, Keith H. *Wisdom Sits in Places: Landscape and Language among the Western Apache*. Albuquerque, NM: University of New Mexico Press, 1996.

Battiste, Marie Ann. *Reclaiming Indigenous Voice and Vision*. Vancouver: University of British Columbia Press, 2009.

Beatty, Charles, and Guy Soulliard Klett. *Journals of Charles Beatty, 1762–1769*. Presbyterian Historical Society Publications 3. University Park: Pennsylvania State University Press, 1962.

Beauchamp, William Martin, ed., August Gottlieb Spangenberg, David Zeisberger, John Martin Mack. *Moravian Journals Relating to Central New York, 1745–1766*. Bowie, MD: Heritage Books, 1999.

Becker, John. *Goethe und die Brüdergemeine*. Neudietendorf in Thür, 1922.

Behrendt, Frauke. "GPS Sound Walks, Ecotones and Edge Species; Experiencing Sound within Teri Rueb's Mobile Metaphor." *Soundscape: The Journal of Acoustic Ecology* 12, no. 1 (Winter/Spring 2012): 25–28.

Beneke, Chris, and Christopher S. Grenda, eds. *The First Prejudice: Religious Tolerance and Intolerance in Early America*. Early American Studies. Philadelphia: University of Pennsylvania Press, 2011.

Berger, Anna Maria Busse. *Medieval Music and the Art of Memory*. Berkeley, CA: University of California Press, 2008.

———. *In Search of Medieval Music in Africa and Germany: Scholars, Singers, and Missionaries*. Chicago, IL: University of Chicago Press, forthcoming 2020.

———. "Spreading the Gospel of Singbewegung: An Ethnomusicologist Missionary in Tanganyika of the 1930s." *Journal of the American Musicological Society* 66, no. 2 (2013): 475–522.

Berman, Merrick Lex. "Boundaries or Networks in Historical GIS: Concepts of Measuring Space and Administrative Geography in Chinese History." *Historical Geography* 33 (2005): 118–133.

Bermudez, Egberto. "Urban Musical Life in the European Colonies: Examples from Spanish America, 1530–1650." In *Music and Musicians in Renaissance Cities and Towns*, edited by Fiona Kisby, 167–180. Cambridge: Cambridge University Press, 2006.

Bettermann, Wilhelm. *Theologie und Sprache bei Zinzendorf*. Gotha: L. Klotz, 1935.
Beutler, Ernst. *Essays Um Goethe*. Sammlung Dieterich 101. Bremen: Carl Schunemann Verlag, 1962.
Beyreuther, Erich. *Der Junge Zinzendorf*, vol. 2. Marburg an der Lahn: Verlag der Francke-Buchhandlung, 1957.
———. *Studien Zur Theologie Zinzendorfs; Gesammelte Aufsätze*. Neukirchen-Vluyn: Neukirchener Verlag, 1962.
Beyreuther, Erich, and Gerhard Meyer, eds. *Zinzendorf: Hauptschriften*. Hildesheim: G. Olms, 1962.
Blesser, Barry, and Linda-Ruth Salter. *Spaces Speak, Are You Listening? Experiencing Aural Architecture*. Cambridge, MA: MIT Press, 2009.
Bianchi, Frederick W., and V. J. Manzo. *Environmental Sound Artists: In Their Own Words*. New York, NY: Oxford University Press, 2016.
Bijsterveld, Karin. *Sonic Skills: Listening for Knowledge in Science, Medicine and Engineering (1920s–Present)*. London: Palgrave Macmillan, 2019.
Bloechl, Olivia Ashley. *Native American Song at the Frontiers of Early Modern Music*. New Perspectives in Music History and Criticism. New York: Cambridge University Press, 2008.
Bodenhamer, David J., John Corrigan, and Trevor M. Harris, eds. *Deep Maps and Spatial Narratives*. The Spatial Humanities. Bloomington: Indiana University Press, 2015.
———. *The Spatial Humanities: GIS and the Future of Humanities Scholarship*. The Spatial Humanities. Bloomington: Indiana University Press, 2010.
Bodenhamer, David J., Trevor M. Harris, and John Corrigan. "Deep Mapping and the Spatial Humanities." *Journal of Humanities & Arts Computing: A Journal of Digital Humanities* 7, nos. 1–2 (March 2013): 170–175.
Bohannon, Richard, ed. *Religions and Environments: A Reader in Religion, Nature and Ecology*. New York: Bloomsbury, 2014.
Bossart, Johann Jakob. *Kurze Anweisung Naturalien zu samlen*. Barby, 1774.
Bourne, Russell. *Gods of War, Gods of Peace: How the Meeting of Native and Colonial Religions Shaped Early America*. New York: Harcourt, 2002.
Boutin, Aimee. *City of Noise: Sound and Nineteenth-Century Paris*. Studies in Sensory History. Urbana: University of Illinois Press, 2015.
Braun, Sebastian Felix, ed. *Transforming Ethnohistories: Narrative, Meaning, and Community*. Norman: University of Oklahoma Press, 2013.
Brekus, Catherine A., and W. Clark Gilpin, eds. *American Christianities: A History of Dominance and Diversity*. Chapel Hill: University of North Carolina Press, 2011.
Brooks, Joanna. "Six Hymns by Samson Occom." *Early American Literature* 38, no. 1 (2003): 67–87.
Brooks, Lisa. "Awikhigawôgan Ta Pildowi Ôjmowôgan: Mapping a New History." *The William and Mary Quarterly* 75, no. 2 (May 2018): 259–294.
———. *The Common Pot: The Recovery of Native Space in the Northeast*. Indigenous Americas Series. Minneapolis: University of Minnesota Press, 2008.
———. "Locating an Ethical Native Criticism." In *Reasoning Together: The Native Critics Collective*, edited by Craig S. Womack, Daniel Heath Justice, and Christopher B. Teuton, 234–264. Norman: University of Oklahoma Press, 2008.
———. *Our Beloved Kin: A New History of King Philip's War*. Henry Roe Cloud Series on American Indians and Modernity. New Haven: Yale University Press, 2018.
———. "The Primacy of the Present, the Primacy of Place: Navigating the Spiral of History in the Digital World." *PMLA* 127, no. 2 (2012): 308–316.

Brookes, George S., and Anthony Benezet. *Friend Anthony Benezet (Benezet Letters)*. Philadelphia: University of Pennsylvania Press, 1937.

Brückner, Martin. *Early American Cartographies*. Chapel Hill: University of North Carolina Press, 2011.

———. *The Geographic Revolution in Early America: Maps, Literacy, and National Identity*. Chapel Hill: Published for the Omohundro Institute of Early American History and Culture by the University of North Carolina Press, 2006.

Brunsman, Denver. "Atlantic Passengers." *The William and Mary Quarterly* 72, no. 3 (2015): 509–512.

Bryson, J. Scott. *The West Side of Any Mountain: Place, Space, and Ecopoetry*. Iowa City: University of Iowa Press, 2005.

Burkholder, Jared S. "Neither 'Kriegerisch' nor 'Quäkerisch': Moravians and the Question of Violence in Eighteenth-Century Pennsylvania." *Journal of Moravian History* 12, no. 2 (November 2012): 143–169.

Butler, Toby. "Memoryscape: How Audio Walks Can Deepen Our Sense of Place by Integrating Art, Oral History and Cultural Geography." *Geography Compass* 1, no. 3 (2007): 360–372.

Butt, John. *Playing with History*. Cambridge: Cambridge University Press, 2002.

Caldwell, Alice May. "Music of the Moravian 'Liturgische Gesänge' (1791–1823): From Oral to Written Tradition." PhD dissertation, New York University, 1987.

Callicott, J. Baird, and Michael P. Nelson. *The Great New Wilderness Debate: An Expansive Collection of Writings Defining Wilderness, from John Muir to Gary Snyder*. Athens, GA: University of Georgia Press, 1998.

Calloway, Colin G. *The American Revolution in Indian Country: Crisis and Diversity in Native American Communities*. Cambridge Studies in North American Indian History. Cambridge: Cambridge University Press, 1995.

———, ed. *The World Turned Upside Down: Indian Voices from Early America, A Brief History with Documents*. The Bedford Series in History and Culture. Boston, MA: Bedford/St. Martin's, 2016.

Camporesi, Piero. *Juice of Life: The Symbolic and Magic Significance of Blood*. New York: Continuum, 1995.

Cappon, Lester Jesse, Barbara Bartz Petchenik, and John Hamilton Long, eds. *Atlas of Early American History: The Revolutionary Era, 1760–1790*. Princeton: Published for the Newberry Library and the Institute of Early American History and Culture by the Princeton University Press, 1976.

Caquard, S., G. Brauen, B. Wright, and P. Jasen. "Designing Sound in Cybercartography: From Structured Cinematic Narratives to Unpredictable Sound/Image Interactions." *International Journal of Geographical Information Science* 22, nos. 11–12 (2008): 1219–1245.

Carr, Kurt W., and J. M. Adovasio. *Ice Age Peoples of Pennsylvania*. Harrisburg: PA: Pennsylvania Historical and Museum Commission, in cooperation with the Pennsylvania Archaeological Council, 2002.

Carr, Kurt W., and Roger W. Moeller. *First Pennsylvanians: The Archaeology of Native Americans in Pennsylvania*. Harrisburg: PA: Pennsylvania Historical and Museum Commission, 2015.

Carruthers, Mary J. *The Book of Memory: A Study of Memory in Medieval Culture*. Cambridge: Cambridge University Press, rev. 2014.

Casey, Edward S. *The Fate of Place: A Philosophical History*. Berkeley: University of California Press, rev. 2013.
———. *Representing Place: Landscape Painting and Maps*. Minneapolis: University of Minnesota Press, 2002.
Cayton, Andrew R. L., Fredrika J. Teute, and the Omohundro Institute of Early American History & Culture. *Contact Points: American Frontiers from the Mohawk Valley to the Mississippi, 1750–1830*. Chapel Hill: University of North Carolina Press, 1998.
Chakrabarty, Dipesh. *Provincializing Europe: Postcolonial Thought and Historical Difference*. Princeton Studies in Culture/Power/History. Princeton, NJ: Princeton University Press, 2000.
Chastellux, François Jean. *Travels in North America in the Years 1780, 1781 and 1782*, edited by Howard C. Rice. Institute of Early American History and Culture. Chapel Hill: University of North Carolina Press, 2011.
Chidester, David, and Edward Tabor Linenthal, eds. *American Sacred Space*. Religion in North America. Bloomington: Indiana University Press, 1995.
Cipolla, Craig N. *Becoming Brothertown: Native American Ethnogenesis and Endurance in the Modern World*. Tuscon: University of Arizona Press, 2013.
Colbert, Maile. "Wayback Sound Machine: Sound Through Time, Space, and Place." *Soundscape: The Journal of Acoustic Ecology* 13, no. 1 (Winter/Spring 2013): 21–24.
Constantia, Lady A. (Judith Sargent Murray). "Description of a Journey to Bethlehem, Pennsylvania." *The New York Magazine, or Literary Repository (1790–1797)*. New York: 1790.
Convery, Ian, Gerard Corsane, and Peter Davis, eds. *Making Sense of Place: Multidisciplinary Perspectives*. Heritage Matters 7. Rochester, NY: Boydell & Brewer, 2012.
Cope, Meghan, and Sarah Elwood, eds. *Qualitative GIS: A Mixed Methods Approach*. Los Angeles: SAGE, 2009.
Corbin, Alain. "Identity, Bells, and the Nineteenth-Century French Village." In *Hearing History: A Reader*, edited by Mark M. Smith. Athens: University of Georgia Press, 2004.
———. *Village Bells: Sound and Meaning in the Nineteenth-Century French Countryside*. European Perspectives. New York: Columbia University Press, 1998.
Corrigan, John, ed. *Religion, Space, and the Atlantic World*. Columbia: University of South Carolina Press, 2017.
Corrigan, John, and Lynn S. Neal, eds. *Religious Intolerance in America: A Documentary History*. Chapel Hill: University of North Carolina Press, 2010.
Cotter, John L., Daniel G. Roberts, and Michael Parrington. *The Buried Past: An Archaeological History of Philadelphia*. Philadelphia: University of Pennsylvania Press, 1992.
Crampton, Jeremy W., and John Krygier. "An Introduction to Critical Cartography." *ACME: An International Journal for Critical Geographies* 4, no. 1 (2005): 11–33.
Crang, Mike. "Qualitative Methods: There Is Nothing Outside the Text?" *Progress in Human Geography* 29, no. 2 (April 2005): 225–233.
Cresswell, Tim. *Place: A Short Introduction*. Short Introductions to Geography. Malden, MA: Blackwell, 2004.
Cronon, William. *Changes in the Land: Indians, Colonists, and the Ecology of New England*. New York: Hill and Wang, 1983.
———, ed. *Uncommon Ground: Toward Reinventing Nature*. New York: W. W. Norton, 1995.

Dally Starna, Corinna, and William A. Starna. *Gideon's People: Being a Chronicle of an American Indian Community in Colonial Connecticut and the Moravian Missionaries Who Served There*. The Iroquoians and Their World. Lincoln: University of Nebraska Press, 2009.

Danckaerts, Jasper. *Journal of Jasper Danckaerts, 1679–1680*. New York: Charles Scribner's Sons, 1913.

Daniels, Stephen, et al. *Envisioning Landscapes, Making Worlds*. London: Routledge, 2012.

David, Hans Theodore. *Musical Life in the Pennsylvania Settlements of the Unitas Fratrum*. Winston-Salem, NC: Moravian Music Foundation, 1959.

Davies, Graham I. *The Way of the Wilderness: A Geographical Study of the Wilderness Itineraries in the Old Testament*. New York: Cambridge University Press, 1979.

Delaware Tribe of Indians. Lenape Talking Dictionary. Accessed April 7, 2018, http://talk-lenape.org/stories?id=72#1694.

DeLucia, Christine. "Locating Kickemuit: Springs, Stone Memorials, and Contested Placemaking in the Northeastern Borderlands." *Early American Studies: An Interdisciplinary Journal* 13, no. 2 (April 2015): 467–502.

———. "The Memory Frontier: Uncommon Pursuits of Past and Place in the Northeast after King Philip's War." *The Journal of American History* 98, no. 4 (2012): 975–997.

———. *Memory Lands: King Philip's War and the Place of Violence in the Northeast*. Henry Roe Cloud Series on American Indians and Modernity. New Haven: Yale University Press, 2018.

———. "'The Sound of Violence': Music of King Philip's War and Memories of Settler Colonialism in the American Northeast." *Common-Place: The Journal of Early American Life* 13, no. 2 (Winter 2013). www.common-place-archives.org/vol-13/no-02/delucia/.

———. "Speaking Together: The Brothertown Indian Community and New Directions in Engaged Scholarship." *Early American Literature* 50, no. 1 (February 2015): 167–187.

Del Casino, Vincent J., Jr., and Stephen P. Hanna. "Beyond The 'Binaries': A Methodological Intervention for Interrogating Maps as Representational Practices." *ACME: An International Journal for Critical Geographies* 4, no. 1 (2005): 34–56.

Dell' Antonio, Andrew. *Listening as Spiritual Practice in Early Modern Italy*. Berkeley: University of California Press, 2011.

Deloria, Philip J. "Americans Indians, American Studies and the ASA." *American Quarterly* 55, no. 4 (2003): 669–680.

———. *Indians in Unexpected Places*. Culture America. Lawrence: University Press of Kansas, 2004.

———. *Playing Indian*. Yale Historical Publications. New Haven, CT: Yale University Press, 1998.

———. "What Is the Middle Ground, Anyway?" *The William and Mary Quarterly* 63, no. 1 (2006): 15–22.

DeLyser, Dydia, ed. *The SAGE Handbook of Qualitative Geography*. Thousand Oaks, CA: Sage, 2010.

DeRogatis, Amy. *Moral Geography: Maps, Missionaries, and the American Frontier*. Religion and American Culture. New York: Columbia University Press, 2003.

De Schweinitz, Edmund. *The Life and Times of David Zeisberger, the Western Pioneer and Apostle of the Indians*. Philadelphia: J. B. Lippincott, 1870.

Devall, Bill, and George Sessions. *Deep Ecology: Living as If Nature Mattered*. Salt Lake City: G. M. Smith, 2007.
Deyrup, Marta Mestrovic, ed. *Digital Scholarship*. Routledge Studies in Library and Information Science 6. New York: Routledge, 2009.
Diamond, Beverley. *Native American Music in Eastern North America: Experiencing Music, Expressing Culture*. Global Music Series. New York: Oxford University Press, 2008.
Diamond, Beverley, M. Sam Cronk, and Franziska Von Rosen. *Visions of Sound: Musical Instruments of First Nations Communities in Northeastern America*. Waterloo, Ontario, Canada: Wilfrid Laurier University Press, 1994.
Dickens, Charles. *American Notes and Pictures from Italy*. Oxford Illustrated Dickens. Oxford: Oxford University Press, 1987.
Dodge, Martin, Rob Kitchin, and C. R. Perkins. *Rethinking Maps: New Frontiers in Cartographic Theory*. Routledge Studies in Human Geography 28. New York: Routledge, 2009.
Dohm, Burkhard. "Des Blutes Licht-Tinctur: Alchimistische Konzepte in Herrnhutischer Poesie." In *Künste und Natur in Diskursen der Frühen Neuzeit*, Wolfenbütteler Arbeiten zur Barockforschung, Hartmut Laufhütte et al., eds. Harrassowitz Verlag, 2000.
———. *Poetische Alchimie: Öffnung Zur Sinnlichkeit in Der Hohelied- Und Bibeldichtung von Der Protestantischen Barockmystik Bis Zum Pietismus*. Studien Zur Deutschen Literatur 154. Tübingen: Max Niemeyer Verlag, 2000.
Donehoo, George Patterson. *Indian Villages and Place Names in Pennsylvania*. Baltimore, MD: Gateway, 1977.
Dwyer, Philip G., and Lyndall Ryan, eds. *Theatres of Violence: Massacre, Mass Killing, and Atrocity throughout History*. Studies on War and Genocide 11. New York: Berghahn Books, 2012.
Dwyer, Claire, and Gail Davies. "Qualitative Methods III: Animating Archives, Artful Interventions and Online Environments." *Progress in Human Geography* 34, no. 1 (February 2010): 88–97.
Ekirch, A. Roger. *At Day's Close: Night in Times Past*. New York: Norton, 2005.
Eliade, Mircea. *The Sacred and the Profane: The Nature of Religion*. New York: Harcourt, Brace & World, 1959.
Elder, John. *American Nature Writers*. New York: Simon & Schuster and Prentice Hall International, 1996.
Elwood, Sarah. "Geographic Information Science: Visualization, Visual Methods, and the Geoweb." *Progress in Human Geography* 35, no. 3 (June 2011): 401–408.
Emmerson, Simon, ed. *The Language of Electroacoustic Music*. London: Palgrave Macmillan, 1986.
Engel, Katherine Carté. "Moravians in the Eighteenth-Century Atlantic World." *Journal of Moravian History* 12, no. 1 (May 2012): 1–19.
———. *Religion and Profit: Moravians in Early America*. Philadelphia: University of Pennsylvania Press, 2009.
———. "The Strangers' Store: Moral Capitalism in Moravian Bethlehem, 1753–1775." *Early American Studies: An Interdisciplinary Journal* 1, no. 1 (October 2007): 90–126.
Erb, Peter C., ed. *Pietists: Selected Writings*. The Classics of Western Spirituality. New York: Paulist Press, 1983.

Erbe, Hellmuth. *Bethlehem, Pa. Eine Kommunistische Herrnhuter Kolonie Des 18 Jahrhunderts*. Stuttgart: Ausland und heimat Verlagsaktiengesellschaft, 1929.

Erben, Patrick M. *A Harmony of the Spirits: Translation and the Language of Community in Early Pennsylvania*. Chapel Hill: Published for the Omohundro Institute of Early American History and Culture, Williamsburg, Virginia, by the University of North Carolina Press, 2012.

Erlmann, Veit, ed. *Hearing Cultures: Essays on Sound, Listening, and Modernity*. Wenner-Gren International Symposium Series. New York: Berg, 2004.

Ettwein, John, and John Jordan, ed. "Rev. John Ettwein's Notes of Travel from the North Branch of the Susquehanna to the Beaver River, Pennsylvania, 1772." *The Pennsylvania Magazine of History and Biography* 25, no. 2 (1901): 208–219.

Eyerly, Sarah Justina. "*Der Wille Gottes*: Musical Improvisation in Eighteenth-Century Moravian Communities." In *Self, Community, World, Moravian Education in a Transatlantic World*, edited by Heikki Lempa and Paul Peucker, 201–227. Studies in Eighteenth-Century America and the Atlantic World. Bethlehem, PA: Lehigh University Press, 2010.

———. "Mozart and the Moravians." *Early Music* 47, no. 2 (May 2019): 161–182.

———. "'Singing from the Heart': Memorization and Improvisation in an Eighteenth-Century Utopian Community." PhD dissertation, University of California Davis, 2007.

Faull, Katherine M., ed. *Anthropology and the German Enlightenment: Perspectives on Humanity*. Lewisburg, PA: Bucknell University Press, 1995.

———. "Charting the Colonial Backcountry: Joseph Shippen's Map of the Susquehanna River." *The Pennsylvania Magazine of History and Biography* 136, no. 4 (2012): 461–465.

———. "Faith and Imagination: Nikolaus Ludwig von Zinzendorf's Anti-Enlightenment Philosophy of Self." In *Anthropology and the German Enlightenment: Perspectives on Humanity*, edited by Katherine Faull, 23–56. Lewisburg, PA: Bucknell University Press, 1995.

———. "From Friedenshütten to Wyoming: Johannes Ettwein's Map of the Upper Susquehanna (1768) and an Account of His Journey." *Journal of Moravian History* 11 (2011): 82–96.

———. "Mapping a Mission: The Origins of Golkowsky's 1768 Map of Friedenshütten, Pennsylvania." *Journal of Moravian History* 7 (2009): 107–116.

———, ed. *Masculinity, Senses, Spirit*. Lanham: Rowman & Littlefield, 2011.

———. *Moravian Women's Memoirs: Their Related Lives, 1750–1820*. Women and Gender in North American Religions. Syracuse, NY: Syracuse University Press, 1997.

———. "Speaking and Truth-Telling: *Parrhesia* in the 18th-Century Moravian Church." In *Self, Community, World: Moravian Education in a Transatlantic World*, edited by Heikki Lempa and Paul Peucker, 204–230. Studies in Eighteenth-Century America and the Atlantic World. Bethlehem, PA: Lehigh University Press, 2010.

———. "The Life of Johann Georg Jungmann (1720–1808)." In *The Distinctiveness of Moravian Culture: Essays and Documents in Moravian History in Honor of Vernon H. Nelson on his Seventieth Birthday*, edited by Vernon H. Nelson, Craig D. Atwood, and Peter Vogt. Nazareth, PA: Moravian Historical Society, 2003.

———. "The Experience of the World as the Experience of the Self: Smooth Rocks in a River Archipelago." In *Re-Imagining Nature: Environmental Humanities and Ecosemiotics*, edited by Alfred K. Siewers, 197–214. Lewisburg, PA: Bucknell University Press, 2014.

———. "'You Are the Savior's Widow:' Religion, Sexuality and Bereavement in the Eighteenth-Century Moravian Church." *Journal of Moravian History* 8 (2010): 89–115.
Feld, Steven. *Sound and Sentiment: Birds, Weeping, Poetics, and Song in Kaluli Expression*. Publications of the American Folklore Society 5. Philadelphia: University of Pennsylvania Press, 1982.
Feld, Steven, and Keith H. Basso, eds. *Senses of Place*. School of American Research Advanced Seminar Series. Santa Fe, NM: School of American Research Press; distributed by the University of Washington Press, 1996.
Ferguson, Leland. "What Means 'Gottes Acker'? Leading and Misleading Translations of Salem Records." *Journal of Moravian History* 5 (2008): 68–87.
Fisher, Alexander J. *Music, Piety, and Propaganda: The Soundscapes of Counter-Reformation Bavaria*. The New Cultural History of Music. New York: Oxford University Press, 2014.
Fisher, Kyle. "After Gnadenhütten: The Moravian Indian Mission in the Old Northwest, 1782–1812." *Journal of Moravian History* 17, no. 1 (2017): 27–57.
Fisher, Linford D. "'I Believe They Are Papists!': Natives, Moravians, and the Politics of Conversion in Eighteenth-Century Connecticut." *The New England Quarterly* 81, no. 3 (2008): 410–437.
———. *The Indian Great Awakening: Religion and the Shaping of Native Cultures in Early America*. New York: Oxford University Press, 2012.
Fogleman, Aaron S. "Women on the Trail in Colonial America: A Travel Journal of German Moravians Migrating from Pennsylvania to North Carolina in 1766." *Pennsylvania History: A Journal of Mid-Atlantic Studies* 61, no. 2 (1994): 206–234.
Forsyth, Michael. *Buildings for Music: The Architect, the Musician, and the Listener from the Seventeenth Century to the Present Day*. Cambridge, MA: MIT Press, 1985.
"Forum: Is GIS Changing Historical Scholarship?" *Journal of Humanities & Arts Computing: A Journal of Digital Humanities* 3, nos. 1–2 (March 2009): 1–2.
Foucault, Michel. *Technologies of the Self: A Seminar with Michel Foucault*. Edited by Luther H. Martin, Huck Gutman, and Patrick H. Hutton. Amherst: University of Massachusetts Press, 1988.
Fowler, Michael. "Mapping Sound-Space: The Japanese Garden as Auditory Model." *Arq: Architectural Research Quarterly* 14, no. 1 (March 2010): 63–70.
Franklin, Benjamin. *A Narrative of the Late Massacres, in Lancaster County, of a Number of Indians, Friends of This Province, by Persons Unknown. With Some Observations on the Same*. Philadelphia: Franklin and Hall, 1764.
Franklin, Wayne, and Michael Steiner. *Mapping American Culture* 1. The American Land & Life Series. Iowa City: University of Iowa Press, 1992.
Funchion, John, Keri Holt, and Edward Watts, eds. *Mapping Region in Early American Writing*. Athens, GA: The University of Georgia Press, 2015.
Fur, Gunlög Maria. *A Nation of Women: Gender and Colonial Encounters among the Delaware Indians*. Early American Studies. Philadelphia: University of Pennsylvania Press, 2009.
Gaddis, John Lewis. *The Landscape of History: How Historians Map the Past*. New York: Oxford University Press, 2002.
Gandy, Matthew, and B. J. Nilsen. *The Acoustic City*. Berlin: Jovis, 2014.
Gemünden, Gerd, Colin G. Calloway, and Susanne Zantop, eds. *Germans and Indians: Fantasies, Encounters, Projections*. Lincoln: University of Nebraska Press, 2002.

Geoffrey-Schwinden, Rebecca. "Digital Approaches to Historical Acoustemologies: Replication and Reenactment." In *Digital Sound Studies*, edited by Mary Caton Lingold, Darren Mueller, and Whitney Trettien, 231–249. Durham, NC: Duke University Press, 2018.

Gillespie, Michele, and Robert Beachy, eds. *Pious Pursuits: German Moravians in the Atlantic World*. European Expansion & Global Interaction 7. New York: Berghahn Books, 2007.

Gipson, Lawrence Henry. *The Moravian Indian Mission on White River: Diaries and Letters, May 5, 1799, to November 12, 1806*. Indiana Historical Collections 23. Indianapolis: Indiana Historical Bureau, 1938.

Glover, Jeffrey, and Paul Chaat Smith, eds. *Colonial Mediascapes: Sensory Worlds of the Early Americas*. Lincoln: University of Nebraska Press, 2014.

Gollin, Gillian Lindt. *Moravians in Two Worlds: A Study of Changing Communities*. New York: Columbia University Press, 1967.

Goodman, Glenda. "American Identities in an Atlantic Musical World: Transhistorical Case Studies." PhD dissertation, Harvard University, 2012.

———. "'But They Differ from Us in Sound': Indian Psalmody and the Soundscape of Colonialism, 1651–75." *The William and Mary Quarterly* 69, no. 4 (2012): 793–822.

Gordon, Scott Paul. "Entangled by the World: William Henry of Lancaster and 'Mixed' Living in Moravian Town and Country Congregations." *Journal of Moravian History* 8 (2010): 7–52.

———. "Patriots and Neighbors: Pennsylvania Moravians in the American Revolution." *Journal of Moravian History* 12, no. 2 (November 2012): 111–142.

———. "The Paxton Boys and the Moravians: Terror and Faith in the Pennsylvania Backcountry." *Journal of Moravian History* 14, no. 2 (November 2014): 119–152.

———. "William Atlee's Description of Bethlehem (1779)." *Journal of Moravian History* 9 (2010): 83–88.

Gordon, Tom. "Found in Translation: The Inuit Voice in Moravian Music." *Newfoundland and Labrador Studies* 22, no. 1 (2007): 287–314.

Gould, Peter, and Rodney R. White. *Mental Maps*. Boston: Allen & Unwin, 1986.

Graber, Linda H. *Wilderness as Sacred Space*. Association of American Geographers Monograph Series 8. Washington: Association of American Geographers, 1976.

Graf, Lanie. "John Frederick Hintz, Eighteenth-Century Moravian Instrument Maker, and the Use of the Cittern in Moravian Worship." *Journal of Moravian History* 5 (2008): 7–39.

Gray, Edward G., and Norman Fiering, eds. *The Language Encounter in the Americas, 1492–1800: A Collection of Essays*. European Expansion and Global Interaction 1. New York: Berghahn Books, 2000.

Gray, Elma E. *Wilderness Christians: The Moravian Mission to the Delaware Indians*. Ithaca, NY: Cornell University Press, 1956.

Greer, Allan. *Property and Dispossession: Natives, Empires and Land in Early Modern North America*. Cambridge Studies in North American Indian History. New York: Cambridge University Press, 2018.

Gregerson, Linda, and Susan Juster, eds. *Empires of God: Religious Encounters in the Early Modern Atlantic*. Philadelphia: University of Pennsylvania Press, 2011.

Gregory, Ian. *A Place in History: A Guide to Using GIS in Historical Research*. History Data Service. Oakville, CT: David Brown, 2003.

Gregory, Ian N., and Alistair Geddes, eds. *Toward Spatial Humanities: Historical GIS and Spatial History*. The Spatial Humanities. Bloomington: Indiana University Press, 2014.

Gregory, Ian N., Andreas Kunz, and David J. Bodenhamer. "A Place in Europe: Enhancing European Collaboration in Historical GIS." *International Journal of Humanities and Arts Computing* 5, no. 1 (March 2011): 23–39.

Gregory, Ian N., and Richard G. Healey. "Historical GIS: Structuring, Mapping and Analysing Geographies of the Past." *Progress in Human Geography* 31, no. 5 (October 2007): 638–653.

Grider, Rufus Alexander. *Historical Notes on Music in Bethlehem, Pennsylvania, from 1741 to 1871*. Bethlehem, PA: J. Hill Martin, 1873.

Griffin, Patrick. *American Leviathan: Empire, Nation, and Revolutionary Frontier*. New York: Hill and Wang, 2007.

Griffiths, Nicholas. *Sacred Dialogues: Christianity and Native Religions in the Colonial Americas, 1492–1700*. England: Lulu Enterprises, 2006.

Griffiths, Nicholas, and Fernando Cervantes. *Spiritual Encounters: Interactions between Christianity and Native Religions in Colonial America*. Birmingham: University of Birmingham Press, 1999.

Gustafson, Sandra M. *Eloquence Is Power: Oratory & Performance in Early America*. Chapel Hill: Published for the Omohundro Institute of Early American History and Culture, Williamsburg, Virginia, by the University of North Carolina Press, 2000.

Halttunen, Karen. "Grounded Histories: Land and Landscape in Early America." *The William and Mary Quarterly* 68, no. 4 (2011): 513–532.

Hamilton, Kenneth G. "John Ettwein and the Moravian Church during the Revolutionary Period." Bethlehem, PA: Times Publishing Company, 1940.

Hanna, Charles A. *The Wilderness Trail: or The Ventures and Adventures of the Pennsylvania Traders on the Allegheny Path*. New York and London: 1911.

Harjo, Joy, and Stephen Strom. *Secrets from the Center of the World*. Sun Tracks 17. Tucson: University of Arizona Press, 1989.

Harley, J. B. "Deconstructing the Map." *Cartographica: The International Journal for Geographic Information and Geovisualization* 26, no. 2 (1989): 1–20.

Harper, Rob. "Looking the Other Way: The Gnadenhütten Massacre and the Contextual Interpretation of Violence." *The William and Mary Quarterly* 64, no. 3 (2007): 621–644.

Harper, Steven Craig. *Promised Land: Penn's Holy Experiment, the Walking Purchase, and the Dispossession of Delawares, 1600–1763*. Bethlehem, PA: Lehigh University Press, 2006.

Hartzell, Lawrence W. "Joshua Jr.: Moravian Indian Musician." *Transactions of the Moravian Historical Society* 26 (1990): 1–19.

Harvey, Oscar Jewell, and Ernest Gray Smith. *A History of Wilkes-Barré, Luzerne County, Pennsylvania from Its First Beginnings to the Present Time*. Wilkes-Barré, PA: Raeder, 1909.

Harvey, Sean P. *Native Tongues: Colonialism and Race from Encounter to the Reservation*. Harvard Historical Studies 184. Cambridge, MA: Harvard University Press, 2015.

Hayden, Rev. Horace Edwin. *Proceedings and Collections of the Wyoming Historical and Geological Society for the Years 1911–1912*. Wilkes-Barré, PA: Printed for the Wyoming Historical and Geological Society by the E. B. Yordy Company, 1912.

Hecker, A. P. *Gespräch eines evangelisch-lutherischen Predigers, mit einem, der über 6 Jahr sich zu der Gemeine der sogenannten Mährischen Brüder gehalten, und nach* erkanten Irrthümern und sündlichem Leben dieser Leute . . . weggegangen: darin unter andern gezeiget wird, unter welchen Ceremonien die Aufnahme zum Bruder in der Herrenhutischen Gemeine geschehe, das Heil. Abendmahl unter ihnen ausgetheilet werde . . . Berlin: Buchhandlung der Realschule, 1751.

Heckewelder, John Gottlieb Ernestus. *A Narrative of the Mission of the United Brethren among the Delaware and Mohegan Indians, from Its Commencement, in the Year 1740, to the Close of the Year 1808, Comprising All the Remarkable Incidents Which Took Place at Their Missionary Stations during That Period, Interspersed with Anecdotes, Historical Facts, Speeches of Indians, and Other Interesting Matter.* Philadelphia, PA: M'Carty & Davis, 1820.

———. *History, Manners, and Customs of the Indian Nations: Who Once Inhabited Pennsylvania and the Neighbouring States.* Memoirs of the Historical Society of Pennsylvania 12. Philadelphia: The Historical Society of Pennsylvania, rev. 1876.

———. *Names Which the Lenni Lenape or Delaware Indians Gave to Rivers, Streams and Localities, within the States of Pennsylvania, New Jersey, Maryland, and Virginia, with Their Significations.* Bethlehem, PA: n.p., 1872.

Helmreich, Stefan, Sophia Roosth, and Michele Friedner. *Sounding the Limits of Life: Essays in the Anthropology of Biology and Beyond.* Princeton, NJ: Princeton University Press, 2016.

Herder, Johann Gottfried von. *Johann Gottfried von Herder's sämmtliche Werke.* Edited by Johann Georg Müller. Tübingen: Cotta, 1828.

Herman, David, Manfred Jahn, and Marie-Laure Ryan, eds. *Routledge Encyclopedia of Narrative Theory.* New York: Routledge, Taylor & Francis Group, 2008.

Hertrampf, Stefan. *Unsere Indianer-Geschwister waren lichte und vegnügt: Die Herrnhuter Missionare bei den Indianern Pennsylvanias, 1745–1765.* Frankfurt am Main: Peter Lang, 1997.

Hickel, Helmut. *Das Abendmahl zu Zinzendorfs Zeiten.* Hamburg: L. Appel, 1956.

Hoffer, Peter Charles. *Sensory Worlds of Early America.* Baltimore, MD: Johns Hopkins University Press, 2003.

Hoffmann, Elizabeth Cobbs, Jon Gjerde, and Thomas G. Paterson, eds. *Major Problems in American History: Documents and Essays* I. Wadsworth: Cengage Learning, 2012.

Holmes, John. *Historical Sketches of the Missions of the United Brethren for Propagating the Gospel among the Heathen, from Their Commencement to the Year 1817.* 2nd ed. London: J. Nisbet, 1827.

Holmes, T. *The American Family of Rev. Obadiah Holmes.* Columbus, OH: 1915.

Hoock, Holger. *Scars of Independence: America's Violent Birth.* Broadway Books, 2018.

Hopkins, Peter, Lily Kong, and Elizabeth Olson, eds. *Religion and Place: Landscape, Politics and Piety.* New York: Springer, 2013.

Horowitz, Seth S. *The Universal Sense: How Hearing Shapes the Mind.* New York: Bloomsbury, 2013.

Howes, David, ed. *Empire of the Senses: The Sensual Culture Reader.* Sensory Formations Series. New York: Berg, 2005.

Hubbard, Phil, Rob Kitchin, and Gill Valentine. *Key Thinkers on Space and Place.* London: Sage, 2004.

Hubert, Archer Butler, and William Nathaniel Schwarze, eds. *David Zeisberger's History of the Northern American Indians.* Publication of the Ohio State Archaeological and Historical Society. Lewisburg, PA: Wennawoods Publishing, 1999.

Huetter, Karen Zerbe. *John Adams' Bethlehem, "a Curious and Remarkable Town."* Bethlehem, PA: Oaks, 1976.
Hutton, J. E. *A History of the Moravian Church.* 2nd ed. London: Moravian Publication Office, 1909.
Ingalls, Monique Marie, Carolyn Landau, and Thomas Wagner, eds. *Christian Congregational Music: Performance, Identity, and Experience.* Surrey, UK: Ashgate, 2013.
Ingram, Jeannine. "Music in American Moravian Communities: Transplanted Traditions in Indigenous Practices." *Communal Societies* 2 (1982): 39–52.
Jacobs, Wilbur R. *The Appalachian Indian Frontier: The Edmond Atkin Report and Plan of 1755.* Lincoln: University of Nebraska Press, 1967.
Järviluoma, Helmi, and R. Murray Schafer. *Acoustic Environments in Change.* Studies in Literature and Culture 14. Tampere: Tampereen ammattikorkeakoulu, 2009.
Jennings, Willie James. *Christian Imagination: Theology and the Origins of Race.* New Haven: Yale University Press, 2011.
Johnson, Frederick C. *Count Zinzendorf and the Moravian and Indian Occupancy of the Wyoming Valley, Pa., 1742–1763.* Wilkes-Barré, PA: 1904.
Jordan, John W. "Spangenberg's Notes of Travel to Onondaga in 1745." *Pennsylvania Magazine of History and Biography* 2, no. 4 (1898): 426.
Justice, Daniel Heath. *Our Fire Survives the Storm: A Cherokee Literary History.* Indigenous Americas. Minneapolis: University of Minnesota Press, 2006.
———. *Why Indigenous Literatures Matter.* Waterloo, Ontario, Canada: Wilfrid Laurier University Press, 2018.
Kahn, Douglas. *Noise, Water, Meat: A History of Sound in the Arts.* Cambridge, MA: MIT Press, 1999.
Kane, Brian. *Sound Unseen: Acousmatic Sound in Theory and Practice.* New York: Oxford University Press, 2016.
Kapchan, Deborah A. *Theorizing Sound Writing.* Middletown, CT: Wesleyan University Press, 2017.
Kauanui, J. Kēhaulani. "'A Structure, Not an Event': Settler Colonialism and Enduring Indigeneity." *Lateral*, June 1, 2016. http://csalateral.org/issue/5-1/forum-alt-humanities-settler-colonialism-enduring-indigeneity-kauanui/.
Keller, Damián. "Compositional Processes from an Ecological Perspective." *Leonardo Music Journal* 10 (2000): 55–60.
Kelman, Ari V. "Rethinking the Soundscape: A Critical Genealogy of a Key Term in Sound Studies." *Senses & Society* 5, no. 2 (July 2010): 212–234.
Kent, Barry C., Janet Rice, and Kakuko Ota. "A Map of 18th Century Indian Towns in Pennsylvania." *Pennsylvania Archaeologist* 51, no. 4 (1981): 1–18.
Keyes, Sarah. "'Like a Roaring Lion': The Overland Trail as a Sonic Conquest." *The Journal of American History* 96, no. 1 (2009): 19–43.
Kidula, Jean Ngoya. "Ethnomusicology, the Music Canon, and African Music: Positions, Tensions, and Resolutions in the African Academy." *Africa Today* 52, no. 3 (2006): 99–113.
———. "Transcending Time, Empowering Space: (Re)iterating a Logooli/Luyia/Kenyan Song." *The World of Music* 51, no. 2 (2009): 119–137.
King, Roberta Rose, Jean Ngoya Kidula, James R. Krabill, and Thomas Oduro. *Music in the Life of the African Church.* Waco, TX: Baylor University Press, 2008.

Kisby, Fiona, ed. *Music and Musicians in Renaissance Cities and Towns*. Cambridge: Cambridge University Press, 2006.

Kitchin, Rob, Justin Gleeson, and Martin Dodge. "Unfolding Mapping Practices: A New Epistemology for Cartography." *Transactions of the Institute of British Geographers* 38, no. 3 (July 2013): 480–496.

Knight, David B. *Landscapes in Music: Space, Place, and Time in the World's Great Music*. Lanham, MD: Rowman & Littlefield, 2006.

Knott, John R. *Imagining the Forest: Narratives of Michigan and the Upper Midwest*. Ann Arbor: University of Michigan Press, 2012.

Knouse, Nola Reed. "Moravian Music: Introduction, Theme, and Variations." *Journal of Moravian History* 2 (2007): 37–54.

———, ed. *The Music of the Moravian Church in America*. Eastman Studies in Music 49. Rochester, NY: University of Rochester Press, 2008.

Knowles, Anne Kelly. *Geographies of the Holocaust*. The Spatial Humanities. Bloomington: Indiana University Press, 2014.

Knowles, Anne Kelly, and Amy Hillier, eds. *Placing History: How Maps, Spatial Data, and GIS Are Changing Historical Scholarship*. Redlands, CA: ESRI, 2008.

Kolar, Miriam A. "Sensing Sonically at Andean Formative Chavín de Huántar, Perú." *Time and Mind* 10, no. 1 (January 2017): 39–59.

Kölbing, Friedrich Ludwig. *Die Gedenktage der erneuerten Brüderkirche*. Gnadau: Nabu, 1821.

Krause, Bernard L. *The Great Animal Orchestra: Finding the Origins of Music in the World's Wild Places*. Boston: Back Bay Books Little Brown, 2013.

———. *Into a Wild Sanctuary: A Life in Music & Natural Sound*. Berkeley, CA.: Heyday, 1998.

Křížová, Markéta. "The Moravian Church and the Society of Jesus: American Mission and American Utopia in the Age of Confessionalization." *Journal of Moravian History* 13, no. 2 (November 2013): 197–226.

Kun, Josh D. "The Aural Border." *Theatre Journal* 52, no. 1 (2000): 1–21.

Kwan, Mei-Po. "Feminist Visualization: Re-Envisioning GIS as a Method in Feminist Geographic Research." *Annals of the Association of American Geographers* 92, no. 4 (2002): 645–661.

Laderman, Gary. *Sacred Remains: American Attitudes Toward Death, 1799–1883*. New Haven, CT: Yale University Press, 1996.

Lake-Thom, Bobby. *Spirits of the Earth: A Guide to Native American Nature Symbols, Stories, and Ceremonies*. New York: Plume, 1997.

Lane, Belden C. *Landscapes of the Sacred: Geography and Narrative in American Spirituality*. Isaac Hecker Studies in Religion and American Culture. New York: Paulist, 1988.

Langner, Ekkehard. "Eine Ortsgemeine um 1800—Die Herrnhuter in Neuwied in Reiseberichten der Zeit." *Unitas Fratrum* 4 (1978): 52–69.

Larson, Paul. "Mahican and Lenape Moravians and Moravian Music." *Unitas Fratrum* 21–22 (1988): 173–188.

Lassiter, Luke E., Clyde Ellis, and Ralph Kotay. *The Jesus Road: Kiowas, Christianity, and Indian Hymns*. Lincoln, NE: University of Nebraska Press, 2002.

Laufhütte, Hartmutt, ed. *Künste und Natur in Diskursen der Frühen Neuzeit*. Wiesbaden: Harrassowitz, 2000.

Leavelle, Tracy Neal. *The Catholic Calumet: Colonial Conversions in French and Indian North America*. Early American Studies. Philadelphia: University of Pennsylvania Press, 2012.

Le Breton, David, and Carmen Ruschiensky. *Sensing the World: An Anthropology of the Senses*. London: Bloomsbury, 2017.
Le Guin, Elisabeth. *Boccherini's Body: An Essay in Carnal Musicology*. Berkeley, CA: University of California Press, 2006.
Lehmkuhl, Ursula. "Good Land–Bad Land: Ecological Knowledge and the Settling of the Old Northwest, 1755–1805." *Settler Colonial Studies* 7, no. 2 (April 2017): 141–163.
Lehmkuhl, Ursula, Hans-Jürgen Lüsebrink, and Laurence McFalls, eds. *Of "Contact Zones" and "Liminal Spaces": Mapping the Everyday Life of Cultural Translation*. Münster: Waxmann, 2015.
Leibert, Eugene. "Wechquetank. A Paper Read before the Moravian Historical Society, Sept. 13, 1900." *Transactions of the Moravian Historical Society* 7, no. 2 (1903): 57–82.
Lempa, Heikki, and Paul Peucker, eds. *Self, Community, World: Moravian Education in a Transatlantic World*. Studies in Eighteenth-Century America and the Atlantic World. Bethlehem, PA: Lehigh University Press, 2010.
Leppert, Richard D. *The Sight of Sound: Music, Representation, and the History of the Body*. Berkeley: University of California Press, 1993.
Levering, Joseph Mortimer. *A History of Bethlehem, Pennsylvania, 1741–1892: with Some Account of Its Founders and Their Early Activity in America*. Bethlehem, PA: Times, 1903.
Lewis, Michael L. *American Wilderness: A New History*. New York: Oxford University Press, 2007.
Libin, Laurence. "What Instrument Did Joshua Make?" In *Proceedings of the Seventh Bethlehem Conference on Moravian Music*, 46–50. Winston-Salem, NC: Moravian Music Foundation, 2007.
Lin, Wen. "The Hearing, the Mapping, and the Web: Investigating Emerging Online Sound Mapping Practices." *Landscape and Urban Planning* 142, Supplement C (October 2015): 187–197.
———. "Situating Performative Neogeography: Tracing, Mapping, and Performing 'Everyone's East Lake.'" *Environment and Planning A* 45, no. 1 (January 2013): 37–54.
Lingold, Mary Caton, Darren Mueller, Whitney Trettien, eds. *Digital Sound Studies*. Durham, NC: Duke University Press, 2018.
Loskiel, George Henry. *Geschichte Der Mission Der Evangelischen Brüder Unter Den Indianern in Nordamerika*. Barby: 1789. Translated as *History of the Mission of the United Brethren among the Indians in North America. In Three Parts. By George Henry Loskiel. Translated from the German by Christian Ignatius La Trobe*. London: printed for the Brethren's Society for the furtherance of the Gospel: sold at No. 10, Nevil's Court, Fetter Lane; and by John Stockdale, Opposite Burlington House, Piccadilly, 1794.
Magowan, Fiona, and Karl Neuenfeldt, eds. *Landscapes of Indigenous Performance: Music, Song and Dance of the Torres Strait and Arnhem Land*. Canberra, Australia: Aboriginal Studies Press, 2005.
Mahr, August C. "Delaware Terms for Plants and Animals in the Eastern Ohio Country: A Study in Semantics." *Anthropological Linguistics* 4, no. 5 (1962): 1–48.
Mahr, August C., and John Gottlieb Ernestus Heckewelder. "A Canoe Journey from the Big Beaver to the Tuscarawas in 1773: A Travel Diary of John Heckewaelder." *Ohio State Archaeological and Historical Quarterly*, 1952.
Mancall, Peter C., and James Hart Merrell, eds. *American Encounters: Natives and Newcomers from European Contact to Indian Removal, 1500–1850*. New York: Routledge, 2007.

Mandell, Daniel R. *Behind the Frontier: Indians in Eighteenth-Century Eastern Massachusetts*. Lincoln: University of Nebraska Press, 1996.

Manuel, Frank Edward, and Fritzie Prigohzy Manuel. *Utopian Thought in the Western World*. Cambridge, MA: Belknap, 1979.

Mann, Kristin Dutcher. *The Power of Song: Music and Dance in the Mission Communities of Northern New Spain, 1590–1810*. Palo Alto, CA: Stanford University Press; Academy of American Franciscan History, 2010.

Mar, Tracey Banivanua, and Penelope Edmonds, eds. *Making Settler Colonial Space: Perspectives on Race, Place and Identity*. New York: Palgrave Macmillan, 2010.

Marini, Stephen. *Hymnody and History: Early American Evangelical Hymns as Sacred Music*. Oxford University Press, 2005.

———. "Hymnody in the Religious Communal Societies of Early America." *Communal Societies* 2 (1982): 1–26.

Martin, Calvin. *Keepers of the Game: Indian-Animal Relationships and the Fur Trade*. Berkeley: University of California Press, 1978.

Martin, Joel W., and Mark A. Nicholas, eds. *Native Americans, Christianity, and the Reshaping of the American Religious Landscape*. Chapel Hill: University of North Carolina Press, 2010.

Martin, John Hill. *Historical Sketch of Bethlehem in Pennsylvania*. 2nd ed. Philadelphia, 1873.

Martinez, David, ed. *The American Indian Intellectual Tradition: An Anthology of Writings from 1772 to 1972*. Ithaca, NY: Cornell University Press, 2011.

Massey, Doreen B. *For Space*. London: Sage, 2015.

Masthay, Carl. *Mahican-Language Hymns, Biblical Prose, and Vocabularies from Moravian Sources, with 11 Mohawk Hymns (Transcription and Translation)*. St. Louis, MO: Masthay, 1980.

Matless, David. "Sonic Geography in a Nature Region." *Social & Cultural Geography* 6, no. 5 (October 2005): 745–766.

Maurer, Maurer (sic). "Music in Wachovia, 1753–1800." *The William and Mary Quarterly* 8, no. 2 (1951): 214–227.

McCarthy, K. "Conversion, Identity, and the Indian Missionary." *Early American Literature* 38 (2001): 353–370.

McConnell, Michael N. *A Country Between: The Upper Ohio Valley and Its Peoples, 1724–1774*. Lincoln: University of Nebraska Press, 1992.

McCorkle, Donald M. "The Moravian Contribution to American Music." *Notes* 13, no. 4 (1956): 597–606.

McLafferty, Sara. "Women and GIS: Geospatial Technologies and Feminist Geographies." *Cartographica: The International Journal for Geographic Information and Geovisualization* 40, no. 4 (2005): 37–45.

McNally, Michael D. "Naming the Legacy." In *Native Americans, Christianity, and the Reshaping of the American Religious Landscape*, Joel W. Martin and Mark A. Nicholas, eds. Chapel Hill: University of North Carolina Press, 2010.

———. *Ojibwe Singers: Hymns, Grief, and a Native Culture in Motion*. Religion in America. New York, NY: Oxford University Press, 2000.

———. "The Practice of Native American Christianity." *Church History* 69, no. 4 (2000): 834–859.

Mels, Tom. *Reanimating Places: A Geography of Rhythms*. Re-Materialising Cultural Geography. Aldershot, UK: Ashgate, 2004.

Mereness, Newton D., ed. *Travels in the American Colonies*. New York: Palgrave Macmillan, 1916.
Merrell, James Hart. *Into the American Woods: Negotiators on the Pennsylvania Frontier*. New York: Norton, 1999.
———. "Shamokin, 'The Very Seat of the Prince of Darkness.'" In *Contact Points: American Frontiers from the Mohawk Valley to the Mississippi, 1750–1830*, edited by Andrew R. L. Cayton, Fredrika J. Teute, and the Omohundro Institute of Early American History & Culture. Chapel Hill: University of North Carolina Press, 1998.
Merritt, Jane T. *At the Crossroads: Indians and Empires on a Mid-Atlantic Frontier, 1700–1763*. Chapel Hill: University of North Carolina Press, 2003.
———. "Cultural Encounters along a Gender Frontier: Mahican, Delaware, and German Women in Eighteenth-Century Pennsylvania." *Pennsylvania History: A Journal of Mid-Atlantic Studies* 67, no. 4 (2000): 502–531.
———. "Dreaming of the Savior's Blood: Moravians and the Indian Great Awakening in Pennsylvania." *The William and Mary Quarterly* 54, no. 4 (1997): 723–746.
———. "Metaphor, Meaning, and Misunderstanding: Language and Power on the Pennsylvania Frontier." In *Contact Points: American Frontiers from the Mohawk Valley to the Mississippi, 1750–1830*, edited by Andrew R. L. Cayton, Fredrika J. Teute, and the Omohundro Institute of Early American History & Culture, 60–87. Chapel Hill: University of North Carolina Press, 1998.
Mettele, Gisela. "Constructions of the Religious Self. Moravian Conversion and Transatlantic Communication." *Journal of Moravian History* 2 (2007): 7–36.
Meyer, Dietrich, Paul Peucker, and Karl-Eugen Langerfeld. *Graf ohne Grenzen, Leben und Werk von Nikolaus Ludwig Graf von Zinzendorf: Ausstellung im Völkerkundemuseum Herrnhut, Aussenstelle des Staatlichen Museums für Völkerkunde Dresden, und im Heimatmuseum der Stadt Herrnhut, vom 26. Mai 2000 bis zum 7. Januar 2001*. Publication of the Völkerkundemuseum Herrnhut and the Heimatmuseum der Stadt Herrnhut. Herrnhut: Unitätsarchiv in Herrnhut im Verlag der Comeniusbuchhandlung, 2000.
Meyer, Henry Herman. *Child Nature and Nurture according to Nicolaus Ludwig von Zinzendorf*. Nashville, TN: Abingdon, 1983.
Miller, Cathleen. "Cash or Credit: Selling the Settlement of Pennsylvania." *Pennsylvania Legacies* 9, no. 2 (2009): 6–11.
Miller, Randall M., William Pencak, and Pennsylvania Historical and Museum Commission, eds. *Pennsylvania: A History of the Commonwealth*. A Keystone Book. University Park, PA: Pennsylvania State University Press; Pennsylvania Historical and Museum Commission, 2002.
Miller, Susan A., and James Riding In, eds. *Native Historians Write Back: Decolonizing American Indian History*. Lubbock: Texas Tech University Press, 2011.
Minderhout, David Jay. *Native Americans in the Susquehanna River Valley, Past and Present*. Lanham, MD: Bucknell University Press, 2013.
Minderhout, David Jay, and Andrea T. Frantz. *Invisible Indians: Native Americans in Pennsylvania*. Amherst, NY: Cambria, 2008.
Mitchell, Peta. *Cartographic Strategies of Postmodernity: The Figure of the Map in Contemporary Theory and Fiction*. Hoboken, NJ: Taylor and Francis, 2013.
Mohr, Rudolf, and Dietrich Meyer, eds. *"Alles Ist Euer, Ihr Aber Seid Christi": Festschrift Für Dietrich Meyer*. Schriftenreihe Des Vereins Für Rheinische Kirchengeschichte 147. Cologne: Rheinland-Verlag, 2000.

Momaday, Natachee Scott. *House Made of Dawn*. New York: Perennial Classics, 1999.

———. *The Names: A Memoir*. Tucson: The University of Arizona Press, 1999.

Moyer, Paul B. *Wild Yankees: The Struggle for Independence along Pennsylvania's Revolutionary Frontier*. Reprint ed. Ithaca, NY: Cornell University Press, 2015.

Mueller, Paul Eugene. "David Zeisberger's Official Diary, Fairfield, 1791–1795." *Transactions of the Moravian Historical Society* 19, no. 1 (1963): 3–229.

Murray, Keat. "John Heckewelder's 'Pieces of Secrecy': Dissimulation and Class in the Writings of a Moravian Missionary." *Journal of the Early Republic* 32, no. 1 (2012): 91–126.

Murtagh, William J. *Moravian Architecture and Town Planning: Bethlehem, Pennsylvania, and Other Eighteenth-Century American Settlements*. Chapel Hill: University of North Carolina Press, 1967.

Mutua, Kagendo, and Beth Blue Swadener, eds. *Decolonizing Research in Cross-Cultural Contexts: Critical Personal Narratives*. Albany: State University of New York Press, 2004.

Myers, James P. "Mapping Pennsylvania's Western Frontier in 1756." *The Pennsylvania Magazine of History and Biography* 123, nos. 1–2 (1999): 3–29.

Nash, Roderick. *Wilderness and the American Mind*. New Haven, CT: Yale University Press, 2001.

Nelson, James David. *Herrnhut: Friedrich Schleiermacher's Spiritual Homeland*. PhD dissertation, University of Chicago, 1963.

Nelson, Lise, and Joni Seager. *A Companion to Feminist Geography*. Malden, MA: Blackwell, 2013.

Nelson, Michael P., and J. Baird Callicott. *The Wilderness Debate Rages On: Continuing the Great New Wilderness Debate*. Athens, GA: The University of Georgia Press, 2008.

Nelson, Vernon H. "The Sun Inn at Bethlehem, Pennsylvania." *Journal of Moravian History*, 9 (2010): 68–82.

Nelson, Vernon H., Craig D. Atwood, and Peter Vogt, eds. *The Distinctiveness of Moravian Culture: Essays and Documents in Moravian History in Honor of Vernon H. Nelson on his Seventieth Birthday*. Nazareth, PA: Moravian Historical Society, 2003.

Newman, Andrew. *On Records: Delaware Indians, Colonists, and the Media of History and Memory*. Lincoln: University of Nebraska Press, 2012.

Nippa, Annegrete, and Staatliches Museum für Völkerkunde (Dresden). *Ethnographie Und Herrnhuter Mission: Völkerkundemuseum Herrnhut: Katalog Zur Ständingen Ausstellung Im Völkerkundemuseum Herrnhut, Aussenstelle Des Staatlichen Museums Für Völkerkunde Dresden*. Dresden: Staatliches Museum für Völkerkunde, 2003.

Norwood, C., and G. Cumming. "Making Maps That Matter: Situating GIS within Community Conversations about Changing Landscapes." *Cartographica: The International Journal for Geographic Information and Geovisualization* 47, no. 1 (2012): 2–17.

O'Brien, Jean M. *Firsting and Lasting: Writing Indians Out of Existence in New England*. Indigenous Americas. Minneapolis: University of Minnesota Press, 2010.

Ochoa Gautier, Ana María. *Aurality: Listening and Knowledge in Nineteenth-Century Colombia*. Durham, NC: Duke University Press, 2014.

O'Connell, Marvin R. *Blaise Pascal: Reasons of the Heart*. Library of Religious Biography. Grand Rapids, MI: W. B. Eerdmans, 1997.

Olmstead, Earl P. *David Zeisberger: A Life among the Indians*. Kent, OH: Kent State University Press, 1997.
Olson, Alison. "The Pamphlet War over the Paxton Boys." *The Pennsylvania Magazine of History and Biography* 123, nos. 1–2 (1999): 31–55.
Olwell, Robert, and Alan Tully, eds. *Cultures and Identities in Colonial British America. Anglo-America in the Transatlantic World*. Baltimore, MD: Johns Hopkins University Press, 2006.
Ostler, Jeffrey. "'To Extirpate the Indians': An Indigenous Consciousness of Genocide in the Ohio Valley and Lower Great Lakes, 1750s–1810." *The William and Mary Quarterly* 72, no. 4 (2015): 587–622.
Parmenter, Jon. *The Edge of the Woods: Iroquoia, 1534–1701*. East Lansing: Michigan State University Press, 2010.
Patterson, Daniel, Roger Thompson, and Scott Bryson. *Early American Nature Writers: A Biographical Encyclopedia*. Westport, CT: Greenwood Press, 2008.
Pavlovskaya, Marianna. "Critical GIS as a Tool for Social Transformation." *The Canadian Geographer/Le Géographe Canadien* 62, no. 1 (2018): 40–54.
———. "Theorizing with GIS: A Tool for Critical Geographies." *Sage Public Administration Abstracts* 34, no. 3 (2007): 2003.
Pavlovskaya, Marianna, and Kevin St. Martin. "Feminism and Geographic Information Systems: From a Missing Object to a Mapping Subject." *Geography Compass* 1, no. 3 (2007): 583–606.
Pencak, William, and Daniel K. Richter, eds. *Friends and Enemies in Penn's Woods: Indians, Colonists, and the Racial Construction of Pennsylvania*. University Park: Pennsylvania State University Press, 2004.
Penn, William, Albert Cook Myers, and John E. Pomfret. *William Penn's Own Account of Lenni Lenape or Delaware Indians*. Moorestown, NJ: Middle Atlantic, rev. 1970.
Pennsylvania Department of Conservation of Natural Resources. "Penn's Woods: A History of Pennsylvania's Forests." Accessed July 28, 2016, www.dcnr.state.pa.us/cs/groups/public/documents/document/dcnr_009325.pdf.
Pennsylvania Historical Commission, ed. *Remember William Penn*. Harrisburg: Commonwealth of Pennsylvania, 1944.
Perea, Jessica Bissett. "The Politics of Inuit Musical Modernities in Alaska." PhD dissertation, University of California, Los Angeles, 2011.
Perea, John-Carlos. *Intertribal Native American Music in the United States: Experiencing Music, Expressing Culture*. Global Music Series. New York: Oxford University Press, 2014.
Perreault, Melanie. "American Wilderness and First Contact." In *American Wilderness: A New History*, edited by Michael L. Lewis, 15–33. New York: Oxford University Press, 2007.
Peucker, Paul. *A Time of Sifting: Mystical Marriage and the Crisis of Moravian Piety in the Eighteenth Century*. Pietist, Moravian, and Anabaptist Studies. University Park: The Pennsylvania State University Press, 2015.
———. "David Nitschmann's Letter to Jonas Weiss in Herrnhaag, on the Beginnings of the Settlement on the Lehigh River, 1741." *Journal of Moravian History* 1 (2006): 65–74.
———. *Herrnhuter Wörterbuch: Kleines Lexicon von brüderischen Begriffen*. Herrnhut: Unitätsarchiv Herrnhut, 2000.

———. "'Inspired by Flames of Love': Homosexuality, Mysticism, and Moravian Brothers around 1750." *Journal of the History of Sexuality* 15, no. 1 (2006): 30–64.
———. "In the Blue Cabinet: Moravians, Marriage, and Sex." *Journal of Moravian History* 10 (2011): 6–37.
———. "The Songs of the Sifting. Understanding the Role of Bridal Mysticism in Moravian Piety during the Late 1740s." *Journal of Moravian History* 3 (2007): 51–87.
Pijanowski, Bryan C., Almo Farina, Stuart H. Gage, Sarah L. Dumyahn, and Bernie L. Krause. "What Is Soundscape Ecology? An Introduction and Overview of an Emerging New Science." *Landscape Ecology* 26, no. 9 (November 2011): 1213–1232.
Pijanowski, Bryan C., Luis J. Villanueva-Rivera, Sarah L. Dumyahn, Almo Farina, Bernie L. Krause, Brian M. Napoletano, Stuart H. Gage, and Nadia Pieretti. "Soundscape Ecology: The Science of Sound in the Landscape." *BioScience* 61, no. 3 (2011): 203–216.
Pijoan, Teresa. *White Wolf Woman: Native American Transformation Myths*. Little Rock, AR: August House, 1992.
Pisani, Michael V. *Imagining Native America in Music*. New Haven, CT: Yale University Press, 2006.
Plane, Ann Marie, Leslie Tuttle, and Anthony F. C. Wallace, eds. *Dreams, Dreamers, and Visions: The Early Modern Atlantic World*. Philadelphia: University of Pennsylvania Press, 2013.
Pleasant, Alyssa Mt., Caroline Wigginton, and Kelly Wisecup. "Materials and Methods in Native American and Indigenous Studies: Completing the Turn." *The William and Mary Quarterly* 75, no. 2 (May 2018): 207–236.
Pointer, Richard W. *Encounters of the Spirit: Native Americans and European Colonial Religion*. Religion in North America. Bloomington: Indiana University Press, 2007.
Pollack, John H. "Native American Words, Early American Texts." PhD dissertation, University of Pennsylvania, 2014.
Porterfield, Amanda, and John Corrigan, eds. *Religion in American History*. Malden, MA: Wiley-Blackwell, 2010.
Powell, Eric. "Toward the Sound of Place." *Soundscape: The Journal of Acoustic Ecology* 14, no. 1 (Winter/Spring 2014): 9–12.
Pownall, Thomas. *A Topographical Description of the Dominions of the United States of America*. Pittsburgh: University of Pittsburgh Press, reprint 1949.
———. *The Administration of the British Colonies*. 6th ed. London: Printed for J. Walter, at Homer's Head, Charing-Cross, 1777.
———. *Scenographia Americana: Or, A Collection of Views in North America and the West Indies. Neatly Engraved by Messrs. Sandby, Grignion, Rooker, Canot, Elliot, and Others; from Drawings Taken on the Spot, by Several Officers of the British Navy and Army. Recueil de Vues de L'Amerique Septentrionale et Des Indes Occidentales. Gravés D'aprš Les Desseins Pris Sur Les Lieux Par Diffŕens Officers Des Troupes et de La Marine Angloises*. London: Robert Sayer, at No. 53, in Fleet-Street; Thomas Jeffery's, at the corner of St. Martin's Lane in the Strand; Carington Bowles, at No. 69, in St. Paul's Churchyard; and Henry Parker, at No. 82, in Cornhill: Printed for John Bowles, at No. 13, in Cornhill, 1768.
Preston, David L. *The Texture of Contact: European and Indian Settler Communities on the Frontiers of Iroquoia, 1667–1783*. The Iroquoians and Their World. Lincoln: University of Nebraska Press, 2009.

Price, Matthew Hunter. "Methodism and Social Capital on the Southern Frontier, 1760–1830." PhD dissertation, Ohio State University, 2014.
Proceedings of the Seventh Bethlehem Conference on Moravian Music. Winston-Salem, NC: Moravian Music Foundation, 2007.
"PROVOKE: Digital Sound Studies." http://soundboxproject.com/.
Pulitano, Elvira. *Toward a Native American Critical Theory.* Lincoln: University of Nebraska Press, 2003.
Raban, Jonathan. *Passage to Juneau: A Sea and Its Meanings.* New York: Pantheon Books, 1999.
Radding, Cynthia. "Borderlands of Knowledge about Nature: Crossing and Creating Boundaries in Early America." *Early American Studies: An Interdisciplinary Journal* 13, no. 2 (April 2015): 503–510.
———. Human Geographies and Landscapes of the Divine in the Northern Mesoamerican Borderlands." In *Re-Imagining Nature: Environmental Humanities and Ecosemiotics*, edited by Alfred K. Siewers. Lewisburg, PA: Bucknell University Press, 2014.
———. *Landscapes of Power and Identity: Comparative Histories in the Sonoran Desert and the Forests of Amazonia from Colony to Republic.* Durham, NC: Duke University Press, 2005.
Rath, Richard Cullen. "Hearing American History." *The Journal of American History* 95, no. 2 (2008): 417–431.
———. *How Early America Sounded.* Ithaca, NY: Cornell University Press, 2003.
Rau, Albert G. "The Autobiography of Johann Christopher Pyrlaeus." *Transactions of the Moravian Historical Society* 12, no. 1 (1938): 18–25.
Rau, Robert. "Sketch of the History of the Moravian Congregation at Gnadenhütten on the Mahoning." *Transactions of the Moravian Historical Society* 2, no. 9 (1886): 399–414.
Reichardt, Johann Friedrich. *Briefe eines aufmerksamen Reisenden die Musik betreffend.* Hildesheim: G. Olms, 1977.
Reichel, Edward H. *Historical Sketch of the Church and Missions of the United Brethren, Commonly Called Moravians.* Bethlehem, PA: J. and W. Held, 1848.
Reichel, Jörn. *Dichtungstheorie udn Sprache bei Zinzendorf.* Bad Homburg: Gehlen, 1969.
Reichel, William Cornelius, ed. *Count Zinzendorf and the Indians, 1742.* Lewisburg, PA: Wennawoods, 2007.
———, ed. *Memorials of the Moravian Church.* Philadelphia: J. B. Lippincott, 1870.
Revill, George. "El Tren Fantasma: Arcs of Sound and the Acoustic Spaces of Landscape." *Transactions of the Institute of British Geographers* 39, no. 3 (July 2014): 333–344.
Revill, George, David Matless, and Andrew Leyshon, eds. *The Place of Music.* Mappings. New York: Guilford Press, 1998.
Rex, Cathy. "Writing Indian Nations: Native Intellectuals and the Politics of Historiography, 1827–1863 (Review)." *Early American Literature* 39, no. 3 (2004): 603–606.
Richter, Daniel K. "A Framework for Pennsylvania Indian History." *Pennsylvania History: A Journal of Mid-Atlantic Studies* 57, no. 3 (1990): 236–261.
———. *Before the Revolution: America's Ancient Pasts.* Cambridge, MA: Belknap Press of Harvard University Press, 2011.
———. *Beyond the Covenant Chain: The Iroquois and Their Neighbors in Indian North America, 1600–1800.* University Park: Pennsylvania State University Press, 2003.

———. *Facing East from Indian Country: A Native History of Early America*. Cambridge, MA: Harvard University Press, 2001.

———. "First Pennsylvanians." In *Pennsylvania, History of the Commonwealth*.

———. *Native Americans' Pennsylvania*. University Park: Pennsylvania State University Press, 2005.

———. *The Ordeal of the Longhouse: The Peoples of the Iroquois League in the Era of European Colonization*. Chapel Hill: Published for the Institute of Early American History and Culture, Williamsburg, Virginia, by the University of North Carolina Press, 1992.

———. *Trade, Land, Power: The Struggle for Eastern North America*. Philadelphia: University of Pennsylvania Press, 2013.

———. "War and Culture: The Iroquois Experience." *The William and Mary Quarterly* 40, no. 4 (1983): 528–559.

———. "Whose Indian History?" *The William and Mary Quarterly* 50, no. 2 (1993): 379–393.

Ridge, Mia, Don Lafreniere, and Scott Nesbit. "Creating Deep Maps and Spatial Narratives through Design." *Journal of Humanities & Arts Computing: A Journal of Digital Humanities* 7, nos. 1–2 (March 2013): 176–189.

Ridner, Judith A. *A Town In-Between: Carlisle, Pennsylvania, and the Early Mid-Atlantic Interior*. Early American Studies. Philadelphia: University of Pennsylvania Press, 2010.

———. "Building Urban Spaces for the Interior: Thomas Penn and the Colonization of Eighteenth-Century Pennsylvania." In *Early American Cartographies*, edited by Martin Brükner, 306–338. Chapel Hill: University of North Carolina Press, 2011.

Rivett, Sarah. *Unscripted America: Indigenous Languages and the Origins of a Literary Nation*. Oxford Studies in American Literary History. New York: Oxford University Press, 2017.

Robertson-DeCarbo, Carol E., ed. *Musical Repercussions of 1492: Encounters in Text and Performance*. Washington: Smithsonian Institution Press, 1992.

Rodaway, Paul. *Sensuous Geographies : Body, Sense, and Place*. London: Routledge, 2011.

Roeber, A. G., ed. *Ethnographies and Exchanges: Native Americans, Moravians, and Catholics in Early North America*. Max Kade German-American Research Institute Series. University Park: Pennsylvania State University Press, 2008.

Rollmann, Hans. *Moravian Beginnings in Labrador: Papers from a Symposium Held in Makkovik and Hopedale*. St. John's, Newfoundland: Newfoundland and Labrador Studies, Faculty of Arts Publications, Memorial University, 2009.

Rubertone, Patricia E. *Archaeologies of Placemaking: Monuments, Memories, and Engagement in Native North America*. New York: Routledge, 2016.

Sachse, Julius Friedrich. *The German Sectarians of Pennsylvania: a Critical and Legendary History of the Ephrata Cloister and the Dunkers*. New York: AMS, 1971.

Salisbury, Neal, and Philip Joseph Deloria, eds. *A Companion to American Indian History*. Blackwell Companions to American History 4. Malden, MA: Blackwell, 2002.

Samuels, David W., Louise Meintjes, Ana Maria Ochoa, and Thomas Porcello. "Soundscapes: Toward a Sounded Anthropology." *Annual Review of Anthropology* 39 (2010): 329–345.

Sanderson, Eleanor. "The Challenge of Placing Spirituality within Geographies of Development." *Geography Compass* 1, no. 3 (2007): 389–404.

Schafer, R. Murray. *Five Village Soundscapes*. The Music of the Environment Series 4. Vancouver, BC: ARC, 1994.

———. "Soundscapes and Earwitnesses." In *Hearing History: A Reader*, edited by Mark M. Smith. Athens: University of Georgia Press, 2004.

———. *The Tuning of the World*. New York: Knopf, 1977.

Schama, Simon. *Landscape and Memory*. New York: A. A. Knopf, 1995.
Schattschneider, David. "Moravians Approach the Indians: Theories and Realities." *Unitas Fratrum* 21, no. 22 (1988): 37–48.
Schatull, Nicole. *Die Liturgie in der Herrnhuter Brüdergemeine Zinzendorfs*. Tübingen: Francke, 2005.
Schmick, Johann Jacob, and Carl Masthay (ed.). *Schmick's Mahican Dictionary*. Philadelphia: American Philosophical Society, 1991.
Schmidt, Leigh Eric. *Hearing Things: Religion, Illusion, and the American Enlightenment*. Cambridge, MA: Harvard University Press, 2000.
Schopenhauer, Johanna. *Ausflucht an den Rhein und dessen nächste Umgebung: im Sommer des ersten friedlichen Jahres*. Leipzig: Brockhaus, 1830.
Schöpf, Johann David. *Travels in the Confederation, 1783–1784: From the German of Johann David Schoepf; Translated and Edited by Alfred J. Morrison*. New York: B. Franklin, 1968.
Schutt, Amy C. *Peoples of the River Valleys: The Odyssey of the Delaware Indians*. Early American Studies. Philadelphia: University of Pennsylvania Press, 2007.
——. "Tribal Identity in the Moravian Missions on the Susquehanna." *Pennsylvania History: A Journal of Mid-Atlantic Studies* 66, no. 3 (1999): 378–398.
Schweinitz, Paul de, and W. H. Rice. "Gnadenhuetten on the Mahoning. Historical and Commemorative. 1746–1755." *Transactions of the Moravian Historical Society* 7, no. 5 (1906): 347–386.
Seeman, Erik R. *Death in the New World: Cross-Cultural Encounters, 1492–1800*. Early American Studies. Philadelphia: University of Pennsylvania Press, 2010.
Sensbach, Jon F. "Race and the Early Moravian Church: A Comparative Perspective." *Transactions of the Moravian Historical Society* 31 (2000): 1–10.
Smith, Bruce R. *The Acoustic World of Early Modern England: Attending to the O-Factor*. Chicago: University of Chicago Press, 1999.
Smith, Claire, and Hans Martin Wobst. *Indigenous Archaeologies: Decolonizing Theory and Practice*. New York: Routledge, 2005.
Smith, Linda Tuhiwai. *Decolonizing Methodologies: Research and Indigenous Peoples*. London: Zed Books, rev. ed. 2012.
Smith, Mark M. "Introduction: Onward to Audible Pasts." In *Hearing History: A Reader*, edited by Mark M. Smith, 417–431. Athens, GA: University of Georgia Press, 2004.
——. "Still Coming to 'Our' Senses: An Introduction." *The Journal of American History* 95, no. 2 (2008): 378–380.
Smith, Mark M., Mitchell Snay, and Bruce R. Smith. "Coda: Talking Sound History." In *Hearing History: A Reader*, edited by Mark M. Smith, 365–404. Athens: University of Georgia Press, 2004.
Spangenberg, August Gottlieb. *The Life of Nicholas Lewis Count Zinzendorf, Bishop and Ordinary of the Church of the United (or Moravian) Brethren*. London: Samuel Holdsworth, 1838.
Sterne, Jonathan. *The Sound Studies Reader*. London: Routledge, 2012.
Stoler, Ann Laura. *Carnal Knowledge and Imperial Power: Race and the Intimate in Colonial Rule*. Berkeley: University of California Press, 2010.
Szego, Kati, Beverley Diamond, and Heather Sparling. "Indigenous Modernities: Introduction." *MUSICultures* 39, no.1 (October 2012).
Shelemay, Kay Kaufman. *Soundscapes: Exploring Music in a Changing World*. 2nd ed. New York: W. W. Norton, 2006.

Shobe, Hunter, and David Banis. "Music Regions and Mental Maps: Teaching Cultural Geography." *Journal of Geography* 109, no. 2 (April 2010): 87–96.

Siewers, Alfred K., ed. *Re-Imagining Nature: Environmental Humanities and Ecosemiotics.* Lewisburg, PA: Bucknell University Press, 2014.

Silver, Peter Rhoads. *Our Savage Neighbors: How Indian War Transformed Early America.* New York: W. W. Norton, 2009.

Silverman, David J. *Faith and Boundaries: Colonists, Christianity, and Community among the Wampanoag Indians of Martha's Vineyard, 1600–1871.* Studies in North American Indian History. New York: Cambridge University Press, 2005.

Slotkin, Richard. *Regeneration through Violence: The Mythology of the American Frontier, 1600–1860.* Middletown, CT: Wesleyan University Press, 1973.

Smaby, Beverly Prior. *The Transformation of Moravian Bethlehem: From Communal Mission to Family Economy.* Philadelphia: University of Pennsylvania Press, 1988.

Smith, Billy G., and Simon Middleton, eds. *Class Matters: Early North America and the Atlantic World.* Early American Studies. Philadelphia: University of Pennsylvania Press, 2008.

Smith, Bruce. *The Acoustic World of Early Modern England: Attending to the O-Factor.* Chicago: University of Chicago Press, 1999.

Smith, Mark M., ed. *Hearing History: A Reader.* Athens, GA: University of Georgia Press, 2004.

Smith, Susan J. "Beyond Geography's Visible Worlds: A Cultural Politics of Music." *Progress in Human Geography* 21, no. 4 (August 1994): 502–529.

———. "Soundscape." *Area* 26, no. 3 (1994): 232–240.

Soderlund, Jean R. *Lenape Country: Delaware Valley Society before William Penn.* Early American Studies. Philadelphia: University of Pennsylvania Press, 2015.

Sommer, Elisabeth. "Fashion Passion: The Rhetoric of Dress within the Eighteenth-Century Moravian Brethren." In *Pious Pursuits: German Moravians in the Atlantic World*, edited by Michele Gillespie and Robert Beachey, 83–96. New York: Berghahn Books, 2007.

Spangenberg, A. G. "Spangenberg's Notes of Travel to Onondaga in 1745." *The Pennsylvania Magazine of History and Biography* 2, no. 4 (1878): 424–432.

———. "Spangenberg's Notes of Travel to Onondaga in 1745 (continued)." *The Pennsylvania Magazine of History and Biography* 3, no. 1 (1879): 56–64.

St. John de Crèvecoeur, J. Hector. *Letters from an American Farmer and Other Essays.* Edited by Dennis Moore. The John Harvard Library. Cambridge, MA: Belknap Press of Harvard University Press, 2013.

Starna, W.A. *From Homeland to New Land: A History of the Mahican Indians, 1600–1830.* Lincoln: University of Nebraska Press, 2013.

Sterne, Jonathan. *The Audible Past: Cultural Origins of Sound Reproduction.* Durham, NC: Duke University Press, 2003.

Steward, Julian Haynes. *Theory of Culture Change: The Methodology of Multilinear Evolution.* Urbana: University of Illinois Press, 1976.

Stievermann, Jan, and Oliver Scheiding, eds. *A Peculiar Mixture: German-Language Cultures and Identities in Eighteenth-Century North America.* Max Kade German-American Research Institute Series. University Park: Pennsylvania State University Press, 2013.

Stocker, Michael. *Hear Where We Are: Sound, Ecology, and Sense of Place.* New York: Springer, 2013.

Strohm, Reinhard. *Music in Late Medieval Bruges*. Oxford: Clarendon Press, 1985.
Swiggers, Pierre. "David Zeisberger's Description of Delaware Morphology (1827)." *Historiographia Linguistica* 36, no. 2 (December, 2009): 325–344.
Taruskin, Richard. *Text and Act: Essays on Music and Performance*. Oxford: Oxford University Press, 1997.
Thomas, Brian W. "Inclusion and Exclusion in the Moravian Settlement in North Carolina, 1770–1790." *Historical Archaeology* 28, no. 3 (1994): 15–29.
Thomas, Keith. *Religion and the Decline of Magic*. New York: Scribner, 1971.
Thoreau, Henry David. *Thoreau on Birds: Notes on New England Birds from the Journals of Henry David Thoreau*. Edited by Francis H. Allen. Boston: Beacon Press, 1993.
Thulin, Samuel. "Sound Maps Matter: Expanding Cartophony." *Social & Cultural Geography* (December 14, 2016): 1–19.
Till, Karen E., Paul C. Adams, and Steven D. Hoelscher, eds. *Textures of Place: Exploring Humanist Geographies*. Minneapolis: University of Minnesota Press, 2001.
Tinker, George E. *Missionary Conquest: The Gospel and Native American Cultural Genocide*. Minneapolis, MN: Fortress, 1993.
Tomlinson, Gary. *Singing of the New World: Indigenous Voice in the Era of European Contact*. New Perspectives in Music History and Criticism. New York: Cambridge University Press, 2009.
Tooker, Elisabeth, ed. *Native North American Spirituality of the Eastern Woodlands: Sacred Myths, Dreams, Visions, Speeches, Healing Formulas, Rituals, and Ceremonials*. Classics of Western Spirituality. New York: Paulist, 1979.
Treat, James. *Native and Christian: Indigenous Voices on Religious Identity in the United States and Canada*. Hoboken, NJ: Taylor and Francis, 2012.
Truax, Barry. *Acoustic Communication*. 2nd edn. Westport, CT: Ablex, 2013.
———. "Composing with Time-Shifted Environmental Sound." *Leonardo Music Journal* 2, no. 1 (1992): 37–40.
———. "Genres and Techniques of Soundscape Composition as Developed at Simon Fraser University." *Organised Sound* 7, no. 1 (April 2002): 5–14.
———. *Handbook for Acoustic Ecology*. Burnaby, BC: Cambridge Street Records, 1999.
———. "Paradigm Shifts and Electroacoustic Music: Some Personal Reflections." *Organised Sound* 20, no. 1 (April 2015): 105–110.
———. "Sound, Listening and Place: The Aesthetic Dilemma." *Organised Sound* 17, no. 3 (December 2012): 193–201.
———. "Soundscape, Acoustic Communication and Environmental Sound Composition." *Contemporary Music Review* 15, nos. 1–2 (1996): 49–65.
———. "Soundscape, Acoustic Communication and Environmental Sound Composition." In *A Poetry of Reality: Composing with Recorded Sound*, edited by Katherine Norman. Reading: Harwood Academic, 1997.
———. "Soundscape Composition as Global Music: Electroacoustic Music as Soundscape." *Organised Sound* 13, no. 2 (August 2008): 103–9.
Tuan, Yi-fu. *Space and Place: The Perspective of Experience*. Minneapolis: University of Minnesota Press, 1977.
———. *Topophilia: A Study of Environmental Perception, Attitudes, and Values*. Englewood Cliffs, NJ: Prentice-Hall, 1974.
Turino, Thomas. *Music as Social Life: The Politics of Participation*. Chicago Studies in Ethnomusicology. Chicago: University of Chicago Press, 2008.

Twiss, Sumner B., and Walter H. Conser, eds. *Religious Diversity and American Religious History: Studies in Traditions and Cultures*. Athens, GA: University of Georgia Press, 1997.
Uttendörfer, Otto. *Zinzendorf und die Mystik*. Berlin: Christlicher Zeitschriften-Verlag, 1952.
———. *Zinzendorfs Gedanken über den Gottesdienst*. Herrnhut: n.p., 1931.
Vanclay, F. M., Matthew Higgins, and Adam Blackshaw, eds. *Making Sense of Place: Exploring Concepts and Expressions of Place through Different Senses and Lenses*. Canberra, Australia: National Museum of Australia Press, 2011.
Vannini, Phillip, and April Vannini. *Wilderness*. London: Taylor & Francis, 2016.
Vaughan, Alden T., and Daniel K. Richter. *Crossing the Cultural Divide: Indians and New Englanders, 1605–1763*. Worcester, MA: American Antiquarian Society, 1980.
Vaux, George. "Extracts from the Diary of Hannah Callender." *The Pennsylvania Magazine of History and Biography (1877–1906)*. Philadelphia: Historical Society of Pennsylvania, 1888.
Vickers, Daniel, ed. *A Companion to Colonial America*. Blackwell Companions to American History. Malden, MA: Blackwell, 2003.
Vimalassery, Manu, Juliana Hu Pegues, and Alyosha Goldstein. "Introduction: On Colonial Unknowing." *Theory & Event* 19, no. 4 (October 2016). https://muse.jhu.edu/article/633283.
Vizenor, Gerald Robert. *Literary Chance: Essays on Native American Survivance*. Valencia, Spain: Universitat de València, Departament de Filologia Anglesa i Alemanya, 2007.
Vizenor, Gerald Robert, ed. *Survivance: Narratives of Native Presence*. Lincoln: University of Nebraska Press, 2008.
Vogt, Peter. "'Everywhere at Home': The Eighteenth-Century Moravian Movement as a Transatlantic Religious Community." *Journal of Moravian History* 1 (2006): 7–29.
———. "'Honor to the Side': The Adoration of the Side Wound of Jesus in Eighteenth-Century Moravian Piety." *Journal of Moravian History* 7 (2009): 83–106.
———. "Listening to 'Festive Stillness': The Sound of Moravian Music according to Descriptions of Non-Moravian Visitors." *Moravian Music Journal* 44, no. 1 (Spring 1999): 15–23.
Von Glahn, Denise. *The Sounds of Place: Music and the American Cultural Landscape*. Boston: Northeastern University Press, 2003.
Waldock, Jacqueline. "Soundmapping: Critiques and Reflections on This New Publicly Engaging Medium." *Journal of Sonic Studies* 2, no. 1 (October, 2011).
———. "Voice, Narrative, Place: Listening to Stories." *Journal of Sonic Studies* 2, no. 1 (May 2012).
Wallace, Paul A. W. *Conrad Weiser, 1696–1760, Friend of Colonist and Mohawk*. Philadelphia: University of Pennsylvania Press, 1945.
———. *Indian Paths of Pennsylvania*. Harrisburg: Commonwealth of Pennsylvania, Pennsylvania Historical and Museum Commission, 2005.
———. "Jacob Eyerly's Journal, 1794: The Survey of Moravian Lands in the Erie Triangle." *Western Pennsylvania Historical Magazine* 45, no. 1 (March 1962): 5–23.
———. *Thirty Thousand Miles with John Heckewelder, Or, Travels among the Indians of Pennsylvania, New York & Ohio in the 18th Century*. The Great Pennsylvania Frontier Series. Lewisburg, PA: Wennawoods, 1998.
Wallace, Paul A. W., and William A. Hunter. *Indians in Pennsylvania*. 2nd ed. Anthropological Series 5. Harrisburg: Pennsylvania Historical and Museum Commission, 2005.

Ward, Matthew C. *Breaking the Backcountry: The Seven Years' War in Virginia and Pennsylvania, 1754–1765*. Pittsburgh: University of Pittsburgh Press, 2003.
Warrior, Robert Allen. *The People and the Word: Reading Native Nonfiction*. Indigenous Americas. Minneapolis: University of Minnesota Press, 2005.
———. *Tribal Secrets: Recovering American Indian Intellectual Traditions*. Minneapolis: University of Minnesota Press, 1995.
Warsh, Molly A. "A Political Ecology in the Early Spanish Caribbean." *The William and Mary Quarterly* 71, no. 4 (2014): 517–548.
Waterman, Stanley. "Geography and Music: Some Introductory Remarks." *GeoJournal* 65, nos. 1–2 (2006): 1–2.
Weber, Julie Tomberlin, and Craig D. Atwood, eds. *A Collection of Sermons from Zinzendorf's Pennsylvania Journey, 1741–42*. Bethlehem, PA: Interprovincial Board of Communication Moravian Church in North America, 2001.
Wehrend, Anja. *Musikanschauung, Musikpraxis, Kantatenkompositionen in Der Herrnhuter Brüdergemeine: Ihre Musikalische Und Theologische Bedeutung Für Das Gemeinleben von 1727 Bis 1760*. Frankfurt am Main: P. Lang, 1995.
Wellenreuthrer, Hermann, and Carola Wessel, eds. *The Moravian Mission Diaries of David Zeisberger, 1771–1781*. University Park, PA: Pennsylvania State University Press, 2005.
Wessel, Carola. *Delaware-Indianer und Herrnhuter Missionare im Upper Ohio Valley*. Halle: Halle Verlag der Franckeschen Stiftungen im Niemeyer-Verlag, 1997.
———. "'We Do Not Want to Introduce Anything New': Transplanting the Communal Life from Herrnhut to the Upper Ohio Valley." In *In Search of Peace and Prosperity: New German Settlements in Eighteenth-Century Europe and America*, edited by Renate Wilson, Hermann Wellenreuther, and Hartmut Lehmann. University Park: Pennsylvania State University Press, 2000.
Westerkamp, Hildegard. "Linking Soundscape Composition and Acoustic Ecology." *Organised Sound* 7, no. 1 (April 2002): 51–56.
Westmeier, Karl-Wilhelm. *The Evacuation of Shekomeko and the Early Moravian Missions to Native North Americans*. Studies in the History of Missions 12. Lewiston, NY: Edwin Mellen, 1994.
Wheeler, Rachel. "An Imagined Mohican-Moravian 'Lebenslauf': Joshua Sr., d. 1775." *Journal of Moravian History* 11 (2011): 29–44.
———. *To Live upon Hope: Mohicans and Missionaries in the Eighteenth-Century Northeast*. Ithaca, NY: Cornell University Press, 2008.
———. "Women and Christian Practice in a Mahican Village." *Religion and American Culture: A Journal of Interpretation* 13, no. 1 (2003): 27–67.
Wheeler, Rachel, and Sarah Eyerly. "Singing Box 331: Re-Sounding Eighteenth-Century Mohican Hymns from the Moravian Archives." *The William and Mary Quarterly* 76, no. 4 (October, 2019): 451–496.
———. "Songs of the Spirit: Hymnody in the Moravian Mohican Missions." *Journal of Moravian History* 17, no. 1 (2017): 1–26.
Wheeler, Rachel, and Thomas Hahn-Bruckart. "On an Eighteenth-Century Trail of Tears: The Travel Diary of Johann Jacob Schmick of the Moravian Indian Congregation's Journey to the Susquehanna, 1765." *Journal of Moravian History* 15, no. 1 (May 2015): 44–88.
Whitridge, Peter. "Landscapes, Houses, Bodies, Things: 'Place' and the Archaeology of Inuit Imaginaries." *Journal of Archaeological Method & Theory* 11, no. 2 (June 2004): 213–250.

———. "The Sound of Contact: Historic Inuit Music-Making in Northern Labrador." *North Atlantic Archaeology* 4 (September, 2015): 17–42.

Williams, Lewis, Rose Alene Roberts, and Alastair McIntosh, eds. *Radical Human Ecology: Intercultural and Indigenous Approaches*. London: Routledge, 2016.

Wilson, Renate, Hermann Wellenreuther, and Hartmut Lehmann, eds. *In Search of Peace and Prosperity: New German Settlements in Eighteenth-Century Europe and America*. University Park: Pennsylvania State University Press, 2000.

Winiarski, Douglas Leo. *Darkness Falls on the Land of Light: Experiencing Religious Awakenings in Eighteenth-Century New England*. Chapel Hill: Published for the Omohundro Institute of Early American History and Culture, Williamsburg, Virginia, by the University of North Carolina Press, 2017.

Withers, Charles W. J. "Place and the 'Spatial Turn' in Geography and in History." *Journal of the History of Ideas* 70, no. 4 (October 2009): 637–658.

Witte, Bernd, ed. *Goethe-Handbuch: In Vier Bänden*. Stuttgart: J. B. Metzler, 1996.

Wohlleben, Peter, and Tim Flannery. *The Hidden Life of Trees: What They Feel, How They Communicate—Discoveries from a Secret World*. Translated by Jane Billinghurst. Vancouver, Canada: Greystone Books, 2016.

Wolf, Eric R. *Europe and the People without History*. Berkeley: University of California Press, 1982.

Wollstadt, Hanns Joachim. *Geordnetes Dienen in der christichen Gemeinde dargestellt an den Lebensformen der Herrnhuter Brüdergemeine in ihren Anfängen. Mit 4 Kunstdrucktafeln*. Göttingen: Vandenhoeck und Ruprecht, 1966.

Womack, Craig S. *Red on Red: Native American Literary Separatism*. Minneapolis: University of Minnesota Press, 1999.

———. "Theorizing American Indian Experience." In *Reasoning Together: The Native Critics Collective*. Norman: University of Oklahoma Press, 2008.

Womack, Craig S., Daniel Heath Justice, and Christopher B. Teuton. *Reasoning Together: The Native Critics Collective*. Norman: University of Oklahoma Press, 2008.

Woodward, Walter. "'Incline Your Second Ear This Way': Song as a Cultural Mediator in Moravian Mission Towns." In *Ethnographies and Exchanges: Native Americans, Moravians, and Catholics in Early North America*, edited by A. G. Roeber, 125–142. Max Kade German-American Research Institute Series. University Park: Pennsylvania State University Press, 2008.

Woude, Joanne van der. "Polyglot Harmony: Moravians among the Indians." In "Towards a Transatlantic Aesthetic: Immigration, Translation, and Mourning in the Seventeenth Century." PhD dissertation, University of Virginia, 2007.

———. "Towards a Transatlantic Aesthetic: Immigration, Translation, and Mourning in the Seventeenth Century." PhD dissertation, University of Virginia, 2007.

Wren, Christopher. "Description of Indian Graves on Bead Hill, Plymouth, Pennsylvania." In *Proceedings and Collections of the Wyoming Historical and Geological Society, 1911–12*, vol. 12. Wilkes-Barre, PA: E. B. Yordy.

Yates, Frances A. *The Rosicrucian Enlightenment*. London: Routledge & Keegan Paul, 1972.

Yonan, Jonathan. "The 1775 Correspondence of John Wesley and Francis Okely." *Journal of Moravian History* 12, no. 1 (May 2012): 93–103.

Zapf, Hubert, and Walter de Gruyter & Co., eds. *Handbook of Ecocriticism and Cultural Ecology*. De Gruyter Reference 2. Berlin: De Gruyter, 2016.

Zeisberger, David. *David Zeisberger's History of the Northern American Indians*. Edited by Archer Butler Hulbert and William Nathaniel Schwarze. Columbus: Ohio State Archaeological and Historical Society, 1910.

———. "Foreword." In *A Collection of Hymns for the Use of the Delaware Christian Indians, of the Mission of the United Brethren in North America*, 2nd ed. Bethlehem, PA: J. and W. Held, 1847.

———. *Schoenbrunn Story; Excerpts from the Diary of the Reverend David Zeisberger, 1772–1777, at Schoenbrunn in the Ohio Country*. Columbus: Ohio Historical Society, 1972.

Zinzendorf, Nicolaus Ludwig Graf von. *Nine Public Lectures on Important Subjects in Religion, Preached in Fetter Lane Chapel in London in the Year 1746*. Iowa City: University of Iowa Press, 1973.

———. *Der Teutsche Socrates, Das ist: Aufrichtige Anzeige verschiedener nicht so wohl unbekannter als vielmehr in Abfall gerathener Haupt-Wahrheiten in den Jahren 1725 und 1726: Anfänglich in der Königl. Residentz-Stadt Dressden, Hernach aber dem gesamten lieben Vaterland teutscher Nation zu einer guten Nachricht nach und nach ausgefertiget, und von dem Autore selbst mit einem kurtzen Inhalt jedes Stücks, nunmehro auch mit verschiedenen Erläuterungen, die sich in der ersten Auflage nicht befinden, und einem Anhange versehen*. Leipzig, 1732.

Ziser, Michael. *Environmental Practice and Early American Literature*. New York: Cambridge University Press, 2013.

INDEX

A
Abraham, 214–215, 216
acoustic ecology, 40n14, 45n56, 45n57, 47n65, 51; of forests, 2, 50, 53–54, 64. *See also* sound maps; sound environments, importance of
Adams, John, 192–193
agency, in adapting religion, 14
agriculture, 195, 196, 220n16
alchemy, 171–174
Allegheny Front, 3, 189, 226; dangers of crossing, 176, 188, 209; Native American path over, 54–55, 106, 187
American Revolution, 7, 10, 31, 210–213, 223
Americans: Gnadenhütten (Ohio) massacre by, 7–11, 214–215, 223; Moravians trusting, 213–214. *See also* settlers
Anderson, Isobel, 38
animals. *See* flora and fauna, of Pennsylvania
animals, of Pennsylvania, 1–2, 54, 56–57, 64, 114, 187, 192
Anishinaabe people, 15–16
Antes, Henry, 140; school on the farm of, 140, 176
Anton, 214–215
art of memory, 163, 181n37
audible history. *See* sonic history; soundscape compositions; sound maps
author's family farm: family's love for, 3, 5, 225; history of, 4–5; loss of, 3–4, 226

B
Baker, Geoffrey, 12–13
Bartram, John, 61–62, 65, 149n87
Bathsheba, 134, 177n1
bells, in sounds of Bethlehem, 35, 115–118, 127, 131, 144n27, 145n35, 146n43
Benezet, Susanna. *See* Susanna Pyrlaeus
Bethlehem, 90, 106, 147n66, 158; beauty of, 192–193; boundary location of, 109, 199;

buildings and layout of, 112, 114–116, 126, 193; daily rituals in, 115–118, 121, 145n35, 183n62; decimation of, 188; diversity of, 133–134, 145n30, 193; economy of, 109–110, 191, 199–200, 224; end of communal economy in, 197, 199–200; Eyerly Jr. in, 1–2; Eyerly Sr. in, 151–152; as first Church-built Moravian mission, 30, 44n39; as hub for missionary work, 30, 109–110; industries of, 109–110, 114–115, 191, 199–200; membership in, 126, 129–131; modern-day Bethlehem, 33–35; Moravian Archives in, 105–106; Native Moravians and, 201–203, 205, 207; population of, 193, 198–201; Pownall on, 107–109, *108*; relations with outsiders, 126–133; religious services in, 116–117, 127–128, 130, 133; remnants of past in, 33–36, 106; singing in, 115, 125–126, 145n35, 146n43, 193; soundscapes of, 35–38, 114–118, 122–123, 144n27 (*see also* soundscapes; sound maps); watchmen of, 118–120, 145n41. *See also* Moravian mission communities
birds. *See* flora and fauna, of Pennsylvania
birds, of Pennsylvania, 1–2, 55–57, 114
bodies, 166–171, *168*, *170*, *172*; Moravians' focus on Christ's suffering, 171–173; transformation of flesh, 169–171
Böhler, Anna, 125
Böhler, Peter, 76–77, 129, 131–132
Böhme, Jakob, 171–172
Britain/British, 7, 92, 198; land and, 52, 209; Moravian missionaries and American Revolution, 211, 212–213; Native Americans and American Revolution, 210–211; Native Moravians and, 211–212, 214; in Seven Years' War, 198–199
Brooks, Lisa, 5, 16
Brüdergemeine (Brethren's Community). *See* Moravian Church

261

Bürstler (Jacob) family, 78, 102n103
Büttner, Gottlob, 132, 160

C
Caldwell, Alice, 200
Cammerhof, Friedrich, 121, 123
Camporesi, Piero, 167–169
Canada, Native Moravians moving to, 31, 218, 224
Captives' Town, 7, 213
Chastellux, Marquis de, 193
Cherokees, 17
children: choir system and, 111, 112; education of, 121, 140, 156–158; hymns and, 117, 140–141, 161, 164, 180n28, 181n37, 204, 214; musical instruction of, 139–140, 164; spiritual seeking by, 157, 165–166, 176
chorale tunes, 25, 72, 122, 140, 141–142, 145n33, 146n51, 163
choir system (Moravian missions' communal housing system), 24–25, 111–112, 143n10, 143n12, 160, 203; children and, 111, 112, 121; hymns and, 121–122, 126, 146n49, 146n51, 166–167; Singstunden and, 161, 173; sound and, 115–118, 126–127, 128; strangers and, 129, 133
Christianity, 16, 52; beliefs of, 166–171, 179n22; Native Moravians' adoption of, 14–15; rituals and practices of, 15, 67–68
Christians: Gnadenhütten (Ohio) massacre by, 10, 214–216; Moravians' status as, 217
colonialism: effects of, 70, 93–94; importance of sounds in, 12–13, 112, 114; mapping and naming new territories in, 81–83, 194, 201; networks of, 13–14
colonization, 16, 97n3, 112–114, 141
conversions, 29, 70, 86, 157, 160
Cooper, James Fenimore, 93
Cooper Township, Pennsylvania, author's family farm in, 3–5, 186, 225–226
Coshocton (Goschachgünk), 209, 211
Creeks, 17
Crèvecoeur, J. Hector St. John de, 196

D
Danish Empire, 14, 17, 20, 29, 198
David, 76–77
death: Esther's, 174; Eyerly Jr.'s, 153; Joshua Jr.'s, 176–177, 185n79; Joshua Sr.'s, 176; Moravians' understanding of, 121, 166, 171, 184, 189; Nathan's, 188–189; Pyrlaeus's, 175; singing at, 152–153, 174–177, 187n79, 216–217; Zinzendorf's, 175, 199
Delaware Moravians, 7. *See also* Native Moravians
Delawares, 56, 176, 187, 188, 198, 202; lands of, 14, 30, 92, 101n97, 209; settlers *vs.*, 8, 10, 198, 211
DePeyster, Arent Schuyler, 211–212
DeRogatis, Amy, 82
de Tocqueville, Alexis, 13
Dickens, Charles, 92–93
Dober, Martin, 157

E
Economy (Moravian missions' communal economic system), 24–25, 109–111, 115, 126, 129, 130, 197; dissolution of, 191, 193, 199–201
education: of children, 121, 140, 156–158; learning improvisation in singing, 161, 163–165; musical, 139–140, 164, 180n28, 181n37; poet and singing schools, 161–163
Einsingen. See death, singing at
Emmaus community, 224
Engel, Katherine Carté, 129–130, 197–198
Erbe, Helmut, 117
Erickson, Morton, 50
Esther, 174
Ettwein, Johannes, 60, 62, 64, 69, 83; travels of, *84*, 187–189
Europe, Seven Years' War in, 199
European Moravians. *See* Moravians and
Europeans: Pennsylvania landscapes under, 194, 196–197; town planning by, 1–7, 112. *See also* settlers
Eyerly, Christina Elisabeth, 199–200; home of, 220n31
Eyerly, Johann Jacob, Sr.: finances of, 199–200; home of, 220n31; mission work and, 152; Lebenslauf of, 151–152
Eyerly, Johann Jacob, Jr., 1–2, 151–153; death of, 153; journal of, 1–2, 4, 105–106; Lebenslauf of, 106, 187; surveying church lands, 1–2, 106, 187, 223–224

F

Faull, Katherine, 16, 67, 126–127, 201, 218
flora and fauna, of Pennsylvania, 91–92; animals, 1–2, 54, 56–57, 64, 114, 187, 192; birds, 1–2, 55–57, 114; plants, 1–2, 53–54, 192
forests, of Pennsylvania, 53, 58, 83–84; clearing of, 92–93, 191–192, 197; conservation of, 104n129; impressions of, 1–2, 61–62, 64, 93, 192; interpretations of, 55–56, 61–62, 67–68, 71; Moravian missionaries' uses of hymns in, 68–69; navigation through, 1–3, 55–56, 78; sounds of, 1–3, 54, 64; vastness of, 55, 197
Fort Allen, 33, 198. *See also* Gnadenhütten mission, Pennsylvania; Lehighton
France, 52; Seven Years' War and, 198–199, 201
Franklin, Benjamin, 33, 198, 205–206
Fremdenladen. See Strangers' Store
Frey, Andrew, 68
Friedenshütten I (mission community near Bethlehem), 32, 147n66, 188, 224
Friedenshütten II (mission community near Wyalusing), 32, 34, 83, 91, 188, 209, *210*, 224. *See also* Wyalusing
Fröhlich, Christian, 68–69
funerals/graves, 49–50, 112, 122, 128, 145n30, 189, 190, 209, 226

G

Garrison, Nicholas, 147n69
Germany, 30; Moravian missionaries from, 20–21, 91, 217; musical influence of, 22–23, 141–142
Gnadenhütten (Ohio) massacre, 7–11, 187, 194, 213, 218; by Americans, 8–10, 214–216, 223; reactions to, 9–10, 31, 152, 215–216, 223, 224; singing at, 8–11, 214; victims of, 10, 184n77
Gnadenhütten mission, Ohio, 7–10, 46n59, 210, 211, *211*, *212*, 213, 214; missionaries' return to, 7–8; Native Moravians and, 176, 211, 217. *See also* Gnadenhütten (Ohio) massacre
Gnadenhütten mission, Pennsylvania, 46n59, 139, 162; destruction of, 32, 147n66, 188, 201; Eyerly at, 152; Franklin buying site of, 33, 198; massacre at, 147n66, 152, 187, 198; Native Moravians and, 70, 71, 137, 140, 141, 154, 160, 165, 174, 176, 188; rebuilt as Fort Allen, 33, 46n59, 198. *See also* Fort Allen; Lehighton
Goethe, Johann Wolfgang von, 91–92
Golkowsky, George, 147n69
Goodman, Glenda, 12
government: colonial, 21, 201–203; Pennsylvania, 187, 205; US, 223–224
Great Shamokin Path, 20, 34–35, 55, 187–190
Great Warrior's Path, 20, 85
Gregor, Christian, 91–92
Grider, Rufus, 123
Grube, Bernhard Adam, 66, 202, 204, 207

H

Hagen, Johannes, 68, 132, 139
Harris, Captain, 73
Harrison, William Henry, 10
Haudenosaunee, 17, 21, 31, 65, 75, 101n97, 152, 198, 209, 217; lands of, 52, 92
Heckewelder, Johann, 9, 93, 140, 216
Heerendijk community, 109
Heinitz, Johann Friedrich von, 116
Hehl, Mattheus, 81, *82*
Herder, Johann Gottfried, 163, 181n34
Herrnhaag community, 109, 145n33, 182n54
Herrnhut community, 23–24, 45n54, 72, 122–123, 151, 157–158, 199
Herrnhuter. See Moravian Church
Herrnhuter Gesangbuch (Moravian hymnal), 86–87, 163
Hoffer, Peter Charles, 13
Holmes, Obadiah, 9
Horsfield, Timothy, 128, 147n71, 202–203. *See also* Strangers' Store
hunting, 23, 57, 58, 60, 71, 139, 160, 202, 220n16. *See also* spiritual practices around hunting
hymns, 40n13; alchemy and, 171–174; blood and wounds in, 71, 166–171, 183n61; children and, 139–141; choice of, 25; communal singing of, 132; as familiar, tamed soundscape, 71–72; importance to Moravians, 108–109, 217; improvisation of, 25, 154–155, 161–165, 173; Mohican

hymns (*Cont.*)
  musical practices and, 165–166, languages of, 22, 25–29, 66, 141–142, 154; learning and adapting, 139–140, 161–163, 181n37; in Native languages, 29, 132, 134, 137, 141; polyglot singing of, 136–137, *138*; power of, 171–173; in religious services, 116–117, 123; for special occasions, 121–122; uses of, 11, 67–69, 71, 73, 78, 125, 132, 161; by watchmen, 118–120, *119*, 145n41; Zinzendorf composing, 72–77, 85, 86, 96, 125, 146n54; Zinzendorf preparing supplements to hymnal, 86, *89*, 90. *See also* singing

I
identity, 11, 90, 109, 131, 133, 200, 217–218
immigrants, European, 194–195. *See also* settlers
improvisation. *See* education; hymns; singing; Zinzendorf, Nikolaus Ludwig von
Inuit people, 15–17

J
Jacob and Thomas, surviving massacre, 9, 213, 214–215
Jefferson, Thomas, 92
Jephta, 71
Joshua, 57
Joshua Jr., 140, 176, 222; death of, 176–177, 185n79; murder of family, 184n77, 214, 215–216; music of, 140, 203–204, 206
Joshua Sr. (Tassawachamen), 69, 139, 140, 155, 156, 158–160, 178n12, 179n24, 184n77; death of, 176; experience with religion, 159, 179n18; farewell to Pennsylvania governor, 207–208; hymns by, 134, 149n104, 154, 177n1; singing with Pyrlaeus, 154, 160–161, 165; spiritual practices of, 134, 165, 179n18; travels of, 76–77, 188
Judith, 214, 216
Justice, Daniel Heath, 5

K
Kackawatcheky, 85, 96
Keekyuscung, 187
Keyes, Sarah, 13

L
Lakeshore Path, 223
land, 220n16; of Moravian Church, 211–212, 223–224; of Native Americans, 211; settlers and, 223–226
land use, 191, 195, 197, 219n13
languages, 127, 187; diversity of, 67, 133–134, 193; polyglot singing of hymns, 136–137, *138*; of singing, 132, 134, 193
The Last of the Mohicans (Cooper), 93
Le Guin, Elisabeth, 38
Lehigh River, 30, 33, 73, 115, 122, 127, 147n69, 158
Lehigh River valley, 30, 94, 107, 192–193
Lehighton, 33. *See also* Fort Allen; Gnadenhütten mission, Pennsylvania
Lititz community, 110, 224; music in, 140
Little, Juanita, 16
Lorenz, Ruth, 151
Loskiel, George, 57, 60, 70
lots, in Moravian missions, 24–25, 26–28, 102n105, 111, 118, 125, 158, 166, 179–180n26

M
Mack, Jannetje (Rau), 76–77, 85–86
Mack, Johann Martin, 68–70, 76, 85–87
Mann, Kristin Dutcher, 12
maps, 147n69; boundaries and, 201, 211, 224; Europeans', 62, *63*, *81*, 81–82, 96, 194; mental, 55, 59; of missionaries' travels, 78, *79*, *80*, 81–82, *81*, 90, *91*; of Moravian missions, *18*, *210*, *210*, *211*, *212*; naming and, 51–52, 61, 81–83; Native Americans', 53, 58, 60–61; Susquehanna River, *84*; uses of, 82–83, 99n54. *See also* sound maps
Maria, 203
Marie, 86, 96, 103n117
marriages, 125, 133, 143n15, 143n18; effects of, 112, 128–129, 143n15; quarters dependent on, 111, 115
McClure, David, 139
membership, in Moravian communities, 126, 129–131
Menagachsuenk, 30, 115
Meniolagomeka, 30, 72–73
Merrell, James, 16, 67, 197

Index | 265

militia, attacking Gnadenhütten mission, Ohio, 8–9, 213–215
missionaries, 14–16, 82, 159. *See also* Moravian missionaries
Mohawks, 134
Mohican language, 8, 22, 29, 37, 86, 126, 134, 149n97, 154; hymns in, 32, 127, 131–132, 137, 143, 154–155, 225
Mohican Moravians, 7, 9, 76–77, 178n12. *See also* Native Moravians
Mohicans, 23, 176, 225; culture of, 22, 71, 165; mission to, 20, 86
Monocacy Creek, 33, 34, 105, 114, 115, 126, 127, 128, 158, 201
Monocacy Creek valley, 192–193
Montour, Madame, 20, 85, 187
Moravian Archives, 105–106
Moravian Church, 9, 17, 23, 191; colonial governments and, 106, 199; intercessions in, 118, 145n36; land of, 1–2, 14, 30, 197, *211–212*, 223–224; membership in, 126, 129–131, 155, 177n2; network of, 24, 74–75, 141; records kept by, 4, 151; Zinzendorf's hymns in, 72–76, 86–90, 96
Moravian missionaries, 11, 17, 20, 22–23, 29, 30, 158; in American Revolution, 212–213; Bethlehem as hub for, 109–110; on divine deliverance, 69–70, 86–87, 91; gathering natural specimens, 91, 103n126; Native religious life and, 22; Pennsylvania and, 52, 92; relations with Indigenous peoples, 11, 216–217; renaming Indigenous places, 52, 81–83, 177n2, 194, 201; role in destruction of Native Moravians, 216–218; role of, 13–14; uses of music by, 25, 68
Moravian missions, 24, 45n54, 126, 142, 161, 223; boundaries of, 142; church-built, 109; economy of, 24–25, 109–111, 191, 193, 197, 199; financing for, 197–198, 199; first, 22, 109; in Georgia, 29–30; layout of, 24, 109, 114, 145n30; longevity of, 31, 44n39, 223–224; missionaries and home community in, 110–111; multilingualism in, 133; Native Moravians forming, 176, 201–202; networks of, 24, 43n33, 74–75, 108, 110, 141; to North America, *18*; plan for development of, 24, 44n36; planning and layout of, 112, 114; recreation of soundscapes of, 31–37; religious services in, 111–112; sense of community in, 120–121, 166; social life in, 121–122, *124*, 139; social organization of, 24–25, *26–28*, 109–111, 121; soundscapes of, 114–116, 127, 131; travelers and visitors to, 126–129, 205–206; vision for, 24, 111. *See also* Bethlehem
Moravian religion, 23; focus on Christ's suffering, 166–168, *168*, 171–173, *172*, 182n50, 182n53; music in, 25, 29, 141, 142
Moravian Run, 2, 3, 49, 186, 187, 189, 226
Moravians, 15, 126, 210; clothing of, 111–112, *113*, 143n15; efforts to tame wilderness, 71–72, 194; influence of, 4–5; interpretation of nature by, 66–67, 71; maps of, *81*, 82; museums documenting history of, 35; naming and claiming new places, 81–83; return to Gnadenhütten mission, Ohio, 7–8; separateness of, 109; Shikellamy inviting to visit, 75–76; sound and, 38, 108–109, 151; vision of Pennsylvania, 92, 94; wars and conflicts around, 7, 198, 210–211, 213–214, 217–218. *See also* hymns; singing
Moshannon Creek, 2, 3, 5, 6n7, 34, 49–50, 186, 187, 189, 225, 226
mother, author's, 3–5, 225–226
Murray, Judith Sargent, 193
music, 66; German influence on, 141–142; importance of song, 22–23; instruction of children in, 139–140; Joshua's, 154, 160–161, 165, 184n77, 187n79; in Moravian religion, 25–26, 38, 122, 132–133, 157; Native Christians', 15, 165; by Native Moravians, 66, 142, 206; power of, 116–117, 171–172; Pyrlaeus's experience of, 156–157; uses of, 67, 122. *See also* singing

N
Nain community, 33, 110, 126–127, 147n66, 201–202
Nathan, 188–190, 209, 226
Nathan, Andy, 33
Nathanael, 139

Native American communities, 70, 109; conflicts of, 67, 195, 201; languages in, 149n87, 149n105; Moravians in, 17, 22; archaeology of, 224–225; soundscapes of, 65–66; Zinzendorf visiting, 72, 76–77. *See also* Native Moravian communities

Native Americans, 66, *77*; in American Revolution, 210–211; burial ground of, 95–96; conflicts of, 31, 197; distinguishing "friendly" *vs.* "hostile," 202–203; efforts to recognize past of, 94, 104n132; environmental transformations and, 93–94, 159, 194, 202; lands of, 1–2, 5, 195–196, *211*; languages of, 133–134, 176; Moravian missionaries and, 11, 17–18, 23, 147n66, 152, 216–217; relationships between Moravians and settlers and, 217–218; music of, 15–16, 23, 66, 216–217; nativist movements of, 10, 177, 218; Pennsylvania government negotiating with, 187; pushed westward, 94, 197; relations with European settlers, 12–13, 16–17, 65, 94, 196, 205–206, 213, 218; religious life of, 15, 20, 22, 55–56; Seven Years' War and, 198–199; sound's importance to, 12–13, 55–57; vision of Pennsylvania, 92, 94; visiting Bethlehem, 106, 127. *See also* Native Moravians

Native Christians, 20, 23; music and, 15, 23, 139–142; in nearby settlements, 30–31, 126. *See also* Native Moravians

Native languages, 14, 22, 133–134, 139; hymns in, 15, 22, 29, 132, 134, 137, 141. *See also* Mohican language

Native Moravian communities, 147n66; new, 201–202, 209, 224

Native Moravians, 31, 61, 126, 140; Bethlehem and, 193–194, 202–203; British ordering seizure of, 211–212; deaths of, 205–207; decimation of, 188; Franklin defending, 205–206; missing Bethlehem, 205, 207; Moravian missionaries' role in destruction of, 216–218; moving to Ohio, 188, 209; moving westward, 31, 188, 218, 224; music of, 139–142; religion of, 9, 14, 70–71; religious services and, 137, 141, 203–205; stay in Philadelphia, 203–209; repeated relocations of, 193–194, 203–208, 209, 211; violence against, 7–9, 211–215

nature, 69, 104n129; beauty of, 83–84; divine deliverance from, 69–70, 86–87, 91, 103n121; Eyerly documenting, 1–2; gathering specimens of, 91, 103n126, 192; as healing, 70–71; sounds of, 59, 158. *See also* trees, animals, etc.

Nazareth community, 30, 106, 140, 194, 199–200, 203, 224

Nelson, Vernon, 106

Netawatwees, 8, 209, 211

Nicodemus, 71

Nitschmann, Anna, 52, 76, 81, *81*, 82, 85, *88*, 90, 102n105

Nitschmann, Susanna, 152

North America, Moravian Church missions to, 17, *18*, 20, 30–31. *See also* Moravian missions; Moravian missionaries

O

Oeconomie. *See* Economy (Moravian missions' communal living)

Ohio: in American Revolution, 7–8, 210–211; Moravian communities in, 210, 223–224; Native Moravians relocating to, 188, 209, 224

Oley community, 78, 102n103, 44n36

Onondaga, 20, 21, 55, 61, 65, 66, 68, 82, 101n97, 149n97, 209

Orchi, Emanuele, 167–169

Otstonwakin, 20, 78, 85, 187

owls, beliefs about, 56

P

Paxinosa, 66

Paxton Boys, 205–206, 213

Peale, Pennsylvania, 49–50, 186, 187, 189, 225

Penn, Thomas, 82, 83

Penn, William, 51, 194–195, 201

Penn's Creek massacre, 187

Penn's Woods, 2, 51, 52, 92, 94, 107, 194, 224–225

Pennsylvania, 5, 67, 107; archaeological sites in, 96n2; changes in, 103n128, 196–197; early map of, *63*; efforts to recognize Native American past in, 94, 104n132; flora and fauna of, 1–2, 53–54, 55–57, 64, 91–92, 114, 192 (*See also* forests); government of, 21, 187, 201–203, 224; land of, 5, 224; landscapes of, 53, 61–64, 104n129, 194, 196–197; languages, 133–136; Moravian map of, 81, 82; Native Americans and, 5, 94, 176, 188; natural resources of, 194–195; varying visions of, 92, 94; wildness of, 2, 55–57, 196–197; Zinzendorf and, 17–20, 52, 72, 74

Pennsylvania Wilds, 2, 5, 6n6
Petquotting, Moravian mission proposed for, 223
Petquotting, Ohio, 223, 224
Philadelphia, Native Moravians in, 188, 203–209
Pilgerruh community, 109
Pisquetomen, 187
place names, 59, 83, 96; Europeans changing, 52, 194; importance in tracing history, 33–35, 96; Moravians', 52, 81–83, 102n105, 102n106; Native Americans', 51–52
plants. *See* flora and fauna, of Pennsylvania
plants, of Pennsylvania, 1–2, 53–54, 192. *See also* flora and fauna, of Pennsylvania
Plymouth, Pennsylvania. *See* Wyoming
Pointer, Richard, 22
Pontiac's War, 176, 202, 218
Post, Agnes, 187
Post, Christian, 187–188
Post, Rachel (Rahel), 134, 135–137, 187
Post's Island, 50, 186, 187, 189
Pownall, Thomas, 62–63, 107–109, 108, 192, 195–196, 220n16
Price, Matthew Hunter, 13
Prostration. *See* religious services, Moravian
Province Island, 203–205. *See also* Philadelphia, Native Moravians in
Pyrlaeus, Susanna (Benezet), 175
Pyrlaeus, Johann Christoph (Tganniatarecheu): death of, 175; Lebenslauf of, 156–158; Tassawachamen (Joshua Sr.) and, 154, 156, 158, 160–161, 165, 177n2

R
race, growing tensions and, 16, 94, 201–202, 218
Rath, Richard Cullen, 13, 177n5
Rau, Jannetje Mack. *See* Mack, Jannetje

Rauch, Christian Heinrich, 20–23, 30; at Shekomeko, 21–22, 178n12; teachings of, 159–160
recreation, in Moravian communities, 125
Reichardt, Johann Friedrich, 162
religions: Joshua's experience of, 158–160, 176; Native Moravians incorporating differing traditions of, 70–71; Pyrlaeus's experience of, 156–157; Zinzendorf's experience of, 166
religious services, Moravian, 73, 116, 141, 200–201; bodies connected to spirit in, 166–167; at Easter, 122, 132; intercessions in, 74, 145n36; languages of, 127, 132, 139; of Native Moravians, 203–206; Prostration in, 73, 116
Renatus, Christian, 203
Renatus, John, 137
Reuss, Erdmuthe Dorothea (Countess von Zinzendorf), 24, 52, 81, 81, 82, 102n105
Reuter, Christian Gottlieb, 36
Richter, Daniel, 16, 42n18, 96n2, 98n38, 197, 218
Ridner, Judith, 82
Rivett, Sarah, 14
Rollins, Nathan, 8
Roosevelt, Theodore, 215
Roth, Johann, 188, 189

S
Salem mission, 8, 176, 210, 211, 211, 212, 213, 214
Sattelihu (Andrew Montour), 85
Schama, Simon, 53
Schebosch, Joseph, 213, 221n57
Sciuchetti, Mark, 33
Schmick, Johann Jacob, 71, 176, 202, 204, 205, 208
Schönbrunn mission, 8, 176, 210, 211, 211, 212, 213, 221n58
Schopenhauer, Johanna, 162
Schöpf, Johann David, 192–194
Scots-Irish. *See* settlers
Seidel, Anna Johanna, 200
settler communities: conflicts of, 31, 67, 195; European *vs.* Native, 109, 187–188; expansion of, 103n128, 191, 194, 197; relations with Natives, 12–13; soundscapes of, 66, 114

268 | Index

settlers, 52, 66, 197, 220n16; attacks on, 9, 213; forests and, 61, 92–93; Moravians and, 17, 133, 217–218; Native Moravians and, 202–203, 210, 224–225; Natives and, 66, 205–206; relations with Natives, 16–17, 65, 196, 201; violence against Native Moravians, 211, 213, 218

Seven Years' War, 31, 67, 94, 99n54, 118, 197–199; effects of, 176, 191, 197, 198, 201, 218

Shabash (Abraham), 93–94

Shamokin, 20, 30, 55, 66, 68, 75–76, 78, 81, 85, 187, 194

Shamokin Path, 75, 78, 81, 101n97

Shawnee Prophet. See Tenskwatawa

Shawnees, 85, *88*, 96, 198

Shekomeko, 20, 178n12, 188, 203, 225; Moravian missionaries in, 21–22, 30–31, 44n39, 71, 76, 139; Zinzendorf and, 75–76; Joshua Sr. and, 158–160

Shikellamy, 21, 65, 66, 75–76, 85, 100n67, 101n97

singing, 137, 184n73, 193, 200, 214; in Christian communities, 123–124; at death, 152–153, 174–177, 187n79, 216–217; difference of Moravians from other churches, 131–132, 148n82, 148n83, 155, 162–163, 173–174; improvisation in, 155, 161, 163–165, 180n28, 200; languages of, 132, 134, 193; learning, 161–162, 200; in missions' daily rituals, 116–125, 145n35, 146n43, 161; in Moravian identity, 131, 217; Moravians and, 9–11, 22–23, 125–126, 139, 151, 155; Native Americans', 65, 217; by Native Moravians, 206, 208–209; pervasiveness of, 115, 162; polyglot, 136–137, *138*; power of, 183n67; purposes of, 57, 120–121, *124*, 141, 155–156, 166, 172–174; sacred, 116–117, 166; while working, 123–124. See also hymns; music; Singstunden

Singstunden, 25, 29, 125, 149n104, 173, 200; children and, 164; decline of, 200; guide to creating, 146n49, 163–164; in Bethlehem, 76, 101n93, 118, 121, 132, 140; in natural settings, 68, 70, 71, 123; Native Moravians and, 203–204, 206–207, 209; Tassawachamen (Joshua Sr.) and Pyrlaeus and, 154–155, 159–161, 165, 166. See also improvisation; hymns

Six Nations. See Haudenosaunee

snakes, 49, 83, 92; Zinzendorf's incident with, 87, 90, 96, 103n121

sonic history, 11–13, 31–32, 46n56

sound environments, importance of, 11–12. See also acoustic ecology; sounds; soundscapes

sound maps, 31–32; creation of, 35–36, 45n57; importance of, 37–38, 46n61

sounds, 51–52, 53, 106, 171–172, 226; in animal-human communication, 57–58; in colonization, 12–13, 112–114; importance of, 1, 11–12, 38, 55–57, 126, 151; of nature, 158; power of, 108–109; spirituality and, 58–59, 155. See also soundscapes

soundscape compositions, 31–32, 36–37, 45n57, 47n65, 47n67, 48n68

soundscapes, 2, 5, 40n14, 54–55, 59, 64, 127, 131; of Bethlehem, 114–118, 122–123, 126–127; changes in, 71–72, 191, 201, 225; in colonization, 12–13, 14; of Moravian communities, 11–12, 13, 31–37, 114–116, 126–127, 201; of Native communities, 65–66, 141; reconstruction of, 35–38, 47n67; of settler communities, 66, 93, 112, 114; spirituality in, 60, 155, 161, 162. See also acoustic ecology; sounds; soundscape compositions

Spangenberg, August, 21, 61, 81–83, 134, 146n60

Spangenberg, Eva Maria, 134, *135–137*

spatial humanities, 32, 37, 45n56. See also sound maps

spirituality, 155, 177n5, 178n13

spiritual practices: around hunting, 22, 57; around travel, 55–56; baptisms, 11, 15, 22, 25, 31, 86, 177n2, 189, 209; dreams and visions in, 22, 24, 93–94, 98n27, 158, 165–166, 176–177; Native, 134, 159, 165, 177n2. See also soundscapes, in spirituality; spirituality; sounds, spirituality and

Stenton, John, 203

Stockbridge, Massachusetts, 23

Stockbridge-Munsee Reservation, 39n2, 225

strangers. See travelers and visitors

Strangers' Store, 118, 128, 202. See also Timothy Horsfield

T

Tassawachamen. *See* Joshua Sr.
Tatemy, Moses, 72, 93–94
Tecumseh, 10, 95
Teedyuscung, 73, 187, 202
Tenskwatawa (Shawnee Prophet), 176–177
Thomas, 9, 213–215. *See also* Jacob and Thomas, surviving massacre
Tinker, George, 14
Tomlinson, Gary, 11, 183n67
topography, Pennsylvania's, 62
trail networks, 6n1, 52, 55, 59–62, 97n15, 98n38, 106, 186–187, 189
travelers and visitors, to Moravian communities, 126–129, 130, 132, 139, 149n98, 162–163
Treaty of Fort Stanwix, 92, 209
Tulpehocken Valley, 21, 30, 44n36, 59, 73–76, 78, 81, 82, 101n97
Tulpehocken Path, 101n97
Tuscarawas Valley, 7, 176, 209, 210–211, 213, 223

V

van der Woude, Joanne, 25, 134
van Sweringen, M., 9–10
von Heinitz, Johann Friedrich, 116, 162

W

Wajomick. *See* Wyoming
Walking Purchase, 14, 101n97
Wallace, Paul, 105, 187
Washington, George, 106, 223
Washington County militia from, 8–9, 213
watchmen, of Bethlehem, 74, 118–120, 123, 145n41, 161
Watteville, Johannes von, 90, 103n125
Weiser, Conrad, 21, 65, 66, 73, 75–78, 77, 90, 100n67, 149n97, 177n2
Welagameka, 30, 194
Wheeler, Rachel, 12, 22, 23, 94, 158, 178n12, 185n79, 198
Whitefield, George, 29

White River mission, 176–177
Womack, Craig, 16, 38–39
Woodward, Walter, 12
work: after end of communal economy, 199–200; in Bethlehem, 110, 123, 125, 128, 193; daily routines of, 115–118; missionaries, 22, 29, 173; singing and, 123–126, 160–161; Spangenberg on, 146n60
Wyalusing, 34, 188, 208, 209. *See also* Friedenshütten II
Wyandots, 7, 8, 211–212, 213, 214
Wyoming (Wajomick), 20, 59, 60, 78, 85–87, 88, 89, 90, 95–96, 103n125

Z

Zander, J. William, 73
Zander, Johanna Magdalena, 73
Zeisberger, David, 9, 22, 64, 68, 83, 93, 102n113, 177n2, 205, 208, 212–213, 215, 216, 217, 221n58, 223
Zinzendorf, Benigna von (von Watteville), 52, 72, 76, 81, 82, 102n105, 102n106
Zinzendorf, Christian Renatus von, 164
Zinzendorf, Erdmuthe Dorothea von. *See* Reuss, Erdmuthe Dorothea
Zinzendorf, Nikolaus Ludwig von, 30, 52, 111, 152, 157, 181n37; on bodies connected to spirit, 166–168, *170*, 170–173, 182n50, 183n57; on conversion, 29; death of, 175, 199; in development of Moravian Church, 23–24, 29–30, 197–198; diaries of, 52, 73–76, 78, 83, 90; hymns by, 7, 67, 72–75, 85, 87, 89, 90, 96, 101n93, 125, 146n54, 170–171; place names and, 81–82, 96, 102n105; Native name, 72, 76, 90, *91*, 101n90, 177n2; on improvisation in singing, 25, 163–165, 166, 181n37; mission plan, 17, *19*, 20, 30–31; preparing supplements to hymnal, 86, *89*, 90; snake incident of, 87, 103n121; travel diaries of, 73, 78, *79–80*, 83–84, *91*; travels of, 17–20, 90, *91*; visiting Moravian communities, 72–76; visiting Native communities, 44n35, 52, 72, 76–77, *77*, 160; in Wajomick, 85–90, *88*

SARAH EYERLY is Associate Professor of Musicology and Director of the Early Music Program at the Florida State University.

www.ingramcontent.com/pod-product-compliance
Lightning Source LLC
Chambersburg PA
CBHW030611230426
43661CB00053B/1942